MW00534933

Stochastic Finance

A Numeraire Approach

CHAPMAN & HALL/CRC
Financial Mathematics Series

Aims and scope:
The field of financial mathematics forms an ever-expanding slice of the financial sector. This series aims to capture new developments and summarize what is known over the whole spectrum of this field. It will include a broad range of textbooks, reference works and handbooks that are meant to appeal to both academics and practitioners. The inclusion of numerical code and concrete real-world examples is highly encouraged.

Published Titles

American-Style Derivatives; Valuation and Computation, *Jerome Detemple*
Analysis, Geometry, and Modeling in Finance: Advanced Methods in Option Pricing, *Pierre Henry-Labordère*
Credit Risk: Models, Derivatives, and Management, *Niklas Wagner*
Engineering BGM, *Alan Brace*
Financial Modelling with Jump Processes, *Rama Cont and Peter Tankov*
Interest Rate Modeling: Theory and Practice, *Lixin Wu*
Introduction to Credit Risk Modeling, Second Edition, *Christian Bluhm, Ludger Overbeck, and Christoph Wagner*
Introduction to Stochastic Calculus Applied to Finance, Second Edition, *Damien Lamberton and Bernard Lapeyre*
Monte Carlo Methods and Models in Finance and Insurance, *Ralf Korn, Elke Korn, and Gerald Kroisandt*
Numerical Methods for Finance, *John A. D. Appleby, David C. Edelman, and John J. H. Miller*
Portfolio Optimization and Performance Analysis, *Jean-Luc Prigent*
Quantitative Fund Management, *M. A. H. Dempster, Georg Pflug, and Gautam Mitra*
Robust Libor Modelling and Pricing of Derivative Products, *John Schoenmakers*
Stochastic Finance: A Numeraire Approach, *Jan Vecer*
Stochastic Financial Models, *Douglas Kennedy*
Structured Credit Portfolio Analysis, Baskets & CDOs, *Christian Bluhm and Ludger Overbeck*
Understanding Risk: The Theory and Practice of Financial Risk Management, *David Murphy*
Unravelling the Credit Crunch, *David Murphy*

Proposals for the series should be submitted to one of the series editors above or directly to:
CRC Press, Taylor & Francis Group
4th, Floor, Albert House
1-4 Singer Street
London EC2A 4BQ
UK

Chapman & Hall/CRC FINANCIAL MATHEMATICS SERIES

Stochastic Finance

A Numeraire Approach

Jan Vecer

CRC Press is an imprint of the
Taylor & Francis Group, an **informa** business
A CHAPMAN & HALL BOOK

CRC Press
Taylor & Francis Group
6000 Broken Sound Parkway NW, Suite 300
Boca Raton, FL 33487-2742

Printed in the United States of America on acid-free paper
10 9 8 7 6 5 4 3 2 1

International Standard Book Number: 978-1-4398-1250-1 (Hardback)

Library of Congress Cataloging-in-Publication Data

Vecer, Jan.
Stochastic finance : a numeraire approach / Jan Vecer.
p. cm. -- (Chapman & Hall/CRC financial mathematics series)
Includes bibliographical references and index.
ISBN 978-1-4398-1250-1 (hardcover : alk. paper)
1. Finance. 2. Stochastic analysis. I. Title.

HG173.V38 2011
332.01'51922--dc22 2010044425

Visit the Taylor & Francis Web site at
http://www.taylorandfrancis.com

and the CRC Press Web site at
http://www.crcpress.com

Contents

Introduction

This book is based on lecture notes from stochastic finance courses I have been teaching at Columbia University for almost a decade. The students of these courses – graduate students, Wall Street professionals, and aspiring quants – has had a significant impact on this text and on my teaching since they have firsthand feedback from the dynamic world of finance. The content of this book addresses both the needs of practitioners who want to expand their knowledge of stochastic finance, and the needs of students who want to succeed as professionals in this field. Since it also covers relatively advanced techniques of the numeraire change, it can be used as a reference by academics working in the field, and by advanced graduate students.

A typical reader should already have some basic knowledge of stochastic processes (Markov chains, Brownian motion, stochastic integration). Thus the prerequisite material on probability and stochastic calculus appears only in the Appendix, so the reader who wants to review this material should refer to this section first. In addition, most of the students who previously studied this material had also been exposed to some elementary concepts of stochastic finance, so some limited knowledge of the financial markets is assumed in the text. This book revisits some concepts that may be familiar, such as pricing in binomial models, but it presents the material in a new perspective of prices relative to a reference asset.

One of the goals of this book is to present the material in the simplest possible way. For instance, the well-known Black–Scholes formula can be obtained in one line by using the basic principles of finance. I often found that it is quite hard to find the easiest, or the most elegant, solution but certainly a lot of effort has been spent achieving this. The reader should keep in mind that this is a demanding field on the level of the mathematical sophistication, so even the simplest solution may look rather complicated. Nevertheless, most of the ideas presented here rely on intuition, or on basic principles, rather than on technical computations.

This book differs from most of the existing literature in the following way: it treats the price as a number of units of one asset needed for an acquisition of a unit of another asset, rather than expressing prices in dollar terms exclusively. Since the price is a relationship of two assets, we will use a notation that will indicate both assets. The price of an asset X in terms of a reference asset Y at time t will be denoted by $X_Y(t)$. This will allow us to distinguish between the asset X itself, and the price of the asset X_Y. This distinction is important since many financial relationships can be expressed in terms of the assets. The existing literature tends to mix the concept of an asset with the concept of the price of an asset.

The reference asset serves as a choice of coordinates for expressing the prices. The price appears in many different markets, and sometimes it is even not interpreted as a price process. The simplest example is a dollar price of an asset, where a dollar is a reference asset. Dollar prices appear in two major markets: an equity market where the primary assets are stocks, and a foreign exchange market where the primary assets are currencies. The prices in the foreign exchange market are also known as exchange rates.

The foreign exchange market shows that the reference asset that is chosen for pricing can be relative. For instance, information about how many dollars are required to obtain one euro is the same as how many euros are required to obtain one dollar. Since in principle there is nothing special about choosing one or the other currency as a reference asset, it is important to create models of the price processes that treat both assets equivalently. Thus we treat the reference asset as relative, and using an analogy from physics, the theory presented here can be called a *theory of relativity in finance.* It essentially means that the observer – an agent in a given economy – should see a similar type of evolution of prices no matter what reference asset is chosen.

Sometimes a different reference asset than a dollar is used. For instance, when the reference asset is a money market, or a bond, the resulting price is known as a discounted price. An even less obvious example of a price is a forward London Interbank Offered Rate, or LIBOR for short, where the reference asset is a bond. Markets that trade LIBOR are known as fixed income markets. Since the prices in the fixed income markets (in this case known as forward rates) are expressed in terms of bonds, it is strictly suboptimal to use a dollar as a reference asset in this case. This book presents a unified approach that explains how to compute the prices of contingent claims in terms of various reference assets, and the principles presented here apply to different markets.

Using dollars and currencies in general for hedging or investing is problematic since holding money in terms of the banknotes creates an arbitrage opportunity – ability to make a risk free profit – for the issuer of the currency. Stated equivalently, money has time value; a dollar now is more valuable than a dollar tomorrow. We can write $\$_t > \$_{t+1}$. In order not to lose the value with the passage of time, currencies have to be invested in assets that do not lose value with the passage of time, such as bonds, non-dividend paying stocks, interest bearing money market accounts, or precious metals. Note that the currency and the interest bearing money market account are two different assets – the first loses value with time, the second does not. When the asset X keeps that same value with the passage of time, we can write $X_t = X_{t+1}$. This relationship does not mean that the price of such an asset with respect to a reference asset Y would stay the same; the price $X_Y(t)$ can be changing with time. For instance, an ounce of gold is staying physically the same as an asset; the gold today is the same as the gold tomorrow, but the dollar price of the ounce of gold can be changing.

Making a loose connection with physics – money is a choice of a reference asset (or coordinates) that comes with friction. The time value of money is analogous to movement with friction. It is always easier to add friction (money) to the theory of frictionless markets as opposed to removing the friction (say through adding interest on the money market) in the theory of markets inherently built with friction. If one holds a unit of the currency, the unit will keep creating arbitrage opportunities for the issuer of the currency. Money in terms of banknotes is acceptable if we use it as a spot reference asset, but it should not be used for hedging or for investment. Therefore we focus our attention in the following text on reference assets that do not create arbitrage opportunities through time, and develop a frictionless theory of pricing financial contracts.

We call assets that keep the same value with the passage of time as *no-arbitrage assets*, as opposed to *arbitrage assets* that have time value. Note that an arbitrage asset itself, such as a currency, can be bought or sold, but it creates arbitrage opportunities as time elapses. Examples of no-arbitrage assets include interest bearing money market accounts, precious metals, stocks that reinvest dividends, options, or contracts that agree to deliver a unit of a certain asset in the future. The asset to be delivered may not necessarily be a no-arbitrage asset, such as in the case of a zero coupon bond – a contract that delivers a dollar (an arbitrage asset) at some future time. The zero coupon bond itself does not create arbitrage opportunities in time (until expiration), and thus can serve as a no-arbitrage reference asset.

The fundamental principle of the modern finance is the non-existence of any arbitrage opportunity in the markets. Therefore the theory applies only to no-arbitrage assets that do not lose value with the passage of time. The central reason why we can determine the price of a contingent claim is the First Fundamental Theorem of Asset Pricing which underscores the importance of the no-arbitrage principle. This theorem states that when the prices are martingales under the probability measure that corresponds to the reference asset, the model does not admit arbitrage. The existence of such a martingale measure allows us to express the prices of contingent claims as conditional expectations under this measure, giving us a stochastic representation of the prices. However, the First Fundamental Theorem of Asset Pricing applies only to prices expressed in terms of no-arbitrage assets as opposed to dollar values, so only no-arbitrage assets have their own corresponding martingale measure. Arbitrage assets, such as dollars, do not have their own martingale measure, and the prices with respect to arbitrage assets have to be computed from the change of numeraire formula using no-arbitrage assets. The First Fundamental Theorem of Asset Pricing is introduced early in the text, and all the pricing formulas follow from this theorem.

In this book we study financial contracts that are written on other underlying assets. Such contracts are called *derivatives* since they depend on other assets. Sometimes we also call them *contingent claims.* We study the price and the hedge of a derivative contract whose payoff depends on more basic assets. The key idea of pricing and hedging derivative contracts is to identify a portfolio that either matches or at least closely mimics the contract by active trading in the underlying assets. It turns out that such a trading strategy in most cases does not depend on the evolution of the price of the underlying assets, and thus we can to some extent ignore the real price evolution of the basic assets.

Single asset contracts depend on only one underlying asset, which we call X. Such contracts include a contract to deliver a unit of X at some future time T. This is a special case of a forward. A forward is a contract that delivers an asset X for K units of an asset Y. Thus a contract to deliver a unit of X represents a choice of $K = 0$ in the forward contract. When the underlying asset to be delivered is a currency, the contract is known as a bond. A zero coupon bond B^T is a contract that delivers one dollar at time T. Contracts on two assets, say X and Y, include options. An option is a contract that depends on two or more underlying assets that has a nonnegative payoff. This is essentially the right to acquire a certain combination of the underlying assets at the time of maturity of the option contract (European-type options), or any time up to the time of maturity of the contract (American-type options). Contracts written on three or more assets include quantos and most exotic options such as lookback and Asian options.

Assets with a positive price that enter a given contract can be used as reference assets for pricing this financial contract. Such assets are called *numeraires.* Whenever possible, it is desirable to choose a no-arbitrage asset as a reference asset since we can apply the results of the First Fundamental Theorem of Asset Pricing directly. Most existing financial contracts can in fact be expressed only in terms of no-arbitrage assets with one notable exception – American stock options are settled in the stock and the dollar, and there is no way to replace the dollar with a suitable no-arbitrage asset. This makes American options exceptional in terms of pricing, since the price of the option has to be expressed with respect to the dollar, which is an arbitrage asset.

Computation of the dollar prices of contingent claims cannot be done directly by applying the First Fundamental Theorem of Asset Pricing. A widely used approach is to assume a deterministic evolution of the dollar price of the money market account, and relate the dollar value to the money market value by discounting. The First Fundamental Theorem of Asset Pricing applies to the money market account, and so the dollar prices may be computed from this relationship. The martingale measure that is associated with the money market account is also known as the risk neutral measure. This approach has two limitations. The first limitation is that the dollar price of the money market is not typically deterministic due to the stochastic evolution of the interest rate, in which case this method does not apply at all. The second limitation is that for more complex financial products, computation of the price of a contingent claim in terms of a dollar may be unnecessarily complicated when compared to pricing with respect to other reference assets that are more natural to use in a given situation.

Our strategy of computing the dollar prices is different and it applies in general. First, we identify the natural reference no-arbitrage assets which can be used in the First Fundamental Theorem of Asset Pricing. For instance, we will show in the later text that a European stock option has two natural reference no-arbitrage assets: a bond B^T that matures at the time of the maturity of the option, and the stock S itself. We can compute the price of the contingent claim using either the probability measure that comes with the bond B^T (also known as a T-forward measure), or the probability measure that comes with the stock S. Once we have the price of the contingent claim with respect to the bond B^T (or the stock S), we can trivially convert this price to its dollar value by a relationship known as the change of numeraire formula.

The advantage of the numeraire approach described above may not be entirely obvious for a relatively simple financial contract. Its price can be found easily using both methods. However, for more complex products, such as for barrier options, lookback options, quantos, or Asian options, the numeraire approach has clear advantages – it leads to simpler pricing equations. We will also illustrate that the barrier option and the lookback option can be related to a plain vanilla contract. We will also show how to identify the basic assets that enter a given contract; for instance, the lookback option depends on a maximal asset, and the Asian option depends on an average asset.

The understanding of representing prices as a pairwise relationship of two assets is a fundamental concept, but many books treat it as an advanced topic. Our approach has several advantages as it leads to a deeper understanding of derivative contracts. When a given contract depends on several underlying assets, we can compute the price of the contract using all available reference assets. It is often the case that a choice of a particular reference asset leads to a simpler form. We also find some pricing formulas that are model independent.

Examples that admit a simple solution with the approach mentioned in this book include a model independent formula for European call options, a simple method for pricing barrier options, lookback options and Asian options, and a formula for options on LIBOR.

The book has the following structure. The first chapter of this book introduces basic concepts of finance: price, the concept of no arbitrage, portfolio and its evolution, types of financial contracts, the First Fundamental Theorem of Asset Pricing, and the change of numeraire formula. The subsequent chapters apply these general principles for three kinds of models: a binomial model, a diffusion model, and a jump model. The binomial model tends to be too simplistic to be used in practice, and we include it only as an illustration of the concept of the relativity of the reference asset. The novel approach is that the prices of these contracts have two or more natural reference assets, and thus there are two or more equivalent descriptions of the pricing problem. In continuous time, we study both diffusion and jump models of the evolution of the price processes. We study European options, barrier options, lookback options, American options, quantos, Asian options, and term structure models in more detail. The Appendix summarizes basic results from probability and stochastic calculus that are used in the text, and the reader can refer to it while reading the main part of the book.

I am grateful to the audiences of my stochastic finance classes given at Columbia University, the University of Michigan, Kyoto University, and the Frankfurt School of Finance and Management. I have also received valuable feedback from the participants in the seminar talks that I gave at Harvard University, Stanford University, Princeton University, the University of Chicago, Cambridge University, Oxford University, Imperial College, King's College, Carnegie Mellon University, Cornell University, Brown University, the University of Waterloo, the University of California at Santa Barbara, the City University of New York, Humboldt University, LMU Muenchen, Tsukuba University, Osaka University, the University of Wisconsin – Milwaukee, Brigham Young University, Charles University in Prague, CERGE-EI, and the Prague School of Economics. The research on the book was sponsored in part by the Center for Quantitative Finance of the Prague School of Advanced Legal Studies.

I would also like to thank the following people for comments and suggestions that helped to improve this manuscript: Mary Abruzzo, Mario Altenburger, Martin Auer, Jun Kyung Auh, Josh Bissu, Mitch Carpen, Peter Carr, Kan Chen, Ivor Cribben, Emily Doran, Helena Dona Duran, Clemens Feil, Scott Glasgow, Nikhil Gutha, Olympia Hadjiliadis, Adrian Hashizume, Gerardo Hernandez, Amy Herron, Sean Ho, Tomoyuki Ichiba, Karel Janecek, Xiao Jia, Philip Johnston, Armenuhi Khachatryan, David Kim, Thierry Klaa, Sharat Kotikalpudi, Ka-Ho Leung, Jianing Li, Sasha Lv, Rupal Malani, Antonio Medina, Vishal Mistry, Amal Moussa, Daniel Neelson, Petr Novotny, Kimberli Piccolo, Radka Pickova, Dan Porter, Libor Pospisil, Cara Roche, Johannes Ruf, Steven Shreve, Lisa Smith, Li Song, Joyce Yuan Hui Su, Stephen Taylor, Uwe Wystup, Mingxin Xu, Ira Yeung, Wenhua Zou, Hongzhong Zhang, and Ningyao Zhang. The editors and the production team from the CRC Press provided much needed assistance, namely, Sunil Nair, Sarah Morris, Karen Simon, Amber Donley, and Shashi Kumar. The whole project would not be possible without the unconditional support of my family.

Chapter 1

Elements of Finance

Some of the basic concepts of finance are widely understood in broad terms; however this chapter will introduce them from a novel perspective of prices being treated relative to a reference asset. We first show the difference between an asset and the price of an asset. The price of an asset is always expressed in terms of another reference asset. The reference asset is also called a **numeraire.** The numeraire asset should never become worthless so that the price with respect to this asset is well defined. The relationship between prices of an asset expressed with respect to two different reference assets is known as a **change of numeraire.** The concept of price appears in different markets under different names, so it may not be obvious that it is just a particular instance of a more general concept. For instance, an exchange rate is in fact a price representing a pairwise relationship of two currencies. An even less obvious example of a price is a forward London Interbank Offer Rate (LIBOR). By adopting a precise definition of price, we are able to treat various markets (equities, foreign exchange, fixed income) in one single unified framework, which simplifies our analysis.

The second section introduces the concept of arbitrage – the possibility of making a risk free profit. We study models of markets where no agent allows an arbitrage opportunity. One can create an arbitrage opportunity just by holding a single asset such as a banknote. This is known as a time value of money. Thus the concept of no arbitrage splits assets into two groups: no-arbitrage assets – the assets that do not allow any arbitrage opportunities; and arbitrage assets – the assets that do allow arbitrage opportunities. In theory, the market should have only no-arbitrage assets. Financial contracts are typically no-arbitrage assets; they become arbitrage assets only when their holder takes some suboptimal action (such as not exercising the American put option at the optimal exercise time). On the other hand, real markets include arbitrage assets such as currencies.

Currencies, in terms of banknotes, are losing an interest rate when compared to the corresponding bond or money market account. Since the loss of the currency value is typically small, money still serves as a primary reference asset in the economy. However, in order to avoid this loss of value in pricing contingent claims, one should use discounted prices rather than dollar prices of the assets. Discounted prices correspond to either a bond or a money market account as a reference asset. Stocks paying dividends are arbitrage assets when the dividends are taken out, but an asset representing the equity plus the dividends is a no-arbitrage asset. We find a simple relationship between the dividend paying stock and the portfolio of the stock and the dividends.

In the section that follows, we introduce the concept of a portfolio. A portfolio is a combination of several assets, and it is important to realize that it has no numerical value. In fact, one should not confuse the concept of a portfolio (viewed as an asset) with the price of a portfolio (number that represents a pairwise relationship of two assets). It should be noted that a portfolio may be staying physically the same, but the price of this portfolio with respect to some reference asset may be changing. We also introduce the concept of trading. Self-financing trading is exchanging assets that have the same price at a given moment. As a consequence, portfolios may be evolving in time by following a self-financing trading strategy.

When no arbitrage exists in the markets, all prices are martingales with respect to the probability measure that comes with the specific no-arbitrage reference asset. Martingales are processes whose best estimator of the future value is its present value. Mathematically, a process $\mathcal{M}$ that satisfies $\mathbb{E}_s[\mathcal{M}(t)] = \mathcal{M}(s), s \leq t$, is a martingale, where $\mathbb{E}_s[.]$ denotes conditional expectation. The reader should refer to the Appendix for more details about martingales and conditional expectation. The result that prices are martingales under the probability measure that is related to the reference asset is known as the First Fundamental Theorem of Asset Pricing. In particular, every no-arbitrage asset has its own pricing **martingale measure.** Other no-arbitrage assets have different martingale measures. The martingale measure associated with the money market account is known as a **risk-neutral measure.** The martingale measures associated with bonds are known as **T-forward measures.** Stocks have martingale measures known as a **stock measure.** Arbitrage assets, such as currencies, do not have their own martingale measures. In particular, there is no dollar martingale measure.

Many authors do not regard currencies as true arbitrage assets because this arbitrage opportunity is one sided for the issuer of the currency. It is also easy to confuse money (in terms of banknotes) with the money market account. Banknotes deposited in a bank start to earn the interest rate and become a part of the money market account. When borrowing money, the debt is not a currency, but rather the corresponding money market account. The debt earns the interest to the lender, and thus it behaves like the money market account. However, arbitrage pricing theory applies only to no-arbitrage assets, such as the money market account, bonds, or stocks. It does not apply to money in terms of banknotes. No-arbitrage assets have their own martingale measure, while arbitrage assets do not.

An important consequence of the First Fundamental Theorem of Asset Pricing is that the prices are martingales with respect to a probability measure associated with a particular reference asset. Martingales in continuous time models are under some assumptions just combinations of continuous martingales, and purely discontinuous martingales. Moreover, continuous martingales are stochastic integrals with respect to Brownian motion. This limits possible evolutions of the price to this class of stochastic processes since other types of evolutions allow for an existence of arbitrage.

Another related question to the concept of no arbitrage is a possibility of replicating a given financial contract by trading in the underlying primary assets. The martingale measure from the First Fundamental Theorem of Asset Pricing may not necessarily be unique; each reference asset may have infinitely many of such measures. However, each martingale measure under one reference asset has a corresponding martingale measure under a different reference asset that agrees on the prices of the financial contracts. The two measures are linked by a Radon–Nikodým derivative. In particular, when there is a unique martingale measure under one reference asset, the martingale measures that correspond to other reference assets are also unique due to the one-to-one correspondence of the martingale measures.

In the case when the martingale measure is unique, all financial contracts can be perfectly replicated. This result is known as the Second Fundamental Theorem of Asset Pricing. The market is complete essentially in situations when the number of different noise factors does not exceed the number of assets minus one. Thus models with two assets are complete when there is only one noise factor, which is, for instance, the case in the binomial model, in the diffusion model driven by one Brownian motion, or in the jump model with a single jump size. When the market is complete, the financial contracts are in principle redundant since they can be replicated by trading in the underlying primary assets. The replication of the financial contracts is also known as **hedging.**

1.1 Price

This section defines price as a pairwise relationship of two assets.

Price is a number representing how many units of an asset Y are required to obtain a unit of an asset X.

We denote this price at time t by

$$X_Y(t).$$

Here an asset Y serves as a reference asset. The reference asset is known as a **numeraire. Price is always a pairwise relationship of two assets.**

For practical purposes the role of a reference asset is typically played by **money**, a choice of the reference asset Y being a dollar \$. However, the choice of the reference asset is in principle arbitrary as long as the reference asset is not worthless. The reader should note that some financial assets may become worthless at a certain stage (such as options expired out of the money), and such contracts would be a poor choice of the reference asset. There are also some desirable properties that the reference asset should satisfy: it should be sufficiently durable, and there should exist enough identical copies of the asset. From this perspective, consumer goods (such as cars, electronic products, most food products) may be used as a reference asset, but this choice would not be appropriate since the asset itself has time value; it is deteriorating in time.

In practice, a small loss of the value of the reference asset is acceptable. Currencies in particular lose value in time by allowing an arbitrage opportunity with respect to the money market account, and they still play a role of a primary reference asset in the economy. However, when the loss of the value becomes large, for instance in a period of hyperinflation, such currency may no longer be accepted as a reference asset. The property of having sufficient identical copies of the asset ensures that the individuals in the economy can easily acquire the reference asset. The reference asset should be sufficiently liquid. For instance some art works (paintings, sculptures, buildings) have a significant value, but they cannot be easily bought or sold and thus using them as a reference asset would not be a good choice.

Typical choices of a reference asset used in practice are currencies (denoted by \$, €, £, ¥, etc.), bonds (denoted by B^T), a money market (denoted by M), or stocks and stock indices (denoted by S). A **bond** B^T is an asset that delivers one dollar at time T. The **money market** M is an asset that is created by the following procedure. The initial amount equal to one dollar is invested at time $t = 0$ in the bond with the shortest available maturity (ideally in the next infinitesimal instant), and this position is rolled over to the bond with the next shortest maturity once the first bond expires. The resulting asset, the money market M, is a result of an active trading strategy involving a number of these bonds. In principle, there is a counter party risk involved in delivering a unit of a currency at some future time. The counter party may fail to deliver the agreed amount at the specified time. The following text assumes situations when there is no such risk present, as in the case when the delivery of the asset is guaranteed by the government.

The reference asset itself does not need to be a traded asset. As we will see in the chapter on pricing exotic options, some natural reference assets that are useful for pricing complex financial contracts do not exist in real markets. For instance, one can use an asset that represents the running maximum of the price $\max_{0 \leq s \leq t} X_Y(s)$ for pricing lookback options, or one can use an asset that represents the average price for pricing Asian options. A price of a financial contract that is expressed in terms of an asset which is not traded can be easily converted to a price expressed in terms of a traded asset. Thus for practical purposes it does not matter if the reference asset exists or not in real markets.

Let us introduce the following notation. By X_t we mean a unit of an asset X at time t, not its price in terms of a different asset. In principle, an asset X that has no time value stays the same at all times (think of an ounce of gold), so there is really no need to index it with time. However, by adding the time coordinate we express that a particular asset is used at that time for trading, pricing, hedging, or for settling some contract. When there is no ambiguity, we will simply drop the time index, and write only X to stress that the asset in fact stays the same.

Recall that price is a pairwise relationship of two assets denoted by $X_Y(t)$ – the number of units of an asset Y required to obtain one unit of an asset X. The asset Y is known as a reference asset, or as a numeraire. We can write that

$$1 \text{ unit of } X = X_Y(t) \text{ units of } Y,$$

or simply

$$X = X_Y(t) \cdot Y. \tag{1.1}$$

Assets X and Y on their own do not have any numerical value (such as an ounce of gold), and the above equality does not mean that the assets on the left hand side and on the right hand side of the equation are physically the same. Note that we cannot divide by Y in the above equation since Y is an asset.

The relation "=" when used for assets as in Equation (1.1) is an equivalence relation. **We will write $X_t = Y_t$ in the sense of assets when $X_Y(t) = 1$ in the sense of numbers.** Clearly, the relation "=" for assets is reflexive ($X_t = X_t$), symmetric ($X_t = Y_t$ implies $Y_t = X_t$), and transitive ($X_t = Y_t$ and $Y_t = Z_t$ imply $X_t = Z_t$). The assets are also ordered according to their prices. We can write $X_t \geq Y_t$ in terms of assets when $X_Y(t) \geq 1$ in terms of numbers. It should be noted that two assets X and Y with an equal price at time t_1 (meaning $X_Y(t_1) = 1$) may differ in price at some other time t_2 (meaning $X_Y(t_2) \neq 1$). If two assets X and Y have the same price at time t, they can be exchanged for each other at that time. This procedure is known as a **self-financing trade.**

It may not be clear as to why we should adopt notation $X_Y(t)$ for the price, instead of using just a single letter for it, say $S(t)$, which is typically used for the price of a stock in terms of dollars. The following examples illustrate that the concept of price appears in different markets, such as in equity markets, in the foreign exchange markets, or in fixed income markets. By using our notation, we are able to treat these prices in one single framework, rather than studying them separately.

Example 1.1 **Examples of the price**

- The **dollar price** of an asset S, $S_\$(t)$, where the role of the asset X is played by the stock S, and the role of the reference asset Y is played by the dollar **\$**. Most of the current literature writes simply $S(t)$ for the dollar price $S_\$(t)$ of this asset, but we want to avoid in our text confusing the asset S itself with the price of the asset $S_\$(t)$.

- The price of a stock S in terms of the money market M, $S_M(t)$, where the asset X is a stock S, and the asset Y is a money market M with $M_0 = \$_0$. The price $S_M(t)$ is known as a **discounted price** of an asset S.

- The price of a stock S in terms of a zero coupon bond B^T with maturity T, $S_{B^T}(t)$, where the asset X is a stock S, and the asset Y is a bond B^T. This is also a form of a **discounted price** which is more appropriate than S_M for pricing derivative contracts that depend on S and **\$**. Note that we have $S_{B^T}(T) = S_\$(T)$.

- The **exchange rate**, $€_\$(t)$, where X is the foreign currency (**€**), and Y is the domestic currency **\$**. The choice of domestic and foreign currency is relative, and thus $\$_€(t)$ is also an exchange rate.

- **Forward London Interbank Offered Rate**, or forward LIBOR for short,

$$\left[B^T - B^{T+\delta}\right]_{\delta B^{T+\delta}}(t),$$

where the role of the asset X is played by a portfolio of two bonds $[B^T - B^{T+\delta}]$, and the reference asset Y is $\delta \cdot B^{T+\delta}$.

□

We will discuss these examples of price in more detail after introducing the concepts of inverse price, and change of numeraire. Since the assets X and Y considered in the above are arbitrary, it also makes perfect sense to consider the inverse relationship when X is chosen as a reference asset. For instance, one may think about X and Y as two currencies. When $X = €$, and $Y = \$$, we have both the exchange rate $€_{\$}(t)$ – the number of dollars required to obtain a unit of a euro, and the exchange rate $\$_{€}(t)$ – the number of euros required to obtain a unit of a dollar. Thus we can also write

$$1 \text{ unit of } Y = Y_X(t) \text{ units of } X,$$

or simply

$$Y = Y_X(t) \cdot X. \tag{1.2}$$

The price $Y_X(t)$ is the **inverse price** to $X_Y(t)$. Let us show the relationship between $Y_X(t)$ and $X_Y(t)$. Suppose that an agent starts with a unit of an asset Y. He can change it for $Y_X(t)$ units of an asset X. This amount can be split in two parts: $Y_X(t) - X_Y(t)^{-1}$ and $X_Y(t)^{-1}$ units of an asset X. The part of $X_Y(t)^{-1}$ units of an asset X can be exchanged back for a unit Y, which follows from the relationship

$$X = X_Y(t) \cdot Y,$$

which is equivalent to

$$Y = X_Y(t)^{-1} \cdot X.$$

We can rewrite the above trading procedure using the following identities

$$\begin{aligned} Y &= Y_X(t) \cdot X \\ &= (Y_X(t) - X_Y(t)^{-1}) \cdot X + X_Y(t)^{-1} \cdot X \\ &= (Y_X(t) - X_Y(t)^{-1}) \cdot X + Y. \end{aligned}$$

Thus the net result of this exchange is $Y_X(t) - X_Y(t)^{-1}$ units of an asset X, which must be zero in order not to allow a risk-free profit. Therefore the prices $X_Y(t)$ and $Y_X(t)$ are related by the following relationship

$$\boxed{Y_X(t) = \frac{1}{X_Y(t)}.} \tag{1.3}$$

This relationship is valid when $0 < X_Y(t) < \infty$, which is the case that neither the asset X nor the asset Y is worthless. In this case, $X_Y(t)$ and its inverse price $Y_X(t)$ have the same information.

In general, it should not matter which reference asset is chosen, one should observe similar price evolutions. We will use this as a key principle for pricing derivative contracts studied in this book. One can look at it as a **theory of relativity in finance**: how one views prices depends on one's choice of the reference asset.

Given an asset X and two reference assets Y and Z, we can write the price of X with respect to the reference asset Y using

$$X = X_Y(t) \cdot Y. \tag{1.4}$$

Similarly, we can write

$$X = X_Z(t) \cdot Z \tag{1.5}$$

when we use Z as a reference asset. Thus we have

$$\boxed{X = X_Y(t) \cdot Y = X_Z(t) \cdot Z,} \tag{1.6}$$

which is known as a **change of numeraire formula.** The above relationship is written in terms of assets. We can rewrite the above relationship in terms of the price as

$$\boxed{X_Y(t) = X_Z(t) \cdot Z_Y(t).} \tag{1.7}$$

This relationship is valid for assets X, Y, and Z that are not worthless.

Example 1.2 **Foreign Exchange Market**
Let us illustrate the concepts of the inverse price and the change of numeraire on the foreign exchange market. Prices in the real markets satisfy the relationship

$$Y_X(t) = X_Y(t)^{-1}$$

at all times (up to the rounding errors). For instance, on January 8th, 2010, at 8:00PM EST, the exchange rates between **€** and **$** were:

$$€_\$ = 1.4415 \qquad \$_€ = 0.6937.$$

We can easily check that

$$\$_€^{-1} = \frac{1}{0.6937} = 1.441545...$$

Thus the inverse exchange rate $\$_€^{-1}$ matches the first four digits of the exchange rate $€_\$$. The exact match is typically not possible since these exchange rates are quoted in four decimal digits. However, the arbitrage is still not possible due to the difference of the prices offered and asked. An agent who wants to acquire a unit of an asset should be ready to pay more than an agent who wants to sell a unit of the same asset.

More specifically, the market exchange works in the following way: Agents who want to buy a particular asset place their orders on the market exchange, and wait until they find corresponding counter parties that are willing to match their orders. The orders compete according to the price that is quoted; a higher quote has a higher priority of being executed. The highest quote is known as the **best bid.** Similarly, agents who want to sell a particular asset place their orders on the market exchange. A smaller price asked for a unit of a given asset has a higher priority. The smallest price asked is known as the **best ask.** Clearly, the best ask is larger than the best bid. The smallest difference between two possible quoted prices on the exchange is known as a **tick.** In the case of euro/dollar exchange rates, the tick is equal to 0.0001. The difference between the best bid and the best ask is known as a **bid-ask spread.** Bid-ask spreads may be larger than a tick. More liquid assets have smaller bid-ask spreads, the difference between the buying and the selling price being smaller.

From the perspective of having both $X_Y(t)$ and $Y_X(t)$ as prices, there is no absolute direction of up and down in the market. Each trade has two sides, a seller and a buyer. If the market moves in one direction, it is either to the benefit of the seller and at the expense of the buyer, or vice versa. This is another way of saying that when one of the prices $X_Y(t)$ or $Y_X(t)$ goes up, the inverse price must go down.

Exchange rates also serve as an example of the change of numeraire formula. Table 1.1 shows the exchange rate table for four major currencies: dollars, euros, pounds, and yen as seen on January 8th, 2010 at 8:00PM EST. For instance the entry (\$, €) gives the price $\$_{€} = 0.6937$, etc.

TABLE 1.1: Exchange Rate Table.

	$	€	£	¥
$	1	0.6937	0.6238	92.6300
€	1.4415	1	0.8991	133.5260
£	1.6032	1.1122	1	148.5040
¥	0.0108	0.0075	0.0067	1

From the change of numeraire formula, we should also have among other similar relationships

$$\$_{€} = \$_{£} \cdot £_{€}. \tag{1.8}$$

In fact,

$$\$_{£} \cdot £_{€} = 0.6238 \times 1.1122 = 0.693790...$$

This matches the original $\$_{€}$ rate in four decimal digits if we neglect the rounding error in the fourth digit. This match is close enough not to allow for any arbitrage opportunities due to the market imperfections such as the bid-ask spread, or transaction costs. □

REMARK 1.1 The change of numeraire formula (1.7) applies to all assets, with or without time value. Note that Equation (1.8) is an example of the change of numeraire formula for assets with time value. □

Example 1.3 **Forward London Interbank Offered Rate**
The Forward London Interbank Offered Rate, or **LIBOR** for short, is defined as a simple interest rate that corresponds to borrowing money over the time interval T and $T+\delta$ as seen at time $t \leq T$. We denote forward LIBOR by $L(t,T)$. When $t=T$, $L(T,T)$ is known as a spot LIBOR since it corresponds to borrowing money at the present time T.

Suppose that one dollar is borrowed at time T, and assume that $L(t,T)$ is the simple interest rate for the period between T and $T+\delta$. Then the agent should return $1+\delta L(t,T)$ dollars at time $T+\delta$. Thus $L(t,T)$ can be defined by the following relationship:

$$(1+\delta L(t,T)) \cdot B_t^{T+\delta} = B_t^T. \tag{1.9}$$

The right hand side of the above relationship indicates that one dollar will be delivered at time T. The left hand side indicates that $(1+\delta L(t,T))$ dollars will be returned at time $T+\delta$. Therefore

$$\delta L(t,T) \cdot B_t^{T+\delta} = B_t^T - B_t^{T+\delta}. \tag{1.10}$$

We can rewrite this relationship in the following form:

$$\boxed{L(t,T) = \left[B^T - B^{T+\delta}\right]_{\delta \cdot B^{T+\delta}}(t),} \tag{1.11}$$

showing that forward LIBOR $L(t,T)$ is in fact a price, where the asset X is a portfolio $[B^T - B^{T+\delta}]$ (long the B^T bond, and short the $B^{T+\delta}$ bond), and the reference asset Y is δ units of the bond $B^{T+\delta}$.

If we wanted to compute $X_Y(t)$ for two general assets, we can do so from the dollar prices of the assets X and Y:

$$X_Y(t) = X_\$(t) \cdot \$_Y(t) = \frac{X_\$(t)}{Y_\$(t)}, \tag{1.12}$$

where we substitute Z for **\$** in the change of numeraire formula. Using Equation (1.12), we can determine forward LIBOR from dollar prices of bonds by using

$$\begin{aligned} L(t,T) &= \left[B^T - B^{T+\delta}\right]_{\delta B^{T+\delta}}(t) \\ &= \left[B^T - B^{T+\delta}\right]_\$(t) \cdot \$_{\delta B^{T+\delta}}(t) = \frac{B_\$^T(t) - B_\$^{T+\delta}(t)}{\delta B_\$^{T+\delta}(t)}. \end{aligned} \tag{1.13}$$

Here we have used the change of numeraire formula, and linearity of the prices:

$$[aX + bY]_Z(t) = aX_Z(t) + bY_Z(t). \tag{1.14}$$

□

Foreign exchange markets, or fixed income markets that trade on LIBORs, are in fact much larger than the equity markets in terms of the volume traded, and thus the main focus of financial markets is on prices that are not expressed exclusively in dollar terms. It is also not an obvious observation that exchange rates and forward LIBORs are in fact prices. Calling them the exchange rates or forward LIBORs is slightly misleading, and the literature tends to study the asset prices, foreign exchange rates, and forward LIBORs separately. In our approach, they are just special cases of a more general concept of price.

Price is always a pairwise relationship of two assets, and we will use this notation throughout this book to indicate the reference asset. This distinction will help us study derivative contracts later on in the text that are written on more than one underlying asset. The second (or the third asset when applicable in the case of exotic options) asset also serves as a viable reference asset for pricing a given derivative contract. This notation is especially helpful when studying quantos and other exotic options, which represent financial contracts that are written on three underlying assets. The reader should also note here that every contract is settled in units of particular assets (dollars, stocks, bonds) rather than in the price itself – the price indicates only how many units of a particular asset are needed.

1.2 Arbitrage

This section discusses another fundamental concept of finance, namely arbitrage.

Arbitrage is an opportunity to make a risk free profit in the market.

It is important to distinguish an arbitrage opportunity from a profitable trading strategy. Arbitrage means that there is at least one agent that can make money for sure, while a profitable trading strategy simply works on average, meaning that some scenarios may lead to a loss.

An arbitrage opportunity means that one can create a guaranteed profit starting from a portfolio with a zero initial price. It is easy to see that if a portfolio has a zero price with respect to one asset, it has a zero price with respect to any reference asset. A typical example of an arbitrage opportunity is the ability to purchase an asset at a given price and then sell the same asset immediately or some later time for a higher price. The guarantee of a higher price is necessary to make it an arbitrage opportunity, assuring that the portfolio always ends up with more assets than when it started. Such arbitrage trades can happen when a purchase price in one market is less than the selling price in a different market.

Example 1.4

Assume that at time $t = 0$, the price of an asset X with respect to an asset Y is $X_Y(0) = K$. Suppose that at a fixed time $T \geq 0$, the price will be exactly $X_Y(T) = J$ with $J > K$. In such a case one can construct a portfolio, starting at time $t = 0$ with $P_0^0 = 0$, exchange it for the portfolio $P_0^1 = X - K \cdot Y$ that has a zero price (long one unit of X and short K units of Y), and end up with a portfolio $P_T^1 = X - K \cdot Y$ at time T. This portfolio can be exchanged by selling a unit of an asset X for J units of an asset Y for a portfolio with the same price $P_T^2 = (J - K) \cdot Y > 0$. This is clearly an arbitrage opportunity. □

A slightly less obvious arbitrage opportunity is a free lottery ticket. Although in most cases a typical lottery ticket does not win any prize, one is certain not to lose any money and still have a possibility of winning something. That qualifies as an arbitrage opportunity.

Example 1.5

Assume that there is a free lottery ticket L whose price in terms of the dollar \$ is zero: $L_\$(0) = 0$. We have seen in the previous example that having dollars in a portfolio provides an arbitrage opportunity, but let us assume for the purpose of this example that dollars keep their value with respect to bonds in order to illustrate a different kind of arbitrage. The lottery ticket either expires worthless, or it wins N dollars at time T. One can construct the portfolio starting from zero $P_0^0 = 0$, acquiring one zero price lottery ticket, thus creating a portfolio $P_0^1 = L_0$. This portfolio will convert to $P_T^1 = N \cdot \mathbb{I}(\omega = \text{Win}) \cdot \$$, where $\mathbb{I}(\omega = \text{Win})$ is the indicator function of the win. We have that $P_T^1 \geq 0$ for sure, with the possibility of $P_T^1 > 0$. This also constitutes an arbitrage opportunity. □

Another example of an arbitrage opportunity is when the price $X_Y(t)$ of an asset X in terms of an asset Y does not correspond to the price $Y_X(t)$ of an asset Y in terms of an asset X.

***Example 1.6* Arbitrage opportunity when $X_Y(t) \neq Y_X(t)^{-1}$.**
If the relationship

$$X_Y(t) = \frac{1}{Y_X(t)}$$

does not hold, it is possible to realize a risk-free profit. Assume for instance

$$\frac{1}{X_Y(t)} < Y_X(t).$$

In this case, we can start with a unit of an asset Y, and exchange it for $Y_X(t)$ units of an asset X. We can split this position in two parts: $Y_X(t) - X_Y(t)^{-1}$ and $X_Y(t)^{-1}$ units of an asset X. The second part, $X_Y(t)^{-1}$ units of an asset X, can be exchanged back for a unit of an asset Y. This follows from

$$X = X_Y(t) \cdot Y,$$

which is equivalent to

$$Y = X_Y(t)^{-1} \cdot X.$$

Therefore one can generate a certain profit of $Y_X(t) - X_Y(t)^{-1} > 0$ units of an asset X. □

Example 1.7
Assume that $X_Y(t) = 3$, and $Y_X(t) = \frac{1}{2}$. How can one realize a risk free profit? First check that $Y_X(t) = \frac{1}{2} \neq X_Y(t)^{-1}(t) = \frac{1}{3}$. Therefore the prices allow for an arbitrage opportunity. Following the method described in the previous example, we can start with borrowing one unit of Y. Using $Y_X(t) = \frac{1}{2}$, we can immediately exchange the unit of Y for $\frac{1}{2}$ units of X. We can split $\frac{1}{2}$ units of X in two parts, consisting of $\frac{1}{6}$ and $\frac{1}{3}$ units of X. The first part $\frac{1}{6}$ units of X is a net profit from this transaction; the second part can be used for an acquisition and return of a borrowed unit Y using the price relationship $X_Y(t) = 3$. □

Formally, an arbitrage opportunity is defined by:

If one starts with a zero initial portfolio $P_0 = 0$, follows a self-financing strategy, and ends up with $P_T \geq 0$ with probability 1, and has a possible outcome of $P_T > 0$ with positive probability at any given time T, then an arbitrage opportunity is available in the market.

Note that the definition of an arbitrage opportunity does not depend on the choice of the reference asset Y. If $P_Y = 0$ or $P_Y > 0$ for the reference asset Y, then $P_U = 0$ or $P_U > 0$ for any other reference asset U.

1.3 Time Value of Assets, Arbitrage and No-Arbitrage Assets

As stated in the previous section, an asset can either stay the same over time or change over time. In the first case, we say that the asset has **no time value.** Examples of assets that do not change over time include precious metals, a contract to deliver a particular asset in some fixed future time, or a stock that reinvests dividends. One should not confuse the concept of an asset with no time value with the concept of the price of an asset with no time value. For instance an ounce of gold is an asset with no time value, and it does not change over time, but the price of this asset with respect to a dollar may be changing over time.

When the asset is changing over time, we say that the asset has a **time value.** Assets with time value may deteriorate over the passage of time or not. Examples of time value assets that deteriorate over time include currencies, stocks that pay out dividends, and most consumer goods. However, some assets may change over time and not deteriorate, for instance portfolios that actively exchange assets with no time value.

One certainly does not create an arbitrage opportunity by holding an asset that has no time value. On the other hand, assets that have time value may or may not create arbitrage opportunities. It depends if the asset with time value deteriorates (or appreciates) in time or not. If one creates an arbitrage opportunity by holding a given asset, we will call this asset an **arbitrage asset.** If an arbitrage opportunity is not possible by holding a given asset, we call this asset a **no-arbitrage asset.** There is a simple method to determine whether a given asset X is an arbitrage or a no-arbitrage asset. Let V be a contract to deliver a unit of the asset X at some future time T. We can write

$$V_T = X_T.$$

When $V_t = X_t$ at all times $t \leq T$, the asset X is a no-arbitrage asset. When $V_t \neq X_t$ for some $t \leq T$, the asset X is an arbitrage asset.

The identity $V_t = X_t$ means that V, the contract to deliver a unit of an asset X, is identical to the asset X itself. The only way to deliver a no-arbitrage asset is to hold it at all times up to time T. For instance the contract to deliver a stock costs the stock itself, a contract to deliver an ounce of gold costs the ounce of gold (neglecting a possible cost of carry which is close to zero for financial assets). Some hedge funds try to realize arbitrage opportunities even in these primary assets, so it may be hard to tell which asset is a no-arbitrage asset without observing the corresponding contract to deliver. A contract to deliver usually does not exist for a no-arbitrage asset since it coincides with the asset itself, and thus it is completely redundant. However, the nonexistence of the contract to deliver can happen for two reasons: the underlying asset is a no-arbitrage asset, or there is no market for the contract to deliver. This makes it harder to determine whether the asset is a no-arbitrage asset.

Rational investors do not allow any arbitrage opportunities, and thus their portfolios hold only no-arbitrage assets, or arbitrage assets that provide one sided advantage for the investor. If the market has only rational investors, there would be no arbitrage assets at all. For a given asset X, **the contract V to deliver an asset X is always a no-arbitrage asset,** even when the asset X to be delivered is an arbitrage asset. This is easily seen from the following argument. Let U be a contract to deliver the asset V at time T, or in other words, $U_T = V_T$. From the identity $V_T = X_T$, we also have $U_T = X_T$. Thus U is also a contract to deliver X at time T, and therefore U is identical to V. This proves that V, a contract to deliver an asset X at time T, is a no-arbitrage asset. In particular, bonds are no-arbitrage assets.

On the other hand, assets with $V_t \neq X_t$ for some $t < T$ are arbitrage assets. We have either $V_t < X_t$, or $V_t > X_t$. When $V_t < X_t$, it is possible to deliver the asset X at time T at a cheaper price than just holding the asset X itself. The exact procedure to lock the arbitrage opportunity for an arbitrage asset is described in Example 1.8 which follows. When $V_t < X_t$, one should buy a contract to deliver V and sell a corresponding number of units of an asset X.

Arbitrage assets do exist in real markets, mostly representing assets with deteriorating time value (food, consumer goods, banknotes). However, these assets are not typically included in financial portfolios as holding them would create arbitrage opportunities that are not favorable for the holders of such assets. But the arbitrage assets still may appear in the payoffs of financial contracts, such as a contract to deliver a unit of the asset in a fixed future time. We have already seen that a contract to deliver any asset is always a no-arbitrage asset. Such derivative contracts facilitate trading of assets with deteriorating time value. While the underlying asset creates arbitrage opportunities, the contract to deliver does not, and as such it may be included in a financial portfolio that does not deteriorate over time.

Examples of arbitrage assets that appear in such payoffs include certain food products (orange juice, coffee, pork bellies), currencies, or stocks that pay dividends. A stock together with the corresponding dividend payments is a no-arbitrage asset. However, a stock when taken separately without the dividends is an arbitrage asset. Taking away the dividends is an obvious arbitrage opportunity. Another example of an arbitrage asset is an asset that corresponds to a maximum price of an asset X with respect to a reference asset Y defined as $[\max_{0 \leq s \leq t} X_Y(s)] \cdot Y_t$. This asset appears in the payoff of a lookback option, and although it does not exist in the real markets, it can still be used as a reference asset for pricing lookback options.

Arbitrage assets do change over some periods of time; in particular we have

$$\$_t > \$_{t+1}, \tag{1.15}$$

which means that a dollar today is worth more than a dollar tomorrow. Inequality (1.15) is known as the **time value of money.**

Example 1.8 **Arbitrage opportunity created by an arbitrage asset** Let V be a contract that delivers a unit of an asset X at time T, or in other words,

$$V_T = X_T.$$

This equality is written in the sense of two assets, the contract to deliver V has the same price as an asset X at time T. In terms of prices, we can write

$$V_X(T) = 1,$$

which means that the price of the contract to deliver V with respect to the reference asset X is one at time T. When $V_0 < X_0$, we can realize a risk free profit by buying a unit of an asset V, and sell $V_X(0) < 1$ units of an asset X, thus creating a zero price portfolio

$$P_0 = 1 \cdot V - V_X(0) \cdot X.$$

Clearly, $P_X(0) = 1 \cdot V_X(0) - V_X(0) = 0$. This portfolio is kept until time T, when it becomes

$$\begin{aligned} P_T &= 1 \cdot V - V_X(0) \cdot X \\ &= (1 - V_X(0)) \cdot X > 0. \end{aligned}$$

Thus one can get a portfolio with a guaranteed positive price starting from a portfolio with a zero price. □

The most typical examples of arbitrage assets are currencies. Let X be a dollar \$. A contract to deliver a dollar at time T is known as a **bond,** and it is denoted by B^T. The dollar price of the bond is typically less than one ($B^T_\$(0) < 1$), making a dollar an arbitrage asset. In order to lock the risk free profit, one would have to buy a bond B^T, and sell $B^T_\$(0)$ units of a dollar. This means one would have to borrow money to get a short position in dollars, which leads us to the following important remark.

REMARK 1.2 Borrowing money

When one borrows money in terms of a dollar \$, the resulting asset that is owed is not money but rather a money market account M, an interest bearing account. The asset that is borrowed is different from the asset that is owed. In contrast, if one borrows a stock S (in terms of short-selling on the stock exchange), the debt is still the same stock S. The exchange may charge a fee for that, but the asset that is borrowed is the same as the asset that is owed. □

Even governments have to pay interest when borrowing money. The only exception when interest is not paid is when governments issue banknotes. Governments typically have a limited intention to print more banknotes in order to finance their debts, and thus exploration of this arbitrage opportunity is not significant.

1.4 Money Market, Bonds, and Discounting

The fact that currencies have time value means that prices in terms of a dollar may not be consistent in time. This is known as time value of money: A dollar today is worth more than a dollar tomorrow. Thus when one expresses prices of an asset S in terms of a dollar, these prices will have an upward drift component that corresponds to the loss of value of the reference asset.

In order to remove the effect of the depreciation of the reference asset, one can express the price of the asset S in terms of no-arbitrage proxy assets to a dollar, such as a money market M, or a bond B^T. Prices $S_M(t)$ and $S_{B^T}(t)$ are known as **discounted prices** of the asset S.

Recall that the money market M is an asset created by the following procedure. The initial amount equal to one dollar is invested at time $t = 0$ in the bond with the shortest available maturity (ideally in the next infinitesimal instant), and this position is rolled over to the bond with the next shortest maturity once the first bond expires. The resulting no-arbitrage asset, the **money market** M, is a result of an active trading strategy involving a number of these no-arbitrage bonds. The dollar price of the money market is given by

$$\boxed{M_\$(t) = \exp\left(\int_0^t r(u)du\right),} \tag{1.16}$$

where $r(t)$ is a parameter known as the **interest rate**. In practice, the money market asset is replicated as a portfolio of different bonds by banks or investment funds.

Equation (1.16) can be written in a differential form as

$$dM_\$(t) = r(t)M_\$(t)dt. \tag{1.17}$$

The interest rate $r(t)$ can be viewed as a rate of deterioration of an arbitrage asset \$ with respect to a no-arbitrage asset M, the money market account. Since the parameter $r(t)$ is related only to the shortest available bond, in this case B^t, a bond that matures immediately at time t, a simple analog of Equation (1.17) for a bond B^T is not available. Only if we take a simplifying assumption that the interest rate $r(t)$ is **deterministic**, can we also write

$$B_\$^T(t) = \exp\left(-\int_t^T r(u)du\right). \tag{1.18}$$

The reason is that there is only one way to deliver one dollar at time T by investing in the money market account M. If one starts with $\exp\left(-\int_t^T r(u)du\right)$ units of a dollar at time t and invests it in the money market account M, it will be worth

$$\exp\left(-\int_t^T r(u)du\right) \cdot \exp\left(\int_t^T r(u)du\right) = 1$$

unit of a dollar at time T. Therefore the price of the bond B^T at time t must be given by Equation (1.18); otherwise we would have an arbitrage opportunity. In this case, the price of the bond B^T and the price of the money market M are related by the formula

$$B_t^T = \exp\left(-\int_0^T r(u)du\right) \cdot M_t. \tag{1.19}$$

Thus the money market M is just a constant multiple of the bond B^T.

In the case of a deterministic interest rate $r(t)$, we can also write

$$dB_\$^T(t) = r(t)B_\$^T(t)dt, \tag{1.20}$$

which is similar to Equation (1.17). Moreover, when the interest rate is constant, the above relationships lead to

$$M_t = e^{rt} \cdot \$_t, \tag{1.21}$$

$$B_t^T = e^{-r(T-t)} \cdot \$_t, \tag{1.22}$$

and

$$B_t^T = e^{-rT} \cdot M_t. \tag{1.23}$$

The relationship between the money market M and the bond B^T is no longer trivial when the interest rate $r(t)$ is stochastic. In this case, the price of the money market starts at a deterministic value $M_\$(0) = 1$, but at later time t, $M_\$(t)$ will be stochastic in general. On the other hand, the price of the bond $B^T_\$(t)$ is random in general for times $t < T$ before the expiration of the bond, but it becomes one at time T ($B^T_\$(T) = 1$), which is a deterministic value. We study the evolution of bond prices in detail in the chapter on term structure models.

As seen earlier, we can regard both $S_M(t)$ and $S_{B^T}(t)$ as discounted prices of an asset S. When we express the price of S with respect to the money market M using the change of numeraire formula for assets $X = S$, $Y = M$, and $Z = \$$, we get

$$S_M(T) = S_\$(T) \cdot \$_M(T) = \exp\left(-\int_0^T r(u)du\right) \cdot S_\$(T) \le S_\$(T), \tag{1.24}$$

with

$$S_M(0) = S_\$(0) \cdot \$_M(0) = S_\$(0). \tag{1.25}$$

Similarly, when we express the price of S with respect to the bond B^T using the change of numeraire formula for assets $X = S$, $Y = M$, and $Z = \$$, we get

$$S_{B^T}(T) = S_\$(T) \cdot \$_{B^T}(T) = S_\$(T), \tag{1.26}$$

with

$$S_{B^T}(0) = S_\$(0) \cdot \$_{B^T}(0) = \frac{S_\$(0)}{B^T_\$(0)} \ge S_\$(0). \tag{1.27}$$

The two types of discounting are also related by

$$S_{B^T}(t) = S_M(t) \cdot M_{B^T}(t). \tag{1.28}$$

In particular, when the interest rate r is constant, the relation between S_{B^T} and S_M is simply

$$S_{B^T}(t) = e^{rT} \cdot S_M(t). \tag{1.29}$$

The important difference between S_M and S_{B^T} is that the price of S_M agrees with the price $S_\$$ at time $t = 0$, while the price of S_{B^T} agrees with the price $S_\$$ at time T. The reference point for discounting with the money market M is at time $t = 0$, while the reference point for discounting with the bond B^T is at time T. Since typical European-type derivative contracts explained in the next chapter pay off $f(S_\$(T))$ for some function f, discounting with respect to the bond B^T makes more sense as $S_{B^T}(T) = S_\$(T)$.

Bonds usually deliver units of a currency at multiple times until their maturity. However, without loss of generality we consider only bonds with a single delivery time T. A bond B^T that pays one dollar at time T is also known as a zero coupon bond. A bond with multiple delivery times is just a combination of several zero coupon bonds. A zero coupon bond is also a possible choice of a no-arbitrage reference asset.

1.5 Dividends

It is often the case that a stock S pays dividends, making it an arbitrage asset. However, the portfolio $\widetilde{S}$ of the stock and the dividends is a no-arbitrage asset. Let us find the relationship between the dividend-paying stock S and the asset representing the stock plus dividends $\widetilde{S}$.

Consider first the situation when the dividends are paid in discrete times $t_1, t_2, \ldots, t_n$. At the time of the first dividend payment t_1, the stock S splits into two parts; one representing the equity part after the dividend, and one representing the dividend. At the time t_1- immediately before the dividend payment, we have

$$S_{t_1-} = \widetilde{S}_{t_1-}.$$

Assuming that the dividend payment is a fraction $a(t_1) \in (0,1)$ of the stock S taken at time t_1-, we get

$$S_{t_1} = (1 - a(t_1)) \cdot S_{t_1-} = (1 - a(t_1)) \cdot \widetilde{S}_{t_1-} = (1 - a(t_1)) \cdot \widetilde{S}_{t_1}.$$

While the value of the equity S jumps down at the time of the dividend payment, the value of the equity plus dividends $\widetilde{S}$ does not, and thus we have

$$\widetilde{S}_{t_1-} = \widetilde{S}_{t_1}.$$

At the time of the second dividend payment t_2, the dividend is the fraction $a(t_2) \in (0,1)$ of the equity part $S_{t_2-} = (1 - a(t_1)) \cdot \widetilde{S}_{t_2-}$. Thus the stock S satisfies

$$S_{t_2} = (1 - a(t_1)) \cdot (1 - a(t_2)) \cdot \widetilde{S}_{t_2}.$$

Continuing this procedure, we conclude that the stock S and the asset representing the stock plus the dividends $\widetilde{S}$ are related by

$$\boxed{S_{t_n} = \left[\prod_{i=1}^{n} (1 - a(t_i))\right] \cdot \widetilde{S}_{t_n}} \tag{1.30}$$

after n dividend payments.

We can also consider the situation when the stock pays the dividend at the continuous rate. A standard approach is to assume that the relationship between the stock S and the asset representing the stock and the dividends $\widetilde{S}$ is given by

$$\boxed{S_t = \exp\left(-\int_0^t a(s)ds\right) \cdot \widetilde{S}_t,} \tag{1.31}$$

or stated equivalently,

$$d\widetilde{S}_S(t) = a(t)\widetilde{S}_S(t)dt. \tag{1.32}$$

The process $a(t)$ represents the dividend yield.

1.6 Portfolio

This section addresses the following questions: What is a portfolio? What is the price of a portfolio? What is a self-financing trading strategy?

A portfolio is a sum of one's assets

$$\boxed{P_t = \sum_{i=0}^{N} \Delta^i(t) \cdot X^i,} \tag{1.33}$$

where $\Delta^i(t)$ represents how many units of an asset X^i are held at time t.

When $\Delta^i(t) > 0$, we say that the portfolio has a **long position** in the asset X^i. When $\Delta^i(t) < 0$, we say that the portfolio has a **short position** in the asset X^i. When $\Delta^i(t) = 0$, we say that the portfolio has a **neutral position** in the asset X^i.

Note that a portfolio is not a number. A car, a house, paintings, and jewelery are assets that do not take numerical values. Thus a portfolio is a distinct concept from the **price of a portfolio**, the number of units of the reference asset that is required to acquire the entire portfolio. As mentioned earlier, price is relative to the chosen reference asset. If we fix $Y = X^0$ to be the reference asset, the price of a portfolio with respect to the reference asset (numeraire) Y is given by

$$\boxed{P_Y(t) = \sum_{i=0}^{N} \Delta^i(t) \cdot X_Y^i(t).} \tag{1.34}$$

In other words, $P_Y(t)$ is the number of units of the asset Y that one would obtain, should one exchange all assets in one's portfolio for an asset Y at time t.

The individual portfolio position $\Delta^i(t)$ has to be known at time t; it cannot be set in retrospect after observing prices in the future. It is similar to betting in a casino – one first places the stake before observing the outcome of a given game. Mathematically, each $\Delta^i(t)$ has to be a **predictable process,** which means that the portfolio position is set before the market observes the price move. Predictable processes are generated by the processes that have left continuous paths.

A portfolio, P_t, together with prices $X^i_Y(t)$ determine the price of a portfolio $P_Y(t)$. On the other hand, different portfolios may have the same price at a given time t. We assume that one can exchange one's portfolio for any other portfolio that has an equal price at time t. We also assume that all assets in the portfolio are no-arbitrage assets. This procedure of exchanging no-arbitrage assets with equal price is known as a **self-financing trading strategy.** Trading portfolios with equal prices means that no asset is either added or withdrawn from the portfolio without being properly exchanged with a combination of assets of an equal price. Holding only no-arbitrage assets ensures that the resulting portfolio is also a no-arbitrage asset. If the prices of two portfolios are the same with respect to one asset Y, the prices are also the same with respect to any other asset Z. This is easily seen from the change of numeraire formula

$$P_Z(t) = P_Y(t) \cdot Y_Z(t).$$

Since exchanging portfolios with equal price can be done in principle at any given time t, one can have continuously rebalanced portfolios as a result.

Let us give an example of self-financing trading.

Example 1.9 **Self-financing trading**
The portfolio

$$P^1_t = \sum_{i=0}^{N} \Delta^i(t) \cdot X^i$$

can be exchanged for the portfolio

$$P^2_t = \left[\sum_{i=0}^{N} \Delta^i(t) \cdot X^i_Y(t) \right] \cdot Y$$

since the two have the same price. This is easily seen from

$$P^1_Y(t) = \sum_{i=0}^{N} \Delta^i(t) \cdot X^i_Y(t),$$

and

$$P^2_Y(t) = \sum_{i=0}^{N} \Delta^i(t) \cdot X^i_Y(t).$$

Therefore we have

$$P^1_Y(t) = P^2_Y(t).$$

However, the two portfolios are physically different. The first portfolio P^1_t has Δ^i_t units of an asset X^i, for $i = 1, \ldots, N$, while the second portfolio P^2_t has

$\sum_{i=0}^{N} \Delta^i(t) \cdot X_Y^i(t)$ units of an asset Y, and zero positions in the remaining assets. But since they have the same price, they can be exchanged for each other at time t. □

REMARK 1.3 Note that self-financing trading may come with some limitations. For instance in the economy consisting of just two assets X and Y, portfolios of the form

$$P = \Delta^X(t) \cdot X + (P_Y(t) - \Delta^X(t)X_Y(t)) \cdot Y$$

have the same price $P_Y(t)$ with respect to the reference asset Y, where $\Delta^X(t)$ is an arbitrary number. But in reality, one usually cannot take arbitrarily large or arbitrarily small (negative) positions in the underlying assets. These positions are usually bounded. For instance, sometimes it may not be possible to take a short position in a particular asset. The bounds on the portfolio position may depend on a given situation, and they may even be different for different agents (think about credit lines). Therefore it is not clear how to define acceptable portfolio positions in order to reflect the reality of the market. There can be also a physical limit on the number of assets that can be held: some assets are nondivisible, and thus one can have only an integer number of them in a given portfolio.

Another limit is that the price of the portfolio may be required to stay above a certain minimal threshold; otherwise a bankruptcy occurs. An adapted portfolio process $\Delta^i(t)_{i=0}^N$ that guarantees $P_Y(t) \geq L$ for some lower bound L for all t is called **admissible.**

The last concern we mention is continuous trading. The traders in the real markets are allowed to change their portfolio positions rather frequently, but only finitely many times in a given time interval. However, mathematical models in continuous time assume that the portfolio positions can be changed continuously. Such an approach gives realistic results, but one should be careful not to construct portfolios that require an infinite number of trades that are not the result of a limit of discrete trading.

We will not be specific in this text about these limitations since this is not a prime focus of the book, but the reader should be aware of them. □

1.7 Evolution of a Self-Financing Portfolio

Let us discuss how the portfolio can evolve in time, using a self-financing trading strategy. **We also assume that all assets are no-arbitrage as-**

sets; otherwise the portfolio itself is an arbitrage asset. Consider first the discrete time case. Let the portfolio at time k be given by

$$P_k = \sum_{i=0}^{N} \Delta^i(k) \cdot X^i.$$

At time $k+1$, the portfolio will have the same positions Δ_k^i in each asset X^i:

$$P_{k+1} = \sum_{i=0}^{N} \Delta^i(k) \cdot X^i,$$

but since X^i stays the same over time for each $i = 0, 1, \ldots, N$, the portfolios P_k and P_{k+1} are the same, only taken at two different time periods.

While the portfolio remains unchanged, its price with respect to a reference asset may be changing. When we write the difference of the prices of the portfolio taken at two consecutive times k and $k+1$, we get

$$\boxed{P_Y(k+1) - P_Y(k) = \sum_{i=0}^{N} \Delta^i(k) \cdot \left[X_Y^i(k+1) - X_Y^i(k)\right].} \tag{1.35}$$

Note that we can omit the changes in the reference asset $Y = X^0$ since

$$Y_Y(k+1) - Y_Y(k) = 1 - 1 = 0.$$

For example, one ounce of gold in the portfolio will still be one ounce of gold in the portfolio in the next time interval, and its price will stay unchanged if the reference asset is chosen to be gold. Similarly, a particular asset will remain the same in the portfolio, but its price with respect to gold may fluctuate in time.

Equation (1.35) says that the change of the price of the portfolio is explained only by the changes of the prices of individual assets in the portfolio. On the other hand, possible changes in the asset positions Δ_k^i from time k to $k+1$ do not enter this equation. At time $k+1$, the holder of the portfolio is free to exchange his present portfolio for a portfolio that has the same price. If we denote the old portfolio that was inherited from time k by $P_{k+1}^{old} = P_k = \sum_{i=0}^{N} \Delta^i(k) \cdot X^i$, and the newly exchanged portfolio at time $k+1$ by $P_{k+1}^{new} = \sum_{i=0}^{N} \Delta^i(k+1) \cdot X^i$, we have

$$P_Y^{old}(k+1) = P_Y^{new}(k+1).$$

The holder of the portfolio can change his position in the underlying assets X^i from $\Delta^i(k)$ to $\Delta^i(k+1)$ given that the two portfolios under consideration have the same price. It means that

$$\sum_{i=0}^{N} \Delta^i(k) \cdot X_Y^i(k+1) = \sum_{i=0}^{N} \Delta^i(k+1) \cdot X_Y^i(k+1),$$

or in other words,

$$\boxed{\sum_{i=0}^{N} \left[\Delta^i(k+1) - \Delta^i(k)\right] \cdot X_Y^i(k+1) = 0.} \tag{1.36}$$

This is the condition a discretely rebalanced portfolio must satisfy in order to be self-financing. The above identity can be also expressed as

$$\sum_{i=0}^{N} \Big[(\Delta^i(k+1) - \Delta^i(k)) \cdot \left[X_Y^i(k+1) - X_Y^i(k)\right] \\ + (\Delta^i(k+1) - \Delta^i(k)) \cdot X_Y^i(k)\Big] = 0. \tag{1.37}$$

When we consider continuous time models, the above identities will take the following forms. For the evolution of the price of the portfolio, we have

$$\boxed{dP_Y(t) = \sum_{i=0}^{N} \Delta^i(t) \cdot dX_Y^i(t),} \tag{1.38}$$

a continuous analog of Equation (1.35). Similarly, the identity corresponding to Equation (1.37) is

$$\boxed{\sum_{i=0}^{N} \left[(d\Delta^i(t)) \cdot dX_Y^i(t) + (d\Delta^i(t)) \cdot X_Y^i(t)\right] = 0.} \tag{1.39}$$

Indeed, if we applied Ito's formula for the evolution of the price of the portfolio, we would get

$$\begin{aligned} dP_Y(t) &= d\left(\sum_{i=0}^{N} \Delta^i(t) \cdot X_Y^i(t)\right) \\ &= \sum_{i=0}^{N} \left[\Delta^i(t) \cdot dX_Y^i(t) + (d\Delta^i(t)) \cdot dX_Y^i(t) + (d\Delta^i(t)) \cdot X_Y^i(t)\right], \end{aligned}$$

But since the last two terms of the above identity sum to zero from (1.39), we have Equation (1.38).

Example 1.10
Consider a portfolio P that holds $\Delta^X(t) = \left[1 - \frac{t}{T}\right]$ units of an asset X, and $\Delta^Y(t) = \left[\frac{1}{T}\int_0^t X_Y(s)ds\right]$ units of an asset Y at time t, where $t \in [0, T]$. In other words,

$$P_t = \left[1 - \tfrac{t}{T}\right] \cdot X + \left[\tfrac{1}{T}\int_0^t X_Y(s)ds\right] \cdot Y. \qquad (1.40)$$

We can show that this is a self-financing portfolio. The condition of self-financing trading (1.39) reads as

$$(d\Delta^Y(t)) \cdot dY_Y(t) + (d\Delta^Y(t)) \cdot Y_Y(t) \\ + (d\Delta^X(t)) \cdot dX_Y(t) + (d\Delta^X(t)) \cdot X_Y(t) = 0,$$

where we substituted $X^0 = Y$, and $X^1 = X$. Since $Y_Y(t) = 1$, the above relationship simplifies to

$$\boxed{(d\Delta^Y(t)) + (d\Delta^X(t)) \cdot dX_Y(t) + (d\Delta^X(t)) \cdot X_Y(t) = 0.} \qquad (1.41)$$

Note that

$$d\Delta^Y(t) = \tfrac{1}{T}X_Y(t)dt, \\ d\Delta^X(t) = -\tfrac{1}{T}dt,$$

and thus

$$d\Delta^Y(t) + d\Delta^X(t) \cdot dX_Y(t) + d\Delta^X(t) \cdot X_Y(t) = \\ = \tfrac{1}{T}X_Y(t)dt + (-\tfrac{1}{T}dt) \cdot dX_Y(t) + (-\tfrac{1}{T}dt)X_Y(t) = 0.$$

Therefore we have the self-financing evolution of the prices of the portfolio from (1.38). When we choose Y to be the reference asset, we have

$$dP_Y(t) = \Delta^X(t)dX_Y(t) = \left[1 - \tfrac{t}{T}\right] dX_Y(t),$$

when we choose X to be the reference asset, we have

$$dP_X(t) = \Delta^Y(t)dY_X(t) = \left[\tfrac{1}{T}\int_0^t X_Y(s)ds\right] dY_X(t).$$

Note that the portfolio P_t starts with $P_0 = X_0$ and ends with $P_T = \left[\frac{1}{T}\int_0^T X_Y(s)ds\right] \cdot Y_T$. Therefore the above described self-financing strategy delivers $\left[\frac{1}{T}\int_0^T X_Y(s)ds\right]$ units of Y at time T. The number $\left[\frac{1}{T}\int_0^T X_Y(s)ds\right]$ represents the average price of the asset X in terms of the reference asset Y.
□

The trading strategy described in Example 1.10 does not depend on the evolution of the underlying price $X_Y(t)$. Also, $d\Delta^X(t)$ and $d\Delta^Y(t)$ have only a dt term, so $\Delta^X(t)$ and $\Delta^Y(t)$ are smooth. Because of that, the $(d\Delta^X(t)) \cdot dX_Y(t)$ cross term is zero. However, the positions $\Delta^X(t)$ and $\Delta^Y(t)$ in the underlying assets can be even diffusions, such as in the following example. In that case, the $(d\Delta^X(t)) \cdot dX_Y(t)$ cross term may not disappear. The reader should be familiar with stochastic calculus, or return to this example after reading the Chapter 3 Diffusion Models in order to fully appreciate it.

Example 1.11

Assume that an asset price follows geometric Brownian motion

$$dX_Y(t) = \sigma X_Y(t) dW^Y(t),$$

where X and Y are two no-arbitrage assets. Consider a portfolio P_t which is given by

$$P_t = [N(d_+)] \cdot X + [-KN(d_-)] \cdot Y,$$

where

$$N(x) = \int_{-\infty}^{x} \tfrac{1}{\sqrt{2\pi}} \cdot e^{-\frac{y^2}{2}} dy,$$

and

$$d_\pm = \tfrac{1}{\sigma\sqrt{T-t}} \cdot \log\left(\tfrac{X_Y(t)}{K}\right) \pm \tfrac{1}{2}\sigma\sqrt{T-t}.$$

The portfolio P holds $\Delta^X(t) = N(d_+)$ units of an asset X, and $\Delta^Y(t) = -KN(d_-)$ units of an asset Y. It turns out (Exercise 1.5) that this portfolio is indeed self-financing. The self-financing condition is given by

$$\begin{aligned} d\Delta^Y(t) + d\Delta^X(t) \cdot dX_Y(t) + d\Delta^X(t) \cdot X_Y(t) = \\ = -KdN(d_-) + dN(d_+) \cdot dX_Y(t) + dN(d_+) \cdot X_Y(t). \end{aligned}$$

It is not trivial to show that

$$-KdN(d_-) + dN(d_+) \cdot dX_Y(t) + dN(d_+) \cdot X_Y(t) = 0,$$

but it is true. Thus we have

$$dP_Y(t) = N(d_+) dX_Y(t),$$

and

$$dP_X(t) = -KN(d_-) dY_X(t).$$

The portfolio P_t in this example is in fact a hedging portfolio for a European option with a payoff $(X_T - K \cdot Y_T)^+$ in a geometric Brownian motion model.
□

1.8 Fundamental Theorems of Asset Pricing

The general assumption in finance is that the market does not contain arbitrage. If an arbitrage opportunity appears, the market usually corrects itself in a short time period. On the other hand, profitable trading strategies may exist for long periods. Some profitable trading strategies may even come with a risk of a catastrophic loss.

Obviously, the entire theory depends upon the fact that the assets in the portfolio are no-arbitrage assets to start with; otherwise the portfolio is not arbitrage free. The central result of finance theory is the First Fundamental Theorem of Asset Pricing:

THEOREM 1.1 First Fundamental Theorem of Asset Pricing
If there exists a probability measure $\mathbb{P}^Y$ such that the price processes $X_Y(t)$ are $\mathbb{P}^Y$-martingales, where X is an arbitrary no-arbitrage asset, and Y is an arbitrary no-arbitrage asset with a positive price, then there is no arbitrage in the market.

PROOF Let Y be a fixed reference asset. If there is an arbitrage opportunity, one can start with a zero price portfolio $P_Y(0) = 0$ and obtain a portfolio $P_Y(T)$ in the form $P_Y(T) = \xi(\omega)$, where $\xi(\omega)$ is a non-negative random variable with $\mathbb{P}^Y(\xi(\omega) > 0) > 0$. In this case, $P_Y(T)$ cannot be a martingale since $\mathbb{E}^Y[P_Y(T)] > 0 = P_Y(0)$. $\square$

REMARK 1.4 Market interpretation of $\mathbb{P}^Y$

The probability measure $\mathbb{P}^Y$ associated with a no-arbitrage reference asset Y has the following market interpretation. Let A be an event in $\mathcal{F}_T$, which can be viewed as a set of market scenarios ω that satisfy a condition $\omega \in A$. As an example of such an event, consider $A = \{\omega \in \Omega : X_Y(T, \omega) \geq K\}$. This is a set of scenarios where the market price of X with respect to the reference asset Y exceeds a fixed constant K at time T. Each set A from the information set $\mathcal{F}_T$ has some objective probability $\mathbb{P}(A)$. The probability measure $\mathbb{P}$ is known as the **real probability measure.** However, the real probability measure does not play any role in the First Fundamental Theorem of Asset Pricing, and thus its role in pricing financial contracts is limited.

What is relevant to pricing financial contracts is the probability measure $\mathbb{P}^Y$. Imagine that there is a security V that pays off one unit of the asset Y at time T when the scenario ω is in A; otherwise it pays nothing. In mathematical notation,

$$V_T = \mathbb{I}_A(\omega) \cdot Y_T,$$

where $\mathbb{I}$ denotes an indicator function. We can also rewrite the above equation in terms of the prices as

$$V_Y(T) = \mathbb{I}_A(\omega).$$

The contract V is known as an **Arrow–Debreu security**. If we want to find the price of this contract at time $t = 0$, we can use the fact that $V_Y(t)$ is a martingale under the probability measure $\mathbb{P}^Y$. Therefore

$$V_Y(0) = \mathbb{E}^Y[V_Y(T)] = \mathbb{E}^Y[\mathbb{I}_A(\omega)] = \mathbb{P}^Y(A).$$

In terms of the assets, we have

$$V_0 = \mathbb{P}^Y(A) \cdot Y_0.$$

In other words, $\mathbb{P}^Y(A)$ is the initial market price of the contract V in terms of the asset Y. Clearly, delivering a unit of Y at time T for a set of scenarios in A should cost at most a unit of Y at time $t = 0$. So $\mathbb{P}^Y(A)$ indicates what fraction of Y is required to start with in order to deliver the Arrow–Debreu security at time T. The probability $\mathbb{P}^Y$ does not indicate directly how likely is the event A to occur, but rather how costly it is with respect to the asset Y.

When the number of possible scenarios in Ω is finite, we can consider events with a single scenario only, meaning $A = \{\omega\}$. The price corresponding to the Arrow–Debreu security for this event, $\mathbb{P}^Y(\omega)$, is known as an **Arrow–Debreu state price.** This concept generalizes to a countable number of states. When the number of states is not countable, representing a continuous random variable, the Arrow–Debreu state price can be interpreted as a density:

$$\mathbb{P}^Y(A) = \int_{\omega \in A} d\mathbb{P}^Y(\omega).$$

In this situation, $d\mathbb{P}^Y(\omega)$ is known as an **Arrow–Debreu state price density.**

Note that the probability measure $\mathbb{P}^X$ that is associated with a different no-arbitrage reference asset X is in general different from $\mathbb{P}^Y$. The corresponding Arrow–Debreu security U would pay off one unit of an asset X when a scenario ω is in A; otherwise it would pay nothing. In other words,

$$U_T = \mathbb{I}_A(\omega) \cdot X_T.$$

This contract differs from V only in the underlying asset. The initial price of U is given by

$$U_0 = \mathbb{P}^X(A) \cdot X_0.$$

In general, the fraction $\mathbb{P}^Y(A)$ of the asset Y needed for the security V and the fraction $\mathbb{P}^X(A)$ of the asset X needed for the security U will differ.

Consider for instance a geometric Brownian motion model for an asset price

$$X_Y(t) = X_Y(0) \cdot \exp\left(\sigma W^Y(t) - \tfrac{1}{2}\sigma^2 t\right).$$

This model will be studied in detail in Chapter 3 Diffusion Models. Brownian motion and stochastic calculus are also discussed in the Appendix. Let A be the set of scenarios where the terminal price $X_Y(T)$ of the asset ends up below the initial price of the asset $X_Y(0)$, or in other words,

$$A = \{\omega \in \Omega : X_Y(T,\omega) \leq X_Y(0)\}.$$

Let V be the corresponding Arrow–Debreu security that delivers a unit of the asset Y at time T when the asset price $X_Y(T)$ ends up below $X_Y(0)$, and let U be the corresponding Arrow–Debreu security that delivers a unit of the asset X at time T. It turns out that $\mathbb{P}^Y(A) > \frac{1}{2}$, but $\mathbb{P}^X(A) < \frac{1}{2}$. While the mathematical details behind this result are discussed in Chapter 3 on Diffusion Models, intuitively this makes sense. Take as an example X to be a stock market, and Y to be a money market. The set of scenarios in A represents outcomes when the market makes a downturn with respect to the reference asset Y. Should one deliver a unit of Y on the downturn, this happens to cost more than half a unit of Y to start with. But that is not surprising; when the market takes a downturn, the reference asset Y, such as the money market in this case, becomes more expensive to deliver. The reason is that Y has appreciated with respect to X, and thus it takes more than one half units of Y to cover the payoff of the corresponding Arrow–Debreu security. On the other hand, it costs less than half a unit of X to deliver a unit of X on the market downturn. This is also not surprising, since on the downturn, the asset X becomes less valuable, and cheaper to deliver.
□

REMARK 1.5 The inverse statement in the First Fundamental Theorem of Asset Pricing that a no-arbitrage condition implies existence of a martingale measure $\mathbb{P}^Y$ is also true, at least in typical mathematical models. This means no arbitrage implies that prices are martingales with respect to the corresponding probability measure. A proper mathematical statement of this theorem requires a careful definition of an admissible trading strategy. The interested reader should refer to academic literature on this topic. **For practical purposes, it is enough that we start with a martingale evolution of the price.** Furthermore, martingales in continuous time are just combinations of diffusions and jumps, so no other processes (such as a fractional Brownian motion for Hurst index $\neq \frac{1}{2}$) can be considered for a no-arbitrage description of the prices.

We now consider how to determine the probability measure $\mathbb{P}^Y$. One should start with describing the set of possible outcomes Ω that represent the individual scenarios of a price evolution. One can consider discrete time and discrete space models, which are known as tree models (binomial or trinomial tree). Continuous time models can have either continuous paths, which lead to diffusion models, or they can have jumps, which lead to purely discontinuous models. Note that requiring prices to be martingales limits possible types of the price evolution.

A general martingale in continuous time can be written as a sum of a martingale with continuous paths and a purely discontinuous martingale:

$$\mathcal{M}(t) = \mathcal{M}^c(t) + \mathcal{M}^d(t). \tag{1.42}$$

A martingale $\mathcal{M}^d(t)$ is called purely discontinuous if its product with any continuous martingale remains a martingale. For instance, a compensated Poisson process $N(t) - \lambda t$ is a purely discontinuous martingale. Note that a purely discontinuous martingale may have continuous paths. Continuous martingales adapted to a filtration $\mathcal{F}_t^W$ generated by a Brownian motion W are in fact diffusions; they can be represented as stochastic integrals with respect to Brownian motion. Thus

$$\mathcal{M}^c(t) = \mathcal{M}^c(0) + \int_0^t \phi(s)dW(s), \tag{1.43}$$

where $\phi(t)$ is adapted to $\mathcal{F}_t^W$. This result is known as the Martingale Representation Theorem (Theorem A.3). □

The following example lists some possible martingale evolutions of the price.

***Example 1.12* Martingale evolution of the price**

Trinomial Model The price $X_Y(0)$ is assumed to take three possible values in the next time instant: event A – go up to $u \cdot X_Y(0)$ $(u > 1)$, event B – stay the same, or event C – go down to $d \cdot X_Y(0)$ $(d < 1)$.

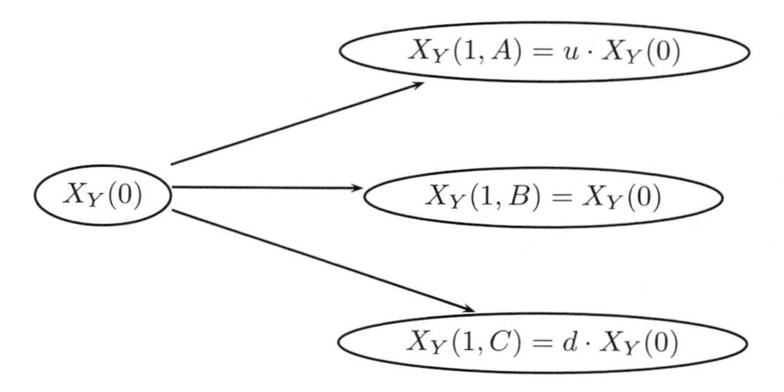

When the probabilities of the events A, B and C are given by

$$\mathbb{P}^{Y,\xi}(A) = \frac{1-d}{u-d} \cdot \xi, \quad \mathbb{P}^{Y,\xi}(B) = 1 - \xi, \quad \mathbb{P}^{Y,\xi}(C) = \frac{u-1}{u-d} \cdot \xi. \quad (1.44)$$

where $\xi \in [0,1]$, the price process $X_Y(n)$ is a martingale. Note that each ξ defines a different probability measure, so in this case there exist infinitely many martingale measures $\mathbb{P}^{Y,\xi}$. One can check that

$$\begin{aligned}\mathbb{E}^{Y,\xi} X_Y(1) &= X_Y(1,A) \cdot \mathbb{P}^{Y,\xi}(A) + X_Y(1,B) \cdot \mathbb{P}^{Y,\xi}(B) \\ &\qquad + X_Y(1,C) \cdot \mathbb{P}^{Y,\xi}(C) \\ &= u \cdot X_Y(0) \cdot \frac{1-d}{u-d} \cdot \xi + X_Y(0) \cdot (1-\xi) \\ &\qquad + d \cdot X_Y(0) \cdot \frac{u-1}{u-d} \cdot \xi \\ &= X_Y(0).\end{aligned}$$

It means that the prices of Arrow–Debreu securities may not be uniquely defined, meaning that there exists a range of the prices when there is no arbitrage present. Consider for instance an Arrow–Debreu security that pays off one unit of Y when the scenario A happens. The initial price of this security is

$$\mathbb{P}^{Y,\xi}(A) = \frac{1-d}{u-d} \cdot \xi,$$

which can be any number in the interval $[0, \frac{1-d}{u-d}]$, depending on the value of the parameter ξ. The market can quote any price in that interval, and there would be no arbitrage opportunity. The question is which martingale measure should one use when there is more than one in order to determine the prices of financial securities? The answer is that it is the market that chooses the martingale measure. For instance, if the market quotes the price of the above mentioned Arrow–Debreu security, it already determines the value of the parameter ξ, thus effectively choosing only one martingale measure.

Binomial Model A binomial model is a special case of a trinomial model with $\xi = 1$. The price either goes up to $u \cdot X_Y(0)$ $(u > 1)$, or goes down to $d \cdot X_Y(0)$ $(0 < d < 1)$.

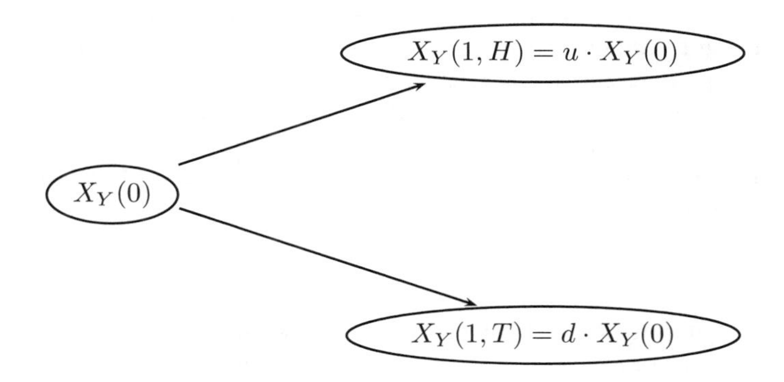

We have a martingale evolution of the price when

$$\mathbb{P}^Y(H) = \frac{1-d}{u-d}, \quad \mathbb{P}^Y(T) = \frac{u-1}{u-d}. \tag{1.45}$$

See Chapter 2 Binomial Models for more details. Note that the martingale measure here is unique.

Geometric Brownian Motion Geometric Brownian motion is a process that satisfies the following stochastic differential equation:

$$dX_Y(t) = \sigma X_Y(t) dW^Y(t). \tag{1.46}$$

The parameter σ is known as the volatility. The above stochastic differential equation admits a closed form solution

$$X_Y(t) = X_Y(0) \cdot \exp\left(\sigma W^Y(t) - \tfrac{1}{2}\sigma^2 t\right), \tag{1.47}$$

which is a martingale. The market noise process, namely Brownian motion $W^Y(t)$, comes with the reference asset Y and determines the martingale measure $\mathbb{P}^Y$. Chapter 3 Diffusion Models studies this model in detail. More information about Brownian motion and stochastic calculus is given in the Appendix.

Geometric Poisson Process Geometric Poisson process satisfies the following stochastic differential equation:

$$dX_Y(t) = (e^\gamma - 1) \cdot X_Y(t-) d(N(t) - \lambda^Y t). \tag{1.48}$$

The price with the above dynamics is given by

$$X_Y(t) = X_Y(0)\exp\left(\gamma \cdot N(t) - (e^{\gamma} - 1)\lambda^Y t\right). \tag{1.49}$$

This is also a martingale process. In contrast to a geometric Brownian motion model, the market noise process $N(t)$ that represents Poisson jumps does not come with a particular asset. However, different assets come with different martingale measures, which is captured by the intensity of jumps λ^Y that comes with a particular reference asset Y. Pricing in the presence of Poisson jumps is studied in Chapter 10 Jump Models.

□

Note that when there is more than one asset with a positive price available, any of them can be used as a reference asset. Consider a situation when both X and Y are no-arbitrage assets with a positive price, and let V be an arbitrary no-arbitrage asset. Then we have that $V_Y(t)$ is a $\mathbb{P}^Y$ martingale, but also $V_X(t)$ is a $\mathbb{P}^X$ martingale. The relationship between martingale measures $\mathbb{P}^Y$ and $\mathbb{P}^X$ is explained in detail in the following text. It turns out that an important assumption is that both prices $X_Y(t)$ and $Y_X(t)$ stay positive, which is a reasonable assumption for primary reference assets that are represented by currencies, stocks, or precious metals. It is possible that even such basic assets may become worthless, in which case the worthless asset cannot be used as a numeraire. For instance when $X = 0$, we still have a well-defined price $X_Y(t) = 0$, but $Y_X(t)$ is not well defined. Note that derivative contracts can have in principle any price, they may even take negative values, but in this case they cannot be used as reference assets.

An example of a situation when we have two assets with a positive price is a foreign exchange market, where X stands for a domestic bond and Y stands for a foreign bond. Bonds serve as no-arbitrage proxies to the respective currencies. However, an asset is domestic relative to a location, and thus Y is a domestic asset and X is a foreign asset for somebody else. Therefore it makes sense to consider the price of the foreign asset in terms of the domestic asset $X_Y(t)$, and vice versa, the price of the domestic asset in terms of the foreign asset $Y_X(t)$.

When the underlying asset is a bond B^T with maturity T, the corresponding $\mathbb{P}^T$ measure is known as a **T-forward measure.** The term **risk-neutral measure** is used when the underlying asset is the money market account M. We will denote the risk-neutral measure by $\mathbb{P}^M$. The risk-neutral measure and T-forward measure coincide when the interest rate evolution is deterministic. The reader should note that the natural choice for the pricing measure for contracts that are settled in money is the T-forward measure which works also in situations of random interest rates. The risk-neutral measure can be used for pricing such contracts only when the interest rate is deterministic. There is no martingale measure $\mathbb{P}^{\$}$ that would correspond to a dollar as a reference asset since the dollar is an arbitrage asset. Other no-arbitrage reference assets have their own martingale measure. When the underlying reference asset is a stock S, the corresponding $\mathbb{P}^S$ measure is known as a **stock measure.**

The price of an arbitrary no-arbitrage asset V can be computed from the First Fundamental Theorem of Asset Pricing, which gives us a stochastic representation of the prices. The theorem states that the prices are martingales under a proper probability measure, and thus their expected value does not change with time. We have the following relationship:

$$V_Y(t) = \mathbb{E}_t^Y \left[V_Y(T)\right], \tag{1.50}$$

where V and Y are two no-arbitrage assets. The symbol $\mathbb{E}_t[.]$ denotes conditional expectation. Rewriting the above relationship in terms of assets, we get

$$\boxed{V = \mathbb{E}_t^Y \left[V_Y(T)\right] \cdot Y.} \tag{1.51}$$

This literally means that V is worth $\mathbb{E}_t^Y \left[V_Y(T)\right]$ units of Y at time t. Note that $\mathbb{E}_t^Y \left[V_Y(T)\right]$ is an $\mathcal{F}_t$ measurable random variable that represents the price $V_Y(t)$. Computing this conditional expectation is a key aspect of pricing financial contracts. The computation can be done in the following ways: finding a closed form solution for a particular contract; using Monte Carlo simulation to estimate the expected value; or by using differential methods to compute the price as explained later in the text.

REMARK 1.6 Computing dollar prices
The First Fundamental Theorem of Asset Pricing does not apply when a dollar is used as a reference asset since it is an arbitrage asset. The dollar prices have to be computed from the change of numeraire formula. Consider a contingent claim V with a payoff at a fixed maturity T. The claim will pay $V_{\$}(T)$ units of a dollar \$ at time T. We can use any no-arbitrage asset Y to compute the price of V using the formula

$$V_Y(t) = \mathbb{E}_t^Y[V_Y(T)].$$

From the change of numeraire formula, we can compute the dollar price of the contract by

$$V_{\$}(t) = V_Y(t) \cdot Y_{\$}(t).$$

A natural no-arbitrage asset to use is the bond B^T that matures at time T. In this case we can write

$$V_{B^T}(t) = \mathbb{E}_t^T[V_{B^T}(T)].$$

Converting to dollar prices by the change of numeraire formula and using the fact that $B^T_\$(T) = 1$, we can also write

$$V_\$(t) = V_{B^T}(t) \cdot B^T_\$(t) = \mathbb{E}^T_t[V_\$(T) \cdot \$_{B^T}(T)] \cdot B^T_\$(t) = \mathbb{E}^T_t[V_\$(T)] \cdot B^T_\$(t).$$

Thus we have

$$\boxed{V_\$(t) = B^T_\$(t) \cdot \mathbb{E}^T_t[V_\$(T)].} \tag{1.52}$$

Equation (1.52) is of central importance in the current literature on derivative pricing. The advantage is that one can immediately obtain the dollar value of a given contingent claim by using the corresponding T-forward measure. Note that the interest rate $r(t)$ does not enter the formula. It appears only indirectly in the price of the bond $B^T_\$(t)$ if we assumed some dependence of this price on the interest rate. However, such a step is not needed as we can get the value of $B^T_\$(t)$ directly from the price quoted on the market.

Another possible choice of a no-arbitrage proxy asset to a dollar is the money market M. We can write

$$V_M(t) = \mathbb{E}^M_t[V_M(T)].$$

Converting to dollar prices, we get

$$V_\$(t) = V_M(t) \cdot M_\$(t) = \mathbb{E}^M_t[V_\$(T) \cdot \$_M(T)] \cdot M_\$(t).$$

We have already seen in Equation (1.16) that $M_\$(t)$ is given by $M_\$(t) = \exp\left(\int_0^t r(s)ds\right)$, and thus the above formula simplifies to

$$V_\$(t) = \mathbb{E}^M_t\left[\exp\left(-\int_t^T r(s)ds\right) \cdot V_\$(T)\right]. \tag{1.53}$$

Equation (1.53) says that "the price of a contingent claim V is the expected value of its discounted payoff under the risk-neutral measure." Some authors use this equation as a starting point of pricing financial contracts, but this method can be safely used only in the case of a deterministic interest rate r. When the interest rate process $r(t)$ is stochastic, which is a typical case in real markets, the random variables $\exp\left(-\int_t^T r(s)ds\right)$ and $V_\$(T)$ that show up in the expectation in (1.53) could be correlated, and the problem of pricing a contingent claim V would have to address the joint distribution of $\exp\left(-\int_t^T r(s)ds\right)$ and $V_\$(T)$. This may not be a trivial task, especially when V itself is an interest rate product.

When the interest rate is deterministic, the discount factor $\exp\left(-\int_t^T r(s)ds\right)$ is also deterministic and thus independent of the payoff $V_\$(T)$. Thus it can be factored out from the expectation, and we have

$$V_\$(t) = \exp\left(-\int_t^T r(s)ds\right) \cdot E^M_t[V_\$(T)].$$

However, in the case of a deterministic interest rate we also have $B_{\$}^T(t) = \exp\left(-\int_t^T r(s)ds\right)$, and we can rewrite Equation (1.52) as

$$V_{\$}(t) = \exp\left(-\int_t^T r(s)ds\right) \cdot E_t^T[V_{\$}(T)].$$

This shows that

$$E_t^M[V_{\$}(T)] = E_t^T[V_{\$}(T)]$$

for an arbitrary claim V, and thus the T-forward measure $\mathbb{P}^T$ and the risk-neutral measure $\mathbb{P}^M$ are the same, but only when the interest rate is deterministic.

When the contingent claim V depends on a stock S, we can also choose S as the reference asset. Converting the price to dollar values, we obtain

$$V = \mathbb{E}_t^S\left[V_S(T)\right] \cdot S = \mathbb{E}_t^S\left[V_S(T)\right] \cdot S_{\$}(t) \cdot \$. \tag{1.54}$$

□

Due to the symmetry of the First Fundamental Theorem of Asset Pricing, a similar formula is valid for the choice of a different reference asset X:

$$V_X(t) = \mathbb{E}_t^X\left[V_X(T)\right], \tag{1.55}$$

or in other words,

$$\boxed{V = \mathbb{E}_t^X\left[V_X(T)\right] \cdot X.} \tag{1.56}$$

This means that V is worth $\mathbb{E}_t^X\left[V_X(T)\right]$ units of X at time t.

Let us illustrate the concept of X being a reference asset on a trinomial model. The cases of a binomial model, a geometric Brownian motion model, and a geometric Poisson model are discussed in detail in the corresponding chapters.

Example 1.13 Trinomial model with X as a reference asset

Recall that the trinomial model assumes the following evolution of the price process. The price can take three different values in one time step. When Y was chosen as a reference asset, the price can go up to $X_Y(1,A) = u \cdot X_Y(0)$ for $u > 1$ (event A), it can stay the same $X_Y(1,B) = X_Y(0)$ (event B), or it can go down to $X_Y(1,C) = d \cdot X_Y(0)$ for $0 < d < 1$ (event C). Let us take X as a reference asset, and let us study the inverse price process Y_X. On event A, the price $Y_X(1)$ is equal to $Y_X(1,A) = \frac{1}{u} \cdot Y_X(0)$. This follows from the relationship between the price X_Y and its inverse price Y_X: $Y_X(t) = X_Y(t)^{-1}$. When the price X_Y goes up (such as in the case of event A), the inverse price Y_X goes down, and vice versa. On event B, the price Y_X stays the same: $Y_X(1,B) = Y_X(0)$. On event C, the price Y_X goes up to $Y_X(1,C) = \frac{1}{d} \cdot Y_X(0)$.

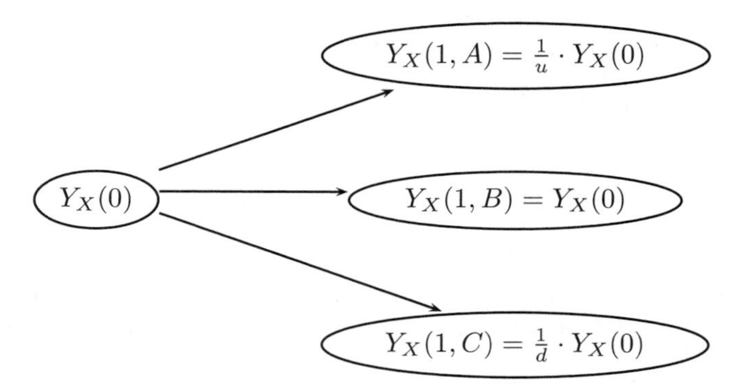

When the probabilities of the events A, B and C are given by

$$\mathbb{P}^{X,\xi}(A) = u \cdot \frac{1-d}{u-d} \cdot \xi, \quad \mathbb{P}^{X,\xi}(B) = 1-\xi, \quad \mathbb{P}^{X,\xi}(C) = d \cdot \frac{u-1}{u-d} \cdot \xi. \tag{1.57}$$

where $\xi \in [0,1]$, the price process $Y_X(n)$ is a martingale. As in the case of Y being a reference asset, we get infinitely many martingale measures $\mathbb{P}^{X,\xi}$, one for each choice of the parameter ξ. One can check that

$$\begin{aligned} \mathbb{E}^{X,\xi} Y_X(1) &= Y_X(1,A) \cdot \mathbb{P}^{X,\xi}(A) + Y_X(1,B) \cdot \mathbb{P}^{X,\xi}(B) + Y_X(1,C) \cdot \mathbb{P}^{X,\xi}(C) \\ &= \tfrac{1}{u} \cdot Y_X(0) \cdot u \cdot \frac{1-d}{u-d} \cdot \xi + Y_X(0) \cdot (1-\xi) \\ &\qquad\qquad + \tfrac{1}{d} \cdot Y_X(0) \cdot d \cdot \frac{u-1}{u-d} \cdot \xi \\ &= Y_X(0). \end{aligned}$$

The probability measure $\mathbb{P}^X$ corresponds to Arrow–Debreu securities that use the asset X as the underlying asset. For instance, a security U that pays off one unit of an asset X when the event A happens has the initial price

$$\mathbb{P}^{X,\xi}(A) = u \cdot \frac{1-d}{u-d} \cdot \xi.$$

The price of U is also not uniquely defined, it can be any number in the interval $[0, u \cdot \frac{1-d}{u-d}]$. □

We have two possible representations of the price of a contract V: it is either $\mathbb{E}_t^Y [V_Y(T)]$ units of an asset Y, or $\mathbb{E}_t^X [V_X(T)]$ units of an asset X. This leads to the following variant of the **change of numeraire formula**

$$\boxed{V = \mathbb{E}_t^Y [V_Y(T)] \cdot Y = \mathbb{E}_t^X [V_X(T)] \cdot X.} \tag{1.58}$$

The reference asset appears in three places in the pricing formula: X – the reference asset; $\mathbb{E}_t^X$ – the conditional expectation that is associated with the reference asset; and X – the discount factor in the payoff function. Thus if one wants to price a contract under a different numeraire Y, one just needs to replace the formula with Y at these three locations.

Note that the probability measure $\mathbb{P}^Y$ in the change of numeraire formula (1.58) may not be unique, and the price $V_Y(t) = \mathbb{E}_t^Y[V_Y(T)]$ of a general contingent claim Y may depend on a particular choice of $\mathbb{P}^Y$. We have seen this situation in the trinomial model. Similarly, the probability measure $\mathbb{P}^X$ may not be unique, and the price $V_X(t) = \mathbb{E}_t^X[V_X(T)]$ may depend on a particular choice of $\mathbb{P}^X$. However, to one particular probability measure $\mathbb{P}^Y$ corresponds one particular probability measure $\mathbb{P}^X$ that agrees on the prices in the sense of the change of numeraire formula (1.58). It turns out that the two measures $\mathbb{P}^Y$ and $\mathbb{P}^X$ are linked by a Radon–Nikodým derivative as we will show in the next section.

Example 1.14 One-to-one correspondence of the probability measures $\mathbb{P}^Y$ and $\mathbb{P}^X$ in the trinomial model

Let us show a one-to-one correspondence of the probability measures $\mathbb{P}^Y$ and $\mathbb{P}^X$ in the trinomial model. Let V be an arbitrary contingent claim. Since the price V_Y is a martingale with respect to $\mathbb{P}^Y$, we can write

$$V_Y(0) = \mathbb{E}^Y[V_Y(1)].$$

This expectation depends on a particular choice of the probability measure $\mathbb{P}^{Y,\xi}$. When we fix a parameter ξ, we get

$$V_Y(0) = \mathbb{E}^{Y,\xi}[V_Y(1)].$$

Note that a different choice of ξ may lead to a different value of $V_Y(0)$. Expanding the expectation, we can also write

$$V_Y(0) = V_Y(1,A)\cdot\mathbb{P}^{Y,\xi}(A) + V_Y(1,B)\cdot\mathbb{P}^{Y,\xi}(B) + V_Y(1,C)\cdot\mathbb{P}^{Y,\xi}(C).$$

Using the change of numeraire formula $V_Y = V_X \cdot X_Y$, the above equality can be rewritten as

$$\begin{aligned} &V_X(0)\cdot X_Y(0) \\ &\quad = V_X(1,A)\cdot X_Y(1,A)\cdot\mathbb{P}^{Y,\xi}(A) + V_X(1,B)\cdot X_Y(1,B)\cdot\mathbb{P}^{Y,\xi}(B) \\ &\qquad\qquad + V_X(1,C)\cdot X_Y(1,C)\cdot\mathbb{P}^{Y,\xi}(C). \end{aligned}$$

After dividing by $X_Y(0)$, we can also write

$$\begin{aligned}
&V_X(0) \\
&\quad = V_X(1,A)\cdot\frac{X_Y(1,A)}{X_Y(0)}\cdot\mathbb{P}^{Y,\xi}(A) + V_X(1,B)\cdot\frac{X_Y(1,B)}{X_Y(0)}\cdot\mathbb{P}^{Y,\xi}(B) \\
&\qquad\qquad + V_X(1,C)\cdot\frac{X_Y(1,C)}{X_Y(0)}\cdot\mathbb{P}^{Y,\xi}(C). \quad (1.59)
\end{aligned}$$

But V_X is a martingale under some probability measure $\mathbb{P}^X$, and thus we have

$$V_X(0) = \mathbb{E}^X[V_X(1)],$$

or

$$V_X(0) = V_X(1,A)\cdot\mathbb{P}^X(A) + V_X(1,B)\cdot\mathbb{P}^X(B) + V_X(1,C)\cdot\mathbb{P}^X(C) \quad (1.60)$$

after expanding the expectation. The prices in (1.59) and (1.60) should agree, so we must have

$$\mathbb{P}^X(\omega) = \frac{X_Y(1,\omega)}{X_Y(0)}\cdot\mathbb{P}^{Y,\xi}(\omega). \quad (1.61)$$

Thus for a particular choice of the martingale measure $\mathbb{P}^{Y,\xi}$ there is a single corresponding measure $\mathbb{P}^X$ given by (1.61) that gives the same prices of contingent claims V. Since

$$\frac{X_Y(1,A)}{X_Y(0)} = u, \qquad \frac{X_Y(1,B)}{X_Y(0)} = 1, \qquad \frac{X_Y(1,C)}{X_Y(0)} = d,$$

the measure $\mathbb{P}^X$ is given by

$$\mathbb{P}^X(A) = u\cdot\mathbb{P}^{Y,\xi}(A), \qquad \mathbb{P}^X(B) = \mathbb{P}^{Y,\xi}(B), \qquad \mathbb{P}^X(C) = d\cdot\mathbb{P}^{Y,\xi}(C). \quad (1.62)$$

It turns out that the probability measure $\mathbb{P}^X$ corresponds to the probability measure $\mathbb{P}^{X,\xi}$ that is given in (1.57). Therefore the price of a contingent claim V would be the same if computed both under $\mathbb{P}^{Y,\xi}$ or under $\mathbb{P}^{X,\xi}$ for a fixed parameter ξ. The relationship between $\mathbb{P}^Y$ and $\mathbb{P}^X$ in a general model is given by the so-called Radon–Nikodým derivative, and it is studied in the next section. □

REMARK 1.7 All martingale measures $\mathbb{P}^Y$ agree on the price of a contract to deliver
A given model of a price evolution may come with infinitely many martingale measures $\mathbb{P}^Y$, and the price of a general contingent claim may depend on the choice of the probability measure $\mathbb{P}^Y$. We have seen this situation in the trinomial model. However, all martingale measures $\mathbb{P}^Y$ agree on a price of a contract to deliver a no-arbitrage asset Y. Let us denote this contract by V,

with $V_T = Y_T$ at the delivery time T. Since V_Y is a martingale, the initial price $V_Y(0)$ is given by

$$V_Y(0) = \mathbb{E}^Y[V_Y(T)] = \mathbb{E}^Y[Y_Y(T)] = \mathbb{E}^Y[1] = 1,$$

and thus $V_0 = Y_0$. This result is independent on the choice of the probability measure $\mathbb{P}^Y$. From the change of numeraire formula (1.58), we would get the same price of the contract V by using any probability measure $\mathbb{P}^X$. Similarly, all martingale measures $\mathbb{P}^X$ and $\mathbb{P}^Y$ agree on the price of a contract to deliver a no-arbitrage asset X. □

In perfectly symmetric situations when the roles of X and Y can be exchanged, it makes sense to study models where $X_Y(t)$ and its inverse price $Y_X(t)$ admit similar evolutions. That would make the observer unable to identify the reference asset just by looking at the price process. More specifically, we can consider the situation when the distribution of the price $\frac{X_Y(T)}{X_Y(0)}$ under the probability measure $\mathbb{P}^Y$ has the same distribution as $\frac{Y_X(T)}{Y_X(0)}$ under the probability measure $\mathbb{P}^X$. When this is the case, we can also write

$$\mathcal{L}^Y\left(\frac{X_Y(T)}{X_Y(0)}\right) = \mathcal{L}^X\left(\frac{Y_X(T)}{Y_X(0)}\right), \tag{1.63}$$

meaning that the laws of the two distributions agree. We will call this principle the **exchangeability of the reference assets.** We show in the following text that it is possible to model the prices of assets in a way that the role of X and Y can be freely exchanged, for instance in the binomial model or in the diffusion model.

Another important question is: when is it possible to replicate a contingent claim V whose payoff depends on underlying assets X^i by trading in these assets? Or in other words, is there a portfolio P of the form

$$P = \sum \Delta^i(t) \cdot X^i$$

such that $P_t = V_t$? We call such a situation a **complete market.** A market is **incomplete** if it is not complete.

THEOREM 1.2 Second Fundamental Theorem of Asset Pricing
A market is complete if and only if the martingale measure $\mathbb{P}^Y$ is unique.

Rule of Thumb: The market is typically complete in situations when the number of different noise factors does not exceed the number of assets minus one asset that serves as a numeraire.

Example 1.15 **Complete models**
Consider a situation when there are just two assets X and Y. The binomial model has one noise factor which can be thought of as a coin toss, and the market is complete. Similarly, the market is complete in a geometric Brownian motion model, where the only source of uncertainty is the Brownian motion. In the case when the asset price has stochastic volatility, there are two noise factors (the original Brownian motion, and stochastic volatility), and the market is incomplete. Jump models are complete only if the jump size takes one single value, such as in a geometric Poisson process which represents the only noise factor. Jump models with multiple jump sizes are incomplete. □

Example 1.16 **Trinomial model**
We have already seen that a trinomial model, with $X_Y(1, A) = u \cdot X_Y(0)$, $X_Y(1, B) = X_Y(0)$, and $X_Y(1, C) = d \cdot X_Y(0)$, where $0 < d < 1 < u$, does not have a unique probability measure $\mathbb{P}^Y$.

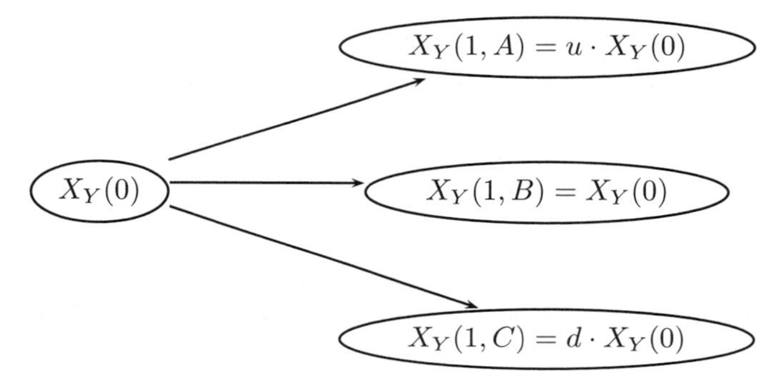

The price process is a martingale when the probability $\mathbb{P}^Y$ is given by

$$\mathbb{P}^{Y,\xi}(A) = \frac{1-d}{u-d} \cdot \xi, \quad \mathbb{P}^{Y,\xi}(B) = 1 - \xi, \quad \mathbb{P}^{Y,\xi}(C) = \frac{u-1}{u-d} \cdot \xi,$$

where $\xi \in [0, 1]$. Here one can think of $\mathbb{P}^Y(A)$ and $\mathbb{P}^Y(B)$ as two sources of uncertainty (or noise factors). The probability of event C, $\mathbb{P}^Y(C)$ is already determined since $\mathbb{P}^Y(C) = 1 - \mathbb{P}^Y(A) - \mathbb{P}^Y(B)$. Let us show that this is indeed an incomplete market. Let V be a contingent claim that pays off $V_Y(1)$ units of Y at time $T = 1$. A hedging portfolio for this claim takes the form

$$P_0 = \Delta^X(0) \cdot X + \Delta^Y(0) \cdot Y. \tag{1.64}$$

If P replicates a contract V, we should have $P_1 = V_1$ for all outcomes A, B, and C. Note that the portfolio P remains unchanged from time $t = 0$ to time $t = 1$, and thus we also have

$$P_1 = \Delta^X(0) \cdot X + \Delta^Y(0) \cdot Y.$$

The identity $P_1 = V_1$ can also be written in terms of the prices as $P_Y(1) = V_Y(1)$. Thus we have three equations, one for each outcome:

$$V_Y(1,A) = \Delta^X(0) \cdot X_Y(1,A) + \Delta^Y(0),$$
$$V_Y(1,B) = \Delta^X(0) \cdot X_Y(1,B) + \Delta^Y(0),$$
$$V_Y(1,C) = \Delta^X(0) \cdot X_Y(1,C) + \Delta^Y(0).$$

However, we have only two unknowns, $\Delta^X(0)$, and $\Delta^Y(0)$ and there is no way to match all three different values of $V_Y(1)$ in general. Since $P_1 = V_1$ cannot be satisfied in general, this model is incomplete.

One way to overcome the incompleteness of the model is to consider more underlying assets that may exist in the real markets, thus completing the model. Let us assume for instance that the market trades an Arrow–Debreu security Z that pays one unit of an asset Y when the outcome A happens. The quote of the price $Z_Y(0)$ already determines the probability measure $\mathbb{P}^{Y,\xi}$ uniquely from the relationship

$$Z_Y(0) = \mathbb{P}^{Y,\xi}(A) = \frac{1-d}{u-d} \cdot \xi,$$

and thus

$$\xi = Z_Y(0) \cdot \frac{u-d}{1-d}.$$

The market becomes complete if we consider a portfolio in the form

$$P_0 = \Delta^X(0) \cdot X + \Delta^Y(0) \cdot Y + \Delta^Z(0) \cdot Z.$$

At time $t = 1$, the portfolio P will remain unchanged. In order to match $P_1 = V_1$ for a general claim V, we must have

$$V_Y(1,A) = \Delta^X(0) \cdot X_Y(1,A) + \Delta^Y(0) + \Delta^Z(0),$$
$$V_Y(1,B) = \Delta^X(0) \cdot X_Y(1,B) + \Delta^Y(0),$$
$$V_Y(1,C) = \Delta^X(0) \cdot X_Y(1,C) + \Delta^Y(0),$$

where we used the fact that $Z_Y(1,\omega) = \mathbb{I}_A(\omega)$. We can always find a solution for $\Delta^X(0)$, $\Delta^Y(0)$, and $\Delta^Z(0)$ that would match the payoff of the contingent claim V.

An alternative way to complete the market with other securities is to change the condition on the hedging portfolio P. Instead of requiring $P_1 = V_1$ which corresponds to a perfect hedge, one may require $P_1 \geq V_1$ which corresponds to a **superhedge.** A superhedging portfolio guarantees that the contractual payoff represented by a claim V is always met, but in some scenarios the resulting portfolio P may have a higher price than V. Unfortunately, it often happens the the superhedging portfolio P has a substantially higher price than the actual claim V. Even the superhedging portfolio that has the smallest initial price $P_Y(0)$ may give unrealistically high prices. For this reason, superhedging is almost never used in practice.

A perfect hedge in the assets X and Y is only possible in two notable situations: either when $V_Y(1) = 1$, or when $V_Y(1) = X_Y(1)$. The first case represents a situation when $V_1 = Y_1$, so the payoff is the asset Y itself. In this case, V becomes a contract to deliver an asset Y, and the corresponding hedge is $\Delta^X(0) = 0$ and $\Delta^Y(0) = 1$. All martingale measures $\mathbb{P}^{Y,\xi}$ do agree that the initial price of this contract is simply $V_0 = Y_0$. The second case represents a situation when $V_1 = X_1$, so the payoff is the asset X itself. The contract V becomes a contract to deliver the asset X, with the initial price $V_0 = X_0$ that is independent of the choice of the martingale measure $\mathbb{P}^{Y,\xi}$ and the corresponding hedge is $\Delta^X(0) = 1$, $\Delta^Y(0) = 0$. ∎

1.9 Change of Measure via Radon–Nikodým Derivative

This section describes the relationship between measures implied by using a different numeraire. Suppose that X is a no-arbitrage reference asset, Y is another no-arbitrage reference asset, and V is a contract to be priced. From the change of numeraire formula, we have

$$V = \mathbb{E}^Y\left[V_Y(T)\right] \cdot Y = \mathbb{E}^X\left[V_X(T)\right] \cdot X. \tag{1.65}$$

Recall that we may have in principle infinitely many different martingale measures $\mathbb{P}^Y$ and $\mathbb{P}^X$, but the change of numeraire formula links one probability measure $\mathbb{P}^Y$ with another probability measure $\mathbb{P}^X$ that agrees with $\mathbb{P}^Y$ on the same prices for an arbitrary claim V.

The two measures $\mathbb{P}^Y$ and $\mathbb{P}^X$ can be also related through a scaling factor $\mathbb{Z}(T)$ in the following sense:

$$\mathbb{E}^Y[V_X(T) \cdot \mathbb{Z}(T)] = \mathbb{E}^X[V_X(T)]. \tag{1.66}$$

Rewriting this equation in integral form

$$\int_\Omega V_X(T,\omega)\mathbb{Z}(T,\omega)d\mathbb{P}^Y(\omega) = \int_\Omega V_X(T,\omega)d\mathbb{P}^X(\omega)$$

which is valid for any integrable random variable $V_X(T,\omega)$, we get the following representation of $\mathbb{Z}(T)$:

$$\mathbb{Z}(T) = \frac{d\mathbb{P}^X}{d\mathbb{P}^Y}. \tag{1.67}$$

In other words,

$$\mathbb{P}^X(A) = \int_A \mathbb{Z}(T,\omega) d\mathbb{P}^Y(\omega), \quad A \in \mathcal{F}. \tag{1.68}$$

Intuitively this represents how much one must increase or decrease the weight placed upon the probability of ω under the $\mathbb{P}^Y$ measure so that one gets the same answer as if one used the $\mathbb{P}^X$ measure to start with. The scaling factor $\mathbb{Z}$ is known as the **Radon–Nikodým derivative**. When the space of outcomes Ω is discrete, Equation (1.68) can be expressed as

$$\mathbb{P}^X(\omega) = \mathbb{Z}(T,\omega) \cdot \mathbb{P}^Y(\omega), \quad \omega \in \Omega. \tag{1.69}$$

We can also consider a reciprocal change of measure

$$\frac{1}{\mathbb{Z}(T)} = \frac{d\mathbb{P}^Y}{d\mathbb{P}^X}, \tag{1.70}$$

meaning that

$$\mathbb{E}^Y[V_Y(T)] = \mathbb{E}^X\left[\frac{V_Y(T)}{\mathbb{Z}(T)}\right]. \tag{1.71}$$

The Radon–Nikodým derivative has the following financial interpretation. We can write

$$\mathbb{E}^X[V_X(T)] \cdot X_0 = \mathbb{E}^Y[V_X(T) \cdot \mathbb{Z}(T)] \cdot X_0 = \mathbb{E}^Y[V_Y(T)] \cdot Y_0,$$

where the first equality results from changing measures, and the second equality comes from the change of numeraire formula. Since this relationship is valid for an arbitrary payoff V, we must have

$$[V_X(T) \cdot \mathbb{Z}(T)] \cdot X_0 = [V_Y(T)] \cdot Y_0,$$

or

$$\boxed{\mathbb{Z}(T) = \frac{d\mathbb{P}^X}{d\mathbb{P}^Y} = \frac{X_Y(T)}{X_Y(0)}.} \tag{1.72}$$

We used that

$$\frac{V_Y(T)}{V_X(T)} = X_Y(T),$$

which follows from the change of numeraire formula. Note that the Radon–Nikodým derivative for the reciprocal change of measure is given by

$$\boxed{\frac{1}{\mathbb{Z}(T)} = \frac{d\mathbb{P}^Y}{d\mathbb{P}^X} = \frac{Y_X(T)}{Y_X(0)},} \tag{1.73}$$

which preserves the symmetry between assets X and Y.

REMARK 1.8 Condition for equivalence of the martingale measures $\mathbb{P}^X$ and $\mathbb{P}^Y$
When both $\mathbb{Z}(T)$ and $\frac{1}{\mathbb{Z}(T)}$ stay positive, the two measures $\mathbb{P}^Y$ and $\mathbb{P}^X$ agree on zero probability events in $\mathcal{F}_T$. When $\mathbb{P}^Y(A) = 0$ for $A \in \mathcal{F}_T$ we also have $\mathbb{P}^X(A) = 0$ and vice versa, $\mathbb{P}^X(A) = 0$ implies $\mathbb{P}^Y(A) = 0$. This follows from the relationships

$$\mathbb{P}^X(A) = \int_A \mathbb{Z}(T, \omega) d\mathbb{P}^Y(\omega),$$

and

$$\mathbb{P}^Y(A) = \int_A \frac{1}{\mathbb{Z}(T, \omega)} d\mathbb{P}^X(\omega).$$

When two probability measures agree on zero probability events in $\mathcal{F}_T$, they are **equivalent.** Thus the probability measures $\mathbb{P}^Y$ and $\mathbb{P}^X$ are equivalent when both prices $X_Y(T)$ and $Y_X(T)$ stay positive. □

REMARK 1.9 The risk-neutral measure $\mathbb{P}^M$ agrees with the T-forward measure $\mathbb{P}^T$ when the interest rate is deterministic
We have already seen that when the interest rate is deterministic, the risk-neutral measure $\mathbb{P}^M$ that comes with the money market account M and the T-forward measure $\mathbb{P}^T$ that comes with the bond B^T that matures at time T give the same prices of contingent claims. This means that the two measures are the same. We can also check this result using the Radon–Nikodým derivative

$$\mathbb{Z}(T) = \frac{d\mathbb{P}^M}{d\mathbb{P}^T} = \frac{M_{B^T}(T)}{M_{B^T}(0)} = \frac{M_\$(T) \cdot \$_{B^T}(T)}{M_\$(0) \cdot \$_{B^T}(0)} = \frac{\exp\left(\int_0^T r(t)dt\right) \cdot 1}{1 \cdot \exp\left(\int_0^T r(t)dt\right)} = 1, \tag{1.74}$$

which implies that

$$\mathbb{P}^M(A) = \int_A \mathbb{Z}(T, \omega) d\mathbb{P}^T(\omega) = \int_A 1 d\mathbb{P}^T(\omega) = \mathbb{P}^T(A), \quad A \in \mathcal{F}. \tag{1.75}$$

Therefore the two martingale measures $\mathbb{P}^T$ and $\mathbb{P}^M$ are the same when the interest rate is deterministic. When the interest rate is stochastic, the Radon–Nikodým derivative becomes

$$\mathbb{Z}(T) = \frac{d\mathbb{P}^M}{d\mathbb{P}^T} = \frac{M_{B^T}(T)}{M_{B^T}(0)} = \frac{M_\$(T) \cdot \$_{B^T}(T)}{M_\$(0) \cdot \$_{B^T}(0)}$$
$$= \frac{\exp\left(\int_0^T r(t)dt\right) \cdot 1}{1 \cdot \$_{B^T}(0)} = \exp\left(\int_0^T r(t)dt\right) \cdot B_\$^T(0),$$

which is no longer one, and the relationship between the risk-neutral measure $\mathbb{P}^M$ and the T-forward measure $\mathbb{P}^T$ is no longer trivial. We study the relationship of $\mathbb{P}^T$ and $\mathbb{P}^M$ in more detail in Chapter 4 Interest Rate Contracts.
□

REMARK 1.10 Radon–Nikodým derivative for conditional expectations
The Radon–Nikodým derivative as described in the above text corresponds to changing the measure at time $t = 0$. However, we can generalize this concept to any time $t \leq T$. From the change of numeraire formula, we have

$$\mathbb{E}_t^X [V_X(T)] \cdot X_t = \mathbb{E}_t^Y [V_Y(T)] \cdot Y_t = \mathbb{E}_t^Y [V_X(T) \cdot X_Y(T)] \cdot Y_t.$$

This can be rewritten as

$$\mathbb{E}_t^X [V_X(T)] = \mathbb{E}_t^Y \left[V_X(T) \cdot \frac{X_Y(T)}{X_Y(t)} \right] = \mathbb{E}_t^Y \left[V_X(T) \cdot \frac{\mathbb{Z}(T)}{\mathbb{Z}(t)} \right].$$

Therefore we have

$$\boxed{\mathbb{E}_t^X [V_X(T)] = \frac{1}{\mathbb{Z}(t)} \cdot \mathbb{E}_t^Y [V_X(T) \cdot \mathbb{Z}(T)]} \tag{1.76}$$

in terms of the original Radon–Nikodým derivative $\mathbb{Z}$. This relationship is known as the **Bayes formula.** □

REMARK 1.11 European call option
A **European call option** is a contract that pays off $(X_T - K \cdot Y_T)^+$ at maturity time T, where K is a constant defined by the contract and is known as the **strike.** Let us denote the European call option contract as V. We can assume that both assets X and Y are no-arbitrage assets. If not, we can consider corresponding no-arbitrage assets that deliver a unit of an asset X, or a unit of an asset Y respectively, at time T. We can rewrite the option payoff as

$$V_T = (X_T - K \cdot Y_T)^+ = \mathbb{I}(X_Y(T) \geq K) \cdot X - K \cdot \mathbb{I}(X_Y(T) \geq K) \cdot Y.$$

The above expression suggests that a European option is simply a combination of two Arrow–Debreu securities, one that pays off a unit of an asset X when $X_Y(T) \geq K$, and one that pays off K units of an asset Y on the same event when $X_Y(T) \geq K$. We have already seen in Remark 1.4 that the initial value of the Arrow–Debreu security that pays off a unit of an asset X when event A happens is $\mathbb{P}^X(A)$ units of an asset X. Similarly, the initial value of the Arrow–Debreu security that pays off a unit of an asset Y when event A happens is $\mathbb{P}^Y(A)$ units of an asset Y. If we consider A to be event $X_Y(T) \geq K$, the value of the European call option at time t is simply

$$\boxed{V_t = \mathbb{P}_t^X(X_Y(T) \geq K) \cdot X - K \cdot \mathbb{P}_t^Y(X_Y(T) \geq K) \cdot Y.} \tag{1.77}$$

The above relationship is known as the **Black–Scholes formula.** Note that deriving the Black–Scholes formula in this form does not require any computation. The question how to determine the probabilities $\mathbb{P}_t^X(X_Y(T) \geq K)$ and

$\mathbb{P}_t^Y(X_Y(T) \geq K)$ more explicitly for more specific martingale models of the price evolution is the subject of following chapters.

Note that the choice of the probability measure $\mathbb{P}^Y$ in situations when there is more than one such measure already determines the corresponding probability measure $\mathbb{P}^X$, and vice versa. The two probability measures must agree on the prices of all contingent claims, and thus they are related by the Radon–Nikodým derivative. This follows from

$$\begin{aligned}
V &= \mathbb{E}_t^Y \left[(X - K \cdot Y)_Y^+(T)\right] \cdot Y \\
&= \mathbb{E}_t^Y \left[X_Y(T) \cdot \mathbb{I}(X_Y(T) \geq K)\right] \cdot Y - \mathbb{E}_t^Y \left[K \cdot Y_Y(T) \cdot \mathbb{I}(X_Y(T) \geq K)\right] \cdot Y \\
&= \mathbb{E}_t^X \left[X_X(T) \cdot \mathbb{I}(X_Y(t) \geq K)\right] \cdot X - K \cdot \mathbb{P}_t^Y(X_Y(T) \geq K) \cdot Y \\
&= \mathbb{P}_t^X(X_Y(T) \geq K) \cdot X - K \cdot \mathbb{P}_t^Y(X_Y(T) \geq K) \cdot Y,
\end{aligned}$$

where we have used the change of numeraire formula

$$\mathbb{E}_t^Y \left[X_Y(T) \cdot \mathbb{I}(X_Y(T) \geq K)\right] \cdot Y = \mathbb{E}_t^X \left[X_X(T) \cdot \mathbb{I}(X_Y(T) \geq K)\right] \cdot X.$$

This shows that the probability measures $\mathbb{P}^X$ and $\mathbb{P}^Y$ are indeed linked by the Radon–Nikodým derivative. $\square$

1.10 Leverage: Forwards and Futures

Leverage is one of the most important concepts of finance. It allows investors to magnify their positions in the underlying assets. Let us consider a situation when an investor believes that the price X_Y of a specific asset X will appreciate in the near future. A straightforward way how to realize the potential profit is to buy the asset X now, and sell it at some subsequent time T. The result of this trading is summarized in Table 1.2. At time $t = 0$, the investor has one unit of an asset X that costs him $X_Y(0)$ units of an asset Y. At time $t = T$, the asset X is sold for $X_Y(T)$ units of an asset Y. Therefore at time $t = T$, the position in the asset X is zero, and the position in the asset Y is $X_Y(T) - X_Y(0)$. The net profit or loss of this trading is thus $X_Y(T) - X_Y(0)$ units of an asset Y. When $X_Y(T) - X_Y(0)$ is positive, this trade results in a net profit, when $X_Y(T) - X_Y(0)$ is negative, this trade results in a net loss.

There is an alternative way to realize this profit or loss by trading in contracts to deliver. Instead of buying the asset X at time $t = 0$, one can buy a contract U that delivers the asset X at time T, and pay for it in terms of a contract V that delivers the asset Y at time T. Consider first the case that X and Y are both no-arbitrage assets. We have seen that a contract to deliver a no-arbitrage asset agrees with the asset itself at all times, so $U_t = X_t$, and

TABLE 1.2: Trading in the asset X.

	Time $t = 0$	Time $t = T$
Asset X	1	0
Asset Y	$-X_Y(0)$	$X_Y(T) - X_Y(0)$

$V_t = Y_t$. Thus it may not be obvious why this approach gives any advantage over the case when the investor trades in the primary assets X and Y. Table 1.3 shows the positions in the assets U, V, X and Y. Note that at time $t = 0$, the investor has zero positions in the assets X and Y. The major advantage in this trade is that the investor does not need to have a short position in the reference asset Y. The choice of Y is typically a money market account. In contrast to the previous case, the investors do not need to decrease their position in the money market by paying $X_Y(0)$ units of an asset Y for a unit of an asset X.

TABLE 1.3: Trading in the contracts to deliver U and V.

	Time $t = 0$	Delivery, $t = T$	Sale of X, $t = T$
Asset U	1	0	0
Asset V	$-X_Y(0)$	0	0
Asset X	0	1	0
Asset Y	0	$-X_Y(0)$	$X_Y(T) - X_Y(0)$

The contract U delivers a unit of an asset X at time T. Similarly, $X_Y(0)$ units of the contract V delivers the corresponding number of units of an asset Y to the counter party of this trade at time T. Furthermore, the holder of the asset X may immediately sell it for $X_Y(T)$ units of an asset Y, resulting in the net profit or loss of $X_Y(T) - X_Y(0)$ units of an asset Y. This is the same as in the case when the asset X was bought at time $t = 0$, and sold at time T.

Developing this idea even further, one can introduce a contract that pays off one unit of an asset X for K units of an asset Y at time T:

$$F_T = X_T - K \cdot Y_T. \tag{1.78}$$

The contract F is known as a **forward.** When X and Y are no-arbitrage assets, the price of the forward contract is given by

$$\begin{aligned} F_Y(t) &= \mathbb{E}_t^Y[X_Y(T) - K \cdot Y_Y(T)] \\ &= \mathbb{E}_t^Y[X_Y(T) - K \cdot Y_Y(T)] = X_Y(t) - K. \end{aligned} \tag{1.79}$$

Thus we have

$$F_t = X_t - K \cdot Y_t$$

at all times $t \leq T$. More generally, the forward can be written as

$$F_t = U_t - K \cdot V_t,$$

where U is a contract that delivers a unit of an asset X, and V is a contract that delivers a unit of an asset Y. This relationship is valid in both cases when assets X and Y are arbitrage or no-arbitrage assets.

The forward price $\text{For}(t, T)$ is the value of K that makes the forward contract F have zero price at time t. It is obvious that

$$\text{For}(t, T) = X_Y(t) \tag{1.80}$$

when X and Y are no-arbitrage assets. Table 1.4 shows that one receives $X_Y(T) - X_Y(0)$ units of an asset Y at time T by buying a forward contract F. The forward contract itself has a zero price at time $t = 0$, and entering this contract does not require any change of positions in the assets X and Y. Since the price of the forward contract F is zero, one can potentially enter an unlimited number of forward contracts at a given time. Although the forward contract should formally deliver a unit of the asset X, it is still typically settled entirely in the asset Y. Thus the number of the forward contracts may exceed the total supply of the asset X. This is indeed the case for many typical assets. For instance there are many more contracts to deliver gold or oil than is physically available. However, these contracts are typically settled in money; the asset itself is delivered only in rare cases.

TABLE 1.4: Trading in the forward contract F.

	Time $t = 0$	Time $t = T$
Asset F	1	0
Asset X	0	0
Asset Y	0	$X_Y(T) - X_Y(0)$

Obviously, entering a huge number of forward contracts comes with a significant risk of a bankruptcy. The contractual payoff $X_Y(T) - X_Y(0)$ at time T can be both positive or negative, and having a substantial number of such contracts may lead to a significant gain, or to a significant loss. In order to prevent the situation that one of the contractual parties fails to meet its obligations, one can split the payoff $X_Y(T) - X_Y(0)$ into a series of daily payments that reflect the change of the price of the forward contract.

Splitting the payoff into a series of payments is done in the following way. Let $0 = t_0 < t_1 < \cdots < t_n = T$ be the times of the payments. One can think about them as days if the payments come on a daily basis. At time $t_0 = 0$, one enters a forward contract $F^{t_0} = X - X_Y(t_0) \cdot Y$ that has a zero price. At time t_1, the price of F^{t_0} will change to

$$F_Y^{t_0}(t_1) = \mathbb{E}_{t_1}^Y[X_Y(T) - X_Y(t_0)] = X_Y(t_1) - X_Y(t_0).$$

In order to make F^{t_0} a zero price contract at time t_1, one should subtract $X_Y(t_1) - X_Y(t_0)$ units of the asset Y from it. This technically creates a new forward contract F^{t_1} that has a zero price at time t_1. The relationship between F^{t_1} and F^{t_0} is the following

$$\begin{aligned} F^{t_1} &= F^{t_0} - [X_Y(t_1) - X_Y(t_0)] \cdot Y \\ &= X - X_Y(t_0) \cdot Y - X_Y(t_1) \cdot Y + X_Y(t_0) \cdot Y = X - X_Y(t_1) \cdot Y. \end{aligned}$$

One can continue this procedure for other times t_k. Table 1.5 shows the result of this procedure between times t_{k-1} and t_k.

TABLE 1.5: Splitting the payments.

	Time $t = t_{k-1}$	Time $t = t_k$
Asset $F^{t_{k-1}}$	1	0
Asset F^{t_k}	0	1
Asset X	0	0
Asset Y	0	$X_Y(t_k) - X_Y(t_{k-1})$

In contrast to the forward contract, this procedure does not wait until its expiration T, but rather settles the changes of the price of the forward contract daily. The forward contract $F^{t_{k-1}}$ from the previous time t_{k-1} is replaced by a new forward contract F^{t_k} at time t_k so that F^{t_k} has a zero price. The difference between the prices of $F^{t_{k-1}}$ and F^{t_k} is settled in the asset Y. At the end of this procedure, one would collect

$$\sum_{k=1}^{n} [X_Y(t_k) - X_Y(t_{k-1})] = X_Y(T) - X_Y(0)$$

units of an asset Y. Splitting the payments is a principle of a contract known as **futures.** A futures contract is defined as a series of payments

$$\boxed{\sum_{k=1}^{n} [\text{Fut}(t_k, T) - \text{Fut}(t_{k-1}, T)] \cdot Y_{t_k}} \tag{1.81}$$

that are settled in the asset Y at the corresponding times t_k. The **futures price** $\mathrm{Fut}(t_m, T)$ is a number that makes the series of the remaining payments

$$\sum_{k=m+1}^{n} [\mathrm{Fut}(t_k, T) - \mathrm{Fut}(t_{k-1}, T)] \cdot Y_{t_k}$$

have a zero price at time t_m. At time $t = T$, $\mathrm{Fut}(T, T)$ agrees with the price $X_Y(T)$, the number of units of an asset Y required to obtain a unit of an asset X.

Let us determine $\mathrm{Fut}(t_m, T)$ when X and Y are two no-arbitrage assets. At time t_{n-1}, the futures contract has only one payment left, namely

$$[\mathrm{Fut}(T, T) - \mathrm{Fut}(t_{n-1}, T)] \cdot Y_T = [X_Y(T) - \mathrm{Fut}(t_{n-1}, T)] \cdot Y_T.$$

If the price of this contract be zero at time t_{n-1}, we must have

$$0 = \mathbb{E}^Y_{t_{n-1}}[X_Y(T) - \mathrm{Fut}(t_{n-1}, T)] = X_Y(t_{n-1}) - \mathrm{Fut}(t_{n-1}, T)$$

from the martingale property of $X_Y(t)$. We conclude that

$$\mathrm{Fut}(t_{n-1}, T) = X_Y(t_{n-1}).$$

Repeating this argument, we obtain

$$\mathrm{Fut}(t_m, T) = X_Y(t_m)$$

at all times t_k. Thus in the case when both assets X and Y are no-arbitrage assets, the forward and the futures price agree:

$$\mathrm{Fut}(t, T) = \mathrm{For}(t, T) = X_Y(t),$$

and futures is the same as the forward contract. However, by splitting the payments, one minimizes the default risk of the counter party.

One can avoid the counter party risk completely by trading such contracts on an exchange. Members of the exchange are required to deposit enough funds to cover for all their potential losses that may happen within one day. This deposit is known as a **margin account.** When the funds in the margin account become critically low, the member receives a margin call, a request to add more funds. If the member fails to do so, his positions are closed. Closing the existing positions does not cost anything as the prices of the futures contracts are set to zero continuously.

The most typical futures contracts are settled in currencies, rather than in a no-arbitrage asset. It slightly changes the situation since we also need to take into account the time value of money. Let us assume that the asset X

is a stock S, and the asset Y is a dollar \$. The futures contract is defined in this case as a series of payments of the following form

$$\sum_{k=1}^{n}[\mathrm{Fut}(t_k, T) - \mathrm{Fut}(t_{k-1}, T)] \cdot \$_{t_k}. \tag{1.82}$$

$\mathrm{Fut}(t_m, T)$ is the value that makes the price of the remaining payments

$$\sum_{k=m+1}^{n} [\mathrm{Fut}(t_k, T) - \mathrm{Fut}(t_{k-1}, T)] \cdot \$_{t_k}.$$

to be zero at time t_m. Equation (1.82) is written in terms of an arbitrage asset \$. However, the investor would immediately convert the dollar position into a position in the money market M. Assume that the price of the money market $M_\$(0)$ starts at one, so we have $M_\$(0) = 1$. From the relationship

$$M_\$(t_k) \cdot \$_{t_k} = M_{t_k},$$

we can write

$$\$_{t_k} = \frac{1}{M_\$(t_k)} \cdot M_{t_k}.$$

The dollar \$ at time t_k can be exchanged for $\frac{1}{M_\$(t_k)}$ number of units of the money market M. Thus the payoff of the futures contract can be reexpressed as

$$\sum_{k=1}^{n}[\mathrm{Fut}(t_k, T) - \mathrm{Fut}(t_{k-1}, T)] \cdot \frac{1}{M_\$(t_k)} \cdot M_{t_k}. \tag{1.83}$$

Note that this makes the money market M a natural reference asset for computing the price of the futures contract. Let us determine $\mathrm{Fut}(t_m, T)$. At the terminal time $t_n = T$, $\mathrm{Fut}(T, T)$ agrees with the dollar price of the stock $S_\$(T)$. At time t_{n-1} the futures contract has only a single payment

$$[\mathrm{Fut}(T,T) - \mathrm{Fut}(t_{n-1}, T)] \cdot \frac{1}{M_\$(T)} \cdot M_T = [S_\$(T) - \mathrm{Fut}(t_{n-1}, T)] \cdot \frac{1}{M_\$(T)} \cdot M_T.$$

Should the price of this payment be zero at time t_{n-1}, we must have

$$\begin{aligned} 0 = \mathbb{E}^M_{t_{n-1}} \left[[S_\$(T) - \mathrm{Fut}(t_{n-1}, T)] \cdot \frac{1}{M_\$(T)} \right] \\ = \frac{1}{M_\$(T)} \cdot \left[\mathbb{E}^M_{t_{n-1}}[S_\$(T)] - \mathrm{Fut}(t_{n-1}, T) \right]. \end{aligned}$$

We have used the fact that the price of the money market account $M_\$(T)$ is already known at the prior time t_{n-1}. The reason is that the interest rate that corresponds to the time interval $[t_{n-1}, t_n]$ is set at time t_{n-1}, so the investor knows the price $M_\$(t_n)$ of the money market account one period ahead. Therefore

$$\text{Fut}(t_{n-1}, T) = \mathbb{E}^M_{t_{n-1}}[S_\$(T)].$$

Repeating this procedure for the previous times t, we get

$$\boxed{\text{Fut}(t, T) = \mathbb{E}^M_t[S_\$(T)].} \tag{1.84}$$

Let us compare the futures price $\text{Fut}(t, T)$ with $\text{For}(t, T)$, the price of the corresponding forward contract. The forward contract F when written on a stock S and a dollar \$ pays off

$$F_T = S_T - K \cdot \$_T.$$

This payoff can be rewritten in terms of a bond B^T that delivers a dollar \$ at time T as

$$F_T = S_T - K \cdot B^T_T.$$

The forward price is a number $\text{For}(t, T)$ that corresponds to a choice of K such that the price of the forward contract F is zero at time T. Thus $\text{For}(t, T)$ satisfies the equation

$$0 = \mathbb{E}^T_t[S_{B^T}(T) - \text{For}(t, T)].$$

The natural choice of the reference asset is a bond B^T. Solving for $\text{For}(t, T)$, we get

$$\boxed{\text{For}(t, T) = \mathbb{E}^T_t[S_\$(T)].} \tag{1.85}$$

We used a simple relationship $S_\$(T) = S_{B^T}(T) \cdot B^T_\$(T) = S_{B^T}(T)$.

Both the futures price $\text{Fut}(t, T)$ and the forward price $\text{For}(t, T)$ are expectations of the terminal price of the stock $S_\$(T)$, but under different probability measures. The futures price is computed under the risk-neutral measure $\mathbb{P}^M$, while the forward price is computed under the T-forward measure $\mathbb{P}^T$. We have already seen that when the interest rate $r(t)$ is deterministic, the two measures agree: $\mathbb{P}^M = \mathbb{P}^T$. In this case, the futures price and the forward price agree.

When the interest rate $r(t)$ is stochastic, the two measures $\mathbb{P}^M$ and $\mathbb{P}^T$ are in general different, and the futures price may be different from the forward price. Let us compute the difference between them:

$$\begin{aligned}
\mathrm{Fut}(0,T) - \mathrm{For}(0,T) &= \qquad (1.86)\\
&= \mathbb{E}^M S_\$(T) - \mathbb{E}^T S_\$(T)\\
&= \mathbb{E}^T \left[S_\$(T) \cdot \frac{M_{B^T}(T)}{M_{B^T}(0)} \right] - \frac{\mathbb{E}^T M_{B^T}(T)}{M_{B^T}(0)} \cdot \mathbb{E}^T S_\$(T)\\
&= B^T_\$(0) \Big[\mathbb{E}^T \left[S_\$(T) \cdot M_\$(T) \right] - \mathbb{E}^T \left[S_\$(T) \right] \cdot \mathbb{E}^T \left[M_\$(T) \right] \Big]\\
&= B^T_\$(0) \cdot \mathrm{cov}^T (S_\$(T), M_\$(T))\\
&= B^T_\$(0) \cdot \mathrm{cov}^T \left(S_\$(T), \exp\left(\textstyle\int_0^T r(t)dt \right) \right).
\end{aligned}$$

Thus the difference between $\mathrm{Fut}(0,T)$ and $\mathrm{For}(0,T)$ is proportional to the covariance between the stock price $S_\$(T)$ and the price of the money market account $M_\$(T)$. The covariance is computed in the T-forward measure $\mathbb{P}^T$ that corresponds to the bond B^T as choice of the reference asset. The price of the money market $M_\$(T)$ is directly related to the interest rate $r(t)$: the higher is the interest rate, the higher is the price of the money market.

When the covariance between $S_\$(T)$ and $M_\$(T)$ is positive, the futures price is higher than the forward price. This can be explained by the following argument. In the scenarios when the stock price $S_\$(T)$ ends up above the initial stock price $S_\$(0)$, the corresponding price of the money market $M_\$(T)$ will also tend to increase more than in the scenarios when the stock price $S_\$(T)$ ends up lower than $S_\$(0)$. This follows from the positive correlation of $S_\$(T)$ and $M_\$(T)$. When the price of the stock goes up, the holder of the futures contract will be receiving a positive cash flow, and this cash flow will tend to earn a higher interest rate $r(t)$ on those scenarios. On the other hand, when the stock goes down, the holder of the futures contract will be receiving a negative cash flow, and this cash flow will tend to earn a lower interest rate $r(t)$ on those scenarios. The fact that the resulting cash flow from the futures contracts earns a favorable interest means that the futures price should be higher than the corresponding forward price. In contrast to the futures contract, the forward contract is settled in one single payment at its maturity time, and thus it cannot benefit from varying interest rate.

The reader should check (Exercise 1.7) that the difference between the futures price and the forward price can also be expressed as

$$\begin{aligned}
\mathrm{Fut}(0,T) - \mathrm{For}(0,T) &= -B^T_\$(0) \cdot \mathrm{cov}^M (S_\$(T), \tfrac{1}{M_\$(T)})\\
&= -B^T_\$(0) \cdot \mathrm{cov}^M \left(S_\$(T), \exp\left(- \textstyle\int_0^T r(t)dt \right) \right)
\end{aligned}$$

if we use the risk-neutral measure $\mathbb{P}^M$ that corresponds to the money market M as a reference asset. The idea is to follow the computation in (1.86), but apply the change of measure from $\mathbb{P}^T$ to $\mathbb{P}^M$ in the third line of the equation.

References and Further Reading

The concept of Arrow–Debreu securities traces back to Arrow and Debreu (1954). The idea of pricing financial securities by a no-arbitrage argument was already present in the papers of Black and Scholes (1973) and Merton (1973). However, the theory of pricing under the martingale measure was fully developed only in later papers of Harrison and Kreps (1979) and Harrison and Pliska (1981). Different formulations of the absence of arbitrage and the existence of a martingale measure appear in Delbaen and Schachermayer (1996). An extensive survey of the mathematics of arbitrage appears in the monograph Delbaen and Schachermayer (2006).

The fact that one may use different reference assets for pricing appeared early in the relevant literature. Margrabe (1978) was the first to use a stock measure for pricing an exchange option, a contract written on two stocks. Jamshidian (1989) used bonds as a numeraire in pricing problems of the fixed income markets, introducing the T-forward measure. A more systematic theory of the change of numeraire was developed in Geman et al. (1995). The change of numeraire is now a mainstream technique used in the finance theory, as illustrated in the papers of Gourieroux et al. (1998), Long (1990), Papell and Theodoridis (2001), Brekke (1997), Flemming et al. (1977), Schroder (1999), Platen (2006), Johansson (1998), Karatzas and Kardaras (2007), Platen (2004), and Filipovic (2008). The distinction between the forward and futures contracts was pointed out by Margrabe (1976) and Black (1976).

There are many additional books on quantitative finance that may be useful for the reader who is interested in a more thorough study of the field. An overview of many financial products is given in Hull (2008), which also serves as an introduction to option pricing for practitioners. The book by Baxter and Rennie (1996) serves as a very intuitive introduction to contingent pricing in continuous time. An incomplete list of quantitative finance monographs includes Shreve (2004b), Merton (1992), Bjork (2004), Musiela and Rutkowski (2008), Duffie (2001), Dana and Jeanblanc (2007), Jeanblanc et al. (2009), Karatzas and Shreve (2001), Shiryaev (1999), Joshi (2008), Wilmott (2006), Cerny (2009), or Neftci (2008).

Exercises

1.1 Assume that the change of numeraire formula does not hold at time t, and we have $X_Y(t) < X_Z(t) \cdot Z_Y(t)$. Show how to make a risk-free profit by trading in assets X, Y, and Z.

1.2 Consider a dividend paying stock S in a continuous dividend payment model given by Equation (1.32). Assume that the price the asset representing the stock plus the dividends $\widetilde{S}$ with respect to the money market M follows geometric Brownian motion

$$d\widetilde{S}_M(t) = \sigma \widetilde{S}_M(t) dW^M(t).$$

Determine

$$dS_M(t).$$

Hint: Use $S_M(t) = S_{\widetilde{S}}(t) \cdot \widetilde{S}_M(t)$ and apply the product rule (Remark A.4).

1.3 Let X and Y be two no-arbitrage assets. Determine whether the following portfolios are self-financing or not:
(a) $P_t = [\max_{0 \le s \le t} X_Y(s)] \cdot Y_t$ (portfolio representing the running maximum)
(b) $P_t = \left[\frac{1}{t}\int_0^t X_Y(s)ds\right] \cdot Y_t$ (portfolio representing the running average).

1.4 Show that the price of a discretely rebalanced portfolio $P_k = \sum_{i=1}^N \Delta^i(k) \cdot X^i$ in terms of the reference asset Y is a $\mathbb{P}^Y$ martingale, where X^i are no-arbitrage assets.

1.5 Assume that the asset price follows geometric Brownian motion

$$dX_Y(t) = \sigma X_Y(t) dW^Y(t),$$

where X and Y are two no-arbitrage assets. Show that a portfolio P_t which is given by

$$P_t = N(d_+) \cdot X - KN(d_-) \cdot Y,$$

where

$$N(x) = \int_{-\infty}^{x} \frac{1}{\sqrt{2\pi}} \cdot e^{-\frac{y^2}{2}} dy,$$

and

$$d_\pm = \frac{1}{\sigma\sqrt{T-t}} \cdot \log\left(\frac{X_Y(t)}{K}\right) \pm \frac{1}{2}\sigma\sqrt{T-t},$$

is self-financing. Show that the portfolio P_t is self-financing when

$$X_Y(t) dN(d_+) + dX_Y(t) dN(d_+) - KdN(d_-) = 0,$$

and prove this relationship using the Ito's formula.

1.6 Assume a geometric Brownian motion model for $X_Y(t)$. Determine which of the following portfolios are self-financing:
(a) $P_t = N(d_-) \cdot Y$.
(b) $P_t = \left[\phi(d_-) \cdot \frac{1}{\sigma\sqrt{T-t}} \cdot Y_X(t)\right] \cdot X + \left[N(d_-) - \phi(d_-) \cdot \frac{1}{\sigma\sqrt{T-t}}\right] \cdot Y$.
(c) $P_t = N(d_+) \cdot X$.
(d) $P_t = \left[N(d_+) + \phi(d_+) \cdot \frac{1}{\sigma\sqrt{T-t}}\right] \cdot X + \left[-\phi(d_+) \cdot \frac{1}{\sigma\sqrt{T-t}} \cdot X_Y(t)\right] \cdot Y$.

1.7 Show that the difference between the futures price and the forward price satisfy

$$\begin{aligned}\mathrm{Fut}(0,T) - \mathrm{For}(0,T) &= -B_{\$}^T(0) \cdot \mathrm{cov}^M(S_{\$}(T), \tfrac{1}{M_{\$}(T)}) \\ &= -B_{\$}^T(0) \cdot \mathrm{cov}^M\left(S_{\$}(T), \exp\left(-\int_0^T r(t)dt\right)\right).\end{aligned}$$

Chapter 2

Binomial Models

Binomial models for the price evolution assume that the price of an asset X in terms of a reference asset Y in the next time instant will take only two possible values – an uptick or a downtick. These models are typically too simple to capture market reality. The main reason to include them in this book is to illustrate the fundamental concepts of derivative pricing in a simple model. We will later extend our analysis to more complex models, in particular to diffusions, and to models with jumps. In this chapter we show how to apply the First Fundamental Theorem of Asset Pricing on contracts written on two assets X and Y. Both assets X and Y can be used as reference assets for pricing European option contracts. We show how the pricing martingale measures that come with the assets X and Y are related, using both the basic martingale principles, and their relationship through the Radon–Nikodým derivative. The approach of using both reference assets X and Y is novel; most of the current literature uses only one reference asset, typically represented by a money market account, to price derivative contracts. In particular, we give two alternative characterizations of the price of a contingent claim using both reference assets.

This chapter has two main parts. The first part deals with pricing models with no-arbitrage assets X and Y. This applies to pricing all European-type options. Even when a European option is written on one or two arbitrage assets, such as in the case of stock options, or on foreign exchange options, it is always possible to substitute the arbitrage asset with the corresponding contract to deliver, which is itself a no-arbitrage asset. For example, a stock option is written on a stock and on a dollar, and the dollar can be substituted with a zero coupon bond. Thus we have a martingale evolution of the price to start with, and we can show how to price and hedge European-type contracts with respect to both assets X and Y. We also illustrate via the contract representing the average asset that any no-arbitrage asset has its own martingale measure.

The second part of this chapter studies the case when one of the assets is an arbitrage asset and the contract is an American option. An American option is similar to the European option, but the payoff can be collected at any time up to the maturity of the contract. In this case the arbitrage asset that enters a given American contract cannot be substituted with a no-arbitrage asset. This happens, for instance, for American stock options whose underlying assets are a stock S and the dollar \$. Thus we also have to consider a no-arbitrage proxy asset for the dollar \$ that is used for hedging such an option, a corresponding zero coupon bond B^T. Therefore American option pricing uses three assets: S, B^T and \$.

2.1 Binomial Model for No-Arbitrage Assets

This section applies to a European-type contract V when the payoff is either $V_Y(T)$ units of an asset Y, or equivalently when the payoff is $V_X(T)$ units of an asset X. The prices V_Y and V_X with respect to two different reference assets are linked by the change of numeraire formula $V_Y(t) = V_X(t) \cdot X_Y(t)$. We have already seen that we can assume that both X and Y are no-arbitrage assets. If not, we can replace the arbitrage asset with a corresponding no-arbitrage asset that delivers X or Y at time T. Thus we can directly apply the First Fundamental Theorem of Asset Pricing, using both assets X and Y. Then we have

$$V = \mathbb{E}_t^Y[V_Y(T)] \cdot Y \tag{2.1}$$

if we use Y as a reference asset Y, and

$$V = \mathbb{E}_t^X[V_X(T)] \cdot X \tag{2.2}$$

if we use X as a reference asset for all $0 \leq t \leq T$. This section determines the probability measures $\mathbb{P}^Y$ and $\mathbb{P}^X$ and the hedging portfolio for a general European option. Both measures $\mathbb{P}^Y$ and $\mathbb{P}^X$ are relevant for pricing derivative contracts. For instance the Black–Scholes formula (1.77) is expressed in terms of the two martingale measures that correspond to assets X and Y.

When the contract is a European stock option, the two no-arbitrage assets are a stock S and a bond B^T. Most of the current literature uses a dollar \$ instead of the bond B^T, but this only increases the dimensionality of the problem to three assets, a step that is not necessary for pricing European options. An option must be hedged in no-arbitrage assets only, which are the stock S and the bond B^T in this case, so the dollar \$ becomes an extra asset. The dollar prices of derivative contracts can be obtained simply by the change of numeraire formula from their prices expressed in terms of the stock S or the bond B^T as opposed to computing them from the pricing model that involves all three assets S, B^T, and \$.

Consideration of all three assets S, B^T and \$ is needed only for American options, and such a model will be described in the following section. The reason is that the intrinsic value of the American option is expressed in terms of the dollar \$, and it cannot be replaced by one corresponding no-arbitrage proxy asset.

2.1.1 One-Step Model

Consider the following one step model for the evolution of an asset price where X and Y are two no-arbitrage assets. At time $n = 0$, the price is $X_Y(0)$. At time $n = 1$, the price is either at $X_Y(1,H) = u \cdot X_Y(0)$, or at $X_Y(1,T) = d \cdot X_Y(0)$, where $u > d$.

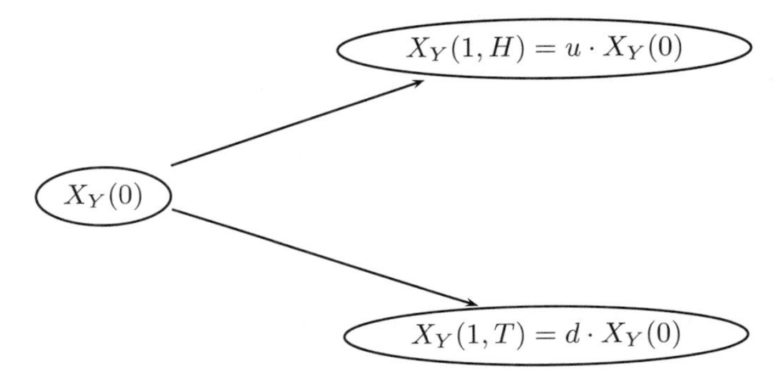

An obvious no-arbitrage condition is given by

$$\boxed{0 < d < 1 < u.}$$

Given that there is no arbitrage, there is a probability measure $\mathbb{P}^Y$ associated with this model such that the price process $X_Y(t)$ is a $\mathbb{P}^Y$-martingale. This is a direct consequence of the First Fundamental Theorem of Asset Pricing. The probability measure $\mathbb{P}^Y$ is determined by $p^Y(u,d) = \mathbb{P}^Y(H)$ and depends on the parameters u and d. Using the martingale property, we must have

$$\begin{aligned} X_Y(0) = \mathbb{E}^Y\left[X_Y(1)\right] &= p^Y(u,d) \cdot X_Y(1,H) + (1 - p^Y(u,d)) \cdot X_Y(1,T) \\ &= p^Y(u,d) \cdot u \cdot X_Y(0) + (1 - p^Y(u,d)) \cdot d \cdot X_Y(0). \end{aligned}$$

Therefore

$$\boxed{\mathbb{P}^Y(H) = p^Y(u,d) = \frac{1-d}{u-d}, \qquad \mathbb{P}^Y(T) = q^Y(u,d) = \frac{u-1}{u-d},} \tag{2.3}$$

where $q^Y = 1 - p^Y = \mathbb{P}^Y(T)$. When $0 < d < 1 < u$, then clearly $0 < p^Y < 1$, so the corresponding $\mathbb{P}^Y$ measure is well defined.

If we want to compute the price of the contingent claim V using Y as a reference asset, we can use the following formula

$$V_Y(0) = \mathbb{E}^Y[V_Y(1)] = V_Y(1,H)\cdot\mathbb{P}^Y(H) + V_Y(1,T)\cdot\mathbb{P}^Y(T) \qquad (2.4)$$
$$= V_Y(1,H)\cdot\frac{1-d}{u-d} + V_Y(1,T)\cdot\frac{u-1}{u-d}. \qquad (2.5)$$

Recall that the probability $\mathbb{P}^Y(A)$ of an event A does not represent the real odds of this event $\mathbb{P}(A)$, but rather how costly the event is in terms of the asset Y. Consider for instance an Arrow–Debreu security V that pays off one unit of Y when $\omega = H$:

$$V_1 = \mathbb{I}(\omega = H)\cdot Y_1. \qquad (2.6)$$

The price of this security is

$$V_Y(0) = \mathbb{E}^Y\left[V_Y(1)\right] = \mathbb{E}^Y\left[\mathbb{I}(\omega = H)\cdot Y_Y(1)\right] = \mathbb{P}^Y(H), \qquad (2.7)$$

where we use the fact that the price process $V_Y(t)$ is a $\mathbb{P}^Y$ martingale. In order to deliver an asset Y on the event $\omega = H$, one should charge $\mathbb{P}^Y(H)$ units of Y to start. Exercise 2.1 asks for one to construct a hedging portfolio for this contract.

Because of the symmetry in the statement of the First Fundamental Theorem of Asset Pricing, we can also use X as a reference asset. In particular, the no arbitrage condition is equivalent to the existence of a probability measure $\mathbb{P}^X$ such that the price process $Y_X(t)$ is a martingale.

Let us study the inverse price Y_X in the binomial model. Assuming the same dynamics as above,

$$Y_X(1,H) = \frac{1}{X_Y(1,H)} = \tfrac{1}{u}\cdot Y_X(0).$$

This is a down tick for Y_X, but an up tick for X_Y. Similarly,

$$Y_X(1,T) = \frac{1}{X_Y(1,T)} = \tfrac{1}{d}\cdot Y_X(0),$$

which is an up tick for Y_X, but a down tick for X_Y.

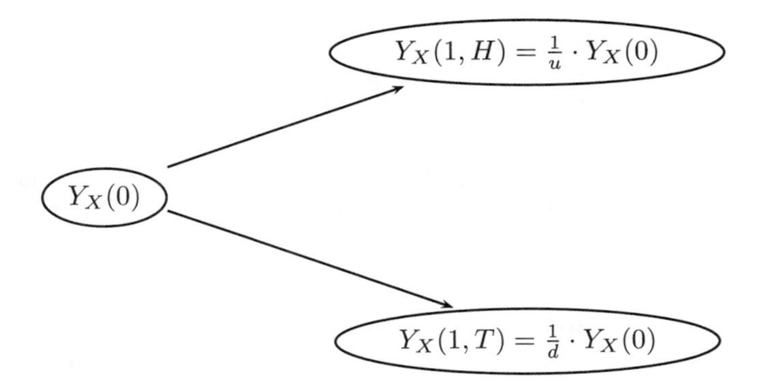

The reference asset X will imply a different probability measure $\mathbb{P}^X$ that is determined by $p^X(u,d) = \mathbb{P}^X(H)$. This probability also depends on the parameters u and d. We can determine $p^X(u,d)$ from the martingale property

$$\begin{aligned} Y_X(0) = \mathbb{E}^X\left[Y_X(1)\right] &= p^X(u,d) \cdot Y_X(1,H) + (1 - p^X(u,d)) \cdot Y_X(1,T) \\ &= p^X(u,d) \cdot \tfrac{1}{u} \cdot Y_X(0) + (1 - p^X(u,d)) \cdot \tfrac{1}{d} \cdot Y_X(0). \end{aligned}$$

Solving for p^X and $q^X = 1 - p^X$, we get

$$\boxed{\mathbb{P}^X(H) = p^X(u,d) = u \cdot \frac{1-d}{u-d}, \qquad \mathbb{P}^X(T) = q^X(u,d) = d \cdot \frac{u-1}{u-d}.} \tag{2.8}$$

Note that the up tick factor u in $X_Y(1,H) = u \cdot X_Y(0)$ has a corresponding down tick factor $\frac{1}{u}$ in $Y_X(1,H) = \frac{1}{u} \cdot Y_X(0)$, and the down tick factor d in $X_Y(1,T) = dX_Y(0)$ has a corresponding up tick factor $\frac{1}{d}$ in $Y_X(1,T) = \frac{1}{d} \cdot Y_X(0)$. Therefore the following symmetry relationship holds:

$$p^X(u,d) = p^Y(\tfrac{1}{u}, \tfrac{1}{d}) = \frac{1 - \frac{1}{d}}{\frac{1}{u} - \frac{1}{d}}, \qquad q^X(u,d) = q^Y(\tfrac{1}{u}, \tfrac{1}{d}) = \frac{\frac{1}{u} - 1}{\frac{1}{u} - \frac{1}{d}}. \tag{2.9}$$

If we want to price a contingent claim V in terms of a reference asset X, we can do so from the following relationship:

$$V_X(0) = \mathbb{E}^X[V_X(1)] = V_X(1,H) \cdot \mathbb{P}^X(H) + V_X(1,T) \cdot \mathbb{P}^X(T) \tag{2.10}$$

$$= V_X(1,H) \cdot u \cdot \frac{1-d}{u-d} + V_X(1,T) \cdot d \cdot \frac{u-1}{u-d}. \tag{2.11}$$

As in the case of the martingale measure $\mathbb{P}^Y$, the martingale measure $\mathbb{P}^X$ represents how costly individual events are in terms of the reference asset X. Consider an Arrow–Debreu security U that pays off one unit of an asset X on the event $\omega = H$:

$$U_1 = \mathbb{I}(\omega = H) \cdot X_1. \tag{2.12}$$

The initial value of this contract is given by

$$U_X(0) = \mathbb{E}^X\left[U_X(1)\right] = \mathbb{E}^X\left[\mathbb{I}(\omega = H) \cdot X_X(1)\right] = \mathbb{P}^X(H). \tag{2.13}$$

In order to deliver a unit of an asset X on the event $\omega = H$, one should charge $\mathbb{P}^X(H)$ units of X to start with. Constructing the hedging portfolio for this claim is the subject of Exercise 2.1.

REMARK 2.1 Radon–Nikodým derivative
Consider a one-step binomial model with tick factors u and d, and corresponding assets X and Y. We have seen that the measure $\mathbb{P}^Y$ that corresponds to the reference asset Y is given by

$$\mathbb{P}^Y(H) = p^Y(u,d) = \frac{1-d}{u-d}, \qquad \mathbb{P}^Y(T) = q^Y(u,d) = \frac{u-1}{u-d}.$$

We can also determine the $\mathbb{P}^X$ measure that corresponds to the reference asset X by using the Radon–Nikodým derivative. From (1.69) we have

$$\mathbb{P}^X(\omega) = \mathbb{Z}(1,\omega) \cdot \mathbb{P}^Y(\omega) \tag{2.14}$$

for $\omega = H, T$. But according to the financial representation of $\mathbb{Z}$ from (1.72), we also have

$$\mathbb{P}^X(\omega) = \mathbb{Z}(1,\omega) \cdot \mathbb{P}^Y(\omega) = \frac{X_Y(1,\omega)}{X_Y(0)} \cdot \mathbb{P}^Y(\omega), \tag{2.15}$$

leading to

$$\mathbb{P}^X(H) = \frac{X_Y(1,H)}{X_Y(0)} \cdot \mathbb{P}^Y(H) = u \cdot \frac{1-d}{u-d}, \tag{2.16}$$

and

$$\mathbb{P}^X(T) = \frac{X_Y(1,T)}{X_Y(0)} \cdot \mathbb{P}^Y(T) = d \cdot \frac{u-1}{u-d}. \tag{2.17}$$

This confirms the formulas for $\mathbb{P}^X(\omega)$ we have previously obtained by using the basic martingale property. □

Note that the probability measure $\mathbb{P}^Y$ under which the price process $X_Y(t)$ is a martingale is inherently tied to the numeraire asset Y. When pricing a contract under the probability measure $\mathbb{P}^Y$, one computes how many units of asset Y are needed in order to settle this contract. Other reference assets imply different probabilities since the number of those assets needed to settle the same contract is, in general, different.

2.1.2 Hedging in the Binomial Model

Let us assume that there is a contingent claim with a given payoff V at time $T = 1$ defined on both outcomes $V(1,H)$ and $V(1,T)$. It is possible to find a portfolio

$$P_0 = \Delta^X(0) \cdot X + \Delta^Y(0) \cdot Y,$$

such that

$$P_Y(0) = V_Y(0),$$

and

$$P_Y(1,H) = V_Y(1,H) \qquad P_Y(1,T) = V_Y(1,T). \tag{2.18}$$

Let us find $\Delta^X(0)$, the number of asset X to be held at time $n = 0$. Since

$$P_1 = \Delta^X(0) \cdot X + \Delta^Y(0) \cdot Y,$$

we also have

$$P_Y(1,H) = \Delta^X(0) \cdot X_Y(1,H) + \Delta^Y(0), \tag{2.19}$$

and

$$P_Y(1,T) = \Delta^X(0) \cdot X_Y(1,T) + \Delta^Y(0). \tag{2.20}$$

Subtracting Equation (2.20) from Equation (2.19), and using the fact that the price of the portfolio P at time $n = 1$ should match the price of the payoff V (Equation (2.18)), we get

$$V_Y(1,H) - V_Y(1,T) = \Delta^X(0) \cdot (X_Y(1,H) - X_Y(1,T)),$$

or in other words,

$$\boxed{\Delta^X(0) = \frac{V_Y(1,H) - V_Y(1,T)}{X_Y(1,H) - X_Y(1,T)}.} \tag{2.21}$$

Similarly, the position $\Delta^Y(0)$ in the asset Y can be determined from the following equations:

$$P_X(1,H) = \Delta^X(0) + \Delta^Y(0) \cdot Y_X(1,H), \tag{2.22}$$

$$P_X(1,T) = \Delta^X(0) + \Delta^Y(0) \cdot Y_X(1,T). \tag{2.23}$$

Subtracting Equation (2.23) from Equation (2.22), and using the fact that the price of the portfolio P at time $n = 1$ should match the price of the payoff V, we get

$$V_X(1,H) - V_X(1,T) = \Delta^Y(0) \cdot (Y_X(1,H) - Y_X(1,T)),$$

or

$$\boxed{\Delta^Y(0) = \frac{V_X(1,H) - V_X(1,T)}{Y_X(1,H) - Y_X(1,T)}.} \tag{2.24}$$

Therefore the replicating portfolio is given by the following formula:

$$\boxed{P_0 = \left(\frac{V_Y(1,H) - V_Y(1,T)}{X_Y(1,H) - X_Y(1,T)}\right) \cdot X + \left(\frac{V_X(1,H) - V_X(1,T)}{Y_X(1,H) - Y_X(1,T)}\right) \cdot Y.} \quad (2.25)$$

REMARK 2.2 The hedging position $\Delta^Y(0)$ in the asset Y can also be computed from the relationship

$$V_Y(1,H) = P_Y(1,H) = \Delta^X(0) \cdot X_Y(1,H) + \Delta^Y(0),$$

or in other words,

$$\Delta^Y(0) = V_Y(1,H) - \frac{V_Y(1,H) - V_Y(1,T)}{X_Y(1,H) - X_Y(1,T)} \cdot X_Y(1,H).$$

After some simplifications, we get

$$\begin{aligned}\Delta^Y(0) &= \frac{[X_Y(1,H) - X_Y(1,T)] \cdot V_Y(1,H)}{X_Y(1,H) - X_Y(1,T)} \\ &\qquad - \frac{[V_Y(1,H) - V_Y(1,T)] \cdot X_Y(1,H)}{X_Y(1,H) - X_Y(1,T)} \\ &= \frac{-V_Y(1,H) \cdot X_Y(1,T) + V_Y(1,T) \cdot X_Y(1,H)}{X_Y(1,H) - X_Y(1,T)} \\ &= \frac{V_X(1,H) - V_X(1,T)}{Y_X(1,H) - Y_X(1,T)},\end{aligned}$$

which confirms the previously obtained formula for $\Delta^Y(0)$. If one wanted to compute $\Delta^Y(0)$ using the prices with respect to the reference asset Y only, one can use the formula

$$\boxed{\Delta^Y(0) = \frac{V_Y(1,T) \cdot X_Y(1,H) - V_Y(1,H) \cdot X_Y(1,T)}{X_Y(1,H) - X_Y(1,T)}.} \quad (2.26)$$

□

2.1.3 Multiperiod Binomial Model

The multiperiod binomial model is a generalization of the one-step binomial model. At time n, the price of $X_Y(n)$ moves either to $X_Y(n+1,H) = u \cdot X_Y(n)$, or to $X_Y(n+1,T) = d \cdot X_Y(n)$. In general, the price X_Y at time n is given by the formula

$$X_Y(n) = X_Y(0) \cdot u^{\#H} \cdot d^{\#T},$$

where $\#H$ is the number of heads, and $\#T$ is number of tails in N trials. Clearly, $\#H + \#T = N$.

Let V be a European option with a payoff V_N at time N. One can determine the price of this contract with respect to the reference asset Y using the martingale property of $V_Y(n)$ under the $\mathbb{P}^Y$ measure:

$$V_Y(n) = \mathbb{E}_n^Y[V_Y(N)].$$

Using the tower property, it is possible to compute this expectation recursively using the relationship

$$\boxed{V_Y(n) = \mathbb{E}_n^Y[V_Y(n+1)] = p^Y \cdot V_Y(n+1,H) + q^Y \cdot V_Y(n+1,T),} \tag{2.27}$$

for $n = N-1, N-2, \ldots, 0$. Similarly, the price of the contract with respect to the reference asset X is given by

$$V_X(n) = \mathbb{E}_n^X[V_X(N)]$$

using the martingale property of $V_X(n)$ under the $\mathbb{P}^X$ measure. The corresponding recursive computation is given by

$$\boxed{V_X(n) = \mathbb{E}_n^X[V_X(n+1)] = p^X \cdot V_X(n+1,H) + q^X \cdot V_X(n+1,T),} \tag{2.28}$$

for $n = N-1, N-2, \ldots, 0$.

The hedging portfolio is analogous to that of the one-step binomial model by the following formulas:

$$\boxed{\Delta^X(n) = \frac{V_Y(n+1,H) - V_Y(n+1,T)}{X_Y(n+1,H) - X_Y(n+1,T)},} \tag{2.29}$$

$$\boxed{\Delta^Y(n) = \frac{V_X(n+1,H) - V_X(n+1,T)}{Y_X(n+1,H) - Y_X(n+1,T)},} \tag{2.30}$$

and thus we can write

$$\begin{aligned} P_n = &\left(\frac{V_Y(n+1,H) - V_Y(n+1,T)}{X_Y(n+1,H) - X_Y(n+1,T)}\right) \cdot X \\ &+ \left(\frac{V_X(n+1,H) - V_X(n+1,T)}{Y_X(n+1,H) - Y_X(n+1,T)}\right) \cdot Y. \end{aligned} \tag{2.31}$$

2.1.4 Numerical Example

Consider pricing a European call option with a payoff $(X_N - K \cdot Y_N)^+$ in a two step ($N = 2$) binomial model with parameters $u = 2$, $d = \frac{1}{2}$, initial price $X_Y(0) = 4$, and strike $K = \frac{14}{5}$. When we take Y as a reference asset, the price $X_Y(n)$ has the following evolution:

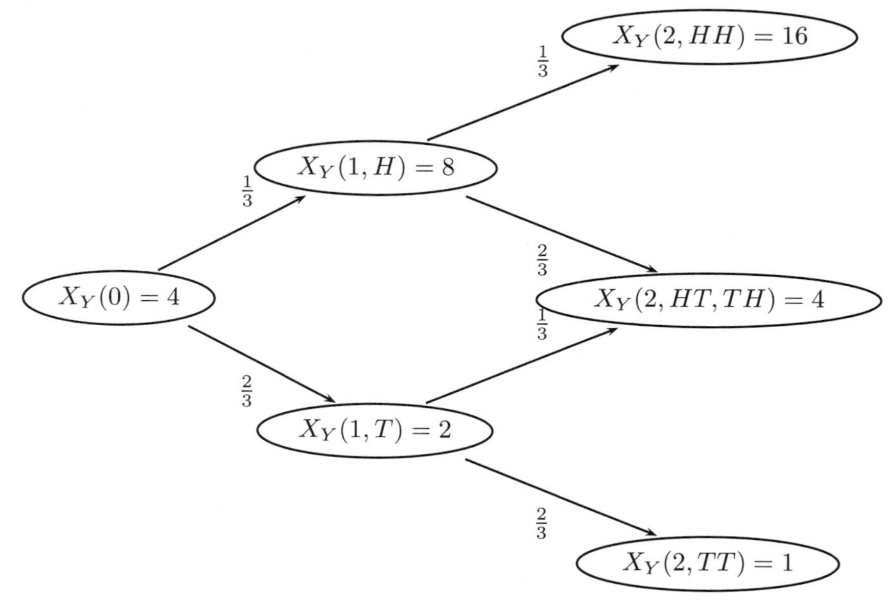

The probabilities $p^Y = \mathbb{P}^Y(H)$ and $q^Y = \mathbb{P}(T)$ are given by

$$p^Y = \frac{1-d}{u-d} = \frac{1-\frac{1}{2}}{2-\frac{1}{2}} = \frac{1}{3}, \quad \text{and} \quad q^Y = \frac{u-1}{u-d} = \frac{2-1}{2-\frac{1}{2}} = \frac{2}{3}.$$

The price process $X_Y(t)$ is a $\mathbb{P}^Y$ martingale, and thus we have

$$X_Y(n,\omega) = \mathbb{E}_n^Y\left[X_Y(n+1)\right](\omega) = p^Y \cdot X_Y(n+1,\omega H) + q^Y \cdot X_Y(n+1,\omega T)$$

for $n = 0, 1$.

Let us denote the European option by V, and first determine the price of V in terms of the asset Y. At the maturity time $N = 2$, the price of V with respect to the reference asset Y is given by

$$V_Y(2) = \left(X_Y(2) - \tfrac{14}{5}\right)^+.$$

The corresponding payoff function is $f^Y(x) = (x - \frac{14}{5})^+$. More specifically,

$$V_Y(2,HH) = \tfrac{66}{5}, \quad V_Y(2,HT) = V_Y(2,TH) = \tfrac{6}{5}, \quad V_Y(2,TT) = 0.$$

From the First Fundamental Theorem of Asset Pricing, the price $V_Y(n)$ is a martingale, and thus

$$V_Y(n,\omega) = \mathbb{E}_n^Y\left[V_Y(n+1)\right](\omega) = p^Y \cdot V_Y(n+1,\omega H) + q^Y \cdot V_Y(n+1,\omega T)$$

for $n = 0, 1$. We first compute $V_Y(1)$ from the known values of $V_Y(2)$ to obtain:

$$\begin{aligned} V_Y(1,H) &= \mathbb{E}_1^Y\left[V_Y(2)\right](H) \\ &= p^Y \cdot V_Y(2,HH) + q^Y \cdot V_Y(2,HT) = \tfrac{1}{3} \cdot \tfrac{66}{5} + \tfrac{2}{3} \cdot \tfrac{6}{5} = \tfrac{26}{5}, \end{aligned}$$

$$\begin{aligned} V_Y(1,T) &= \mathbb{E}_1^Y\left[V_Y(2)\right](T) \\ &= p^Y \cdot V_Y(2,TH) + q^Y \cdot V_Y(2,TT) = \tfrac{1}{3} \cdot \tfrac{6}{5} + \tfrac{2}{3} \cdot 0 = \tfrac{2}{5}. \end{aligned}$$

At time $n = 0$, we have

$$\begin{aligned} V_Y(0) &= \mathbb{E}_0^Y\left[V_Y(1)\right] \\ &= p^Y \cdot V_Y(1,H) + q^Y \cdot V_Y(1,T) = \tfrac{1}{3} \cdot \tfrac{26}{5} + \tfrac{2}{3} \cdot \tfrac{2}{5} = 2. \end{aligned}$$

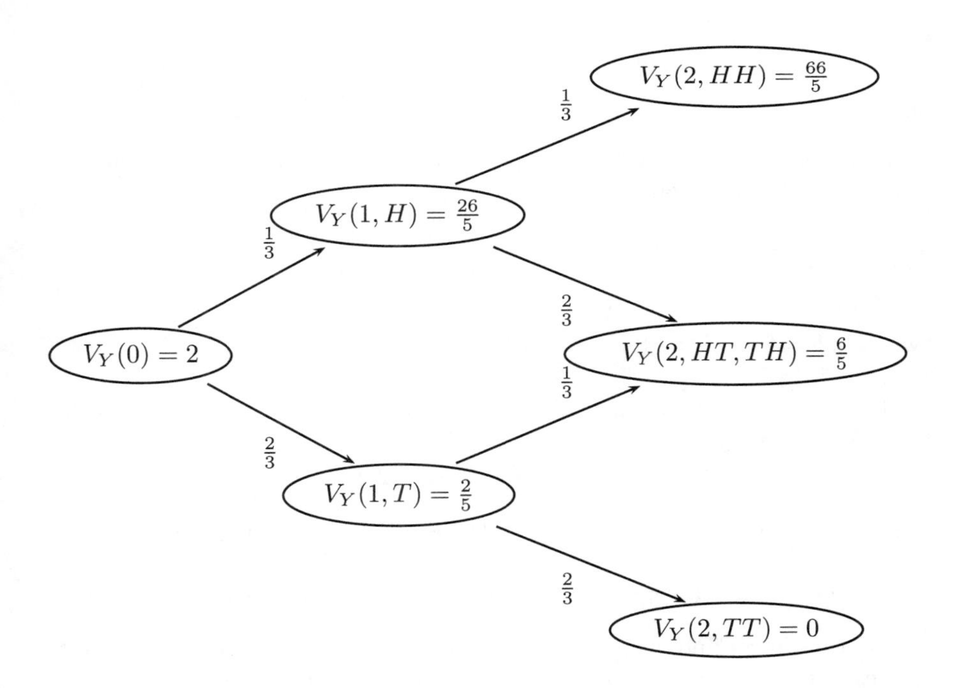

When we take X as a reference asset, the price evolution $Y_X(n)$ is the inverse of $X_Y(n)$:

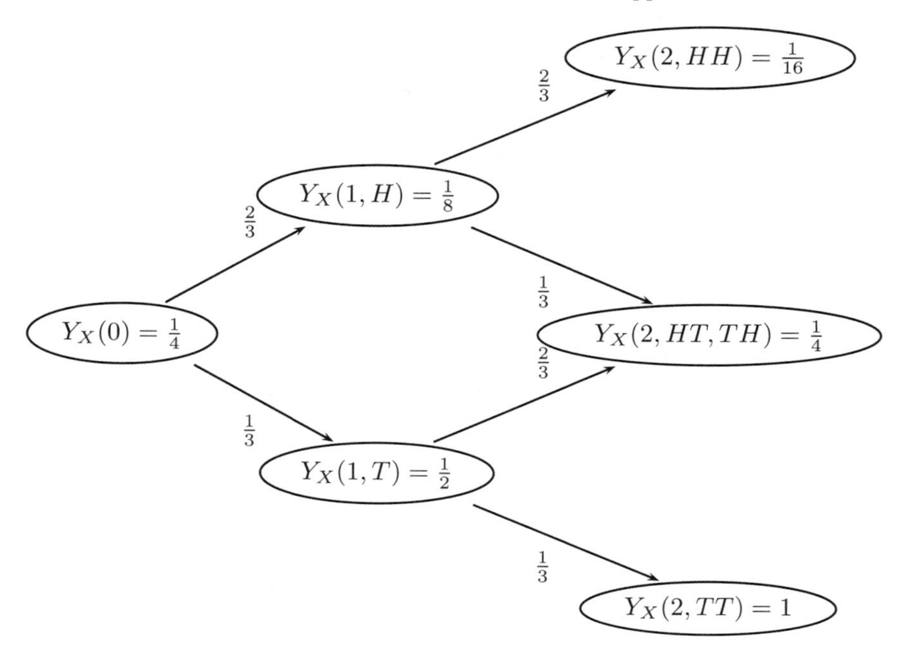

The probability measure $\mathbb{P}^X$ that makes the prices with respect to the reference asset X martingales is given by

$$p^X = u \cdot \frac{1-d}{u-d} = 2 \cdot \frac{1-\frac{1}{2}}{2-\frac{1}{2}} = \frac{2}{3}, \quad \text{and} \quad q^X = d \cdot \frac{u-1}{u-d} = \frac{1}{2} \cdot \frac{2-1}{2-\frac{1}{2}} = \frac{1}{3},$$

where $p^X = \mathbb{P}^X(H)$ and $q^X = \mathbb{P}^X(T)$.

The price of the contract with respect to the reference asset X at time $N = 2$ is given by

$$V_X(2) = \left(1 - \tfrac{14}{5} Y_X(2)\right)^+ .$$

The payoff function is $f^X(x) = \left(1 - \frac{14}{5} \cdot x\right)^+$. This gives us the values

$$V_X(2,HH) = \tfrac{33}{40}, \quad V_X(2,HT) = V_X(2,TH) = \tfrac{3}{10}, \quad V_X(2,TT) = 0.$$

Using the First Fundamental Theorem of Asset Pricing, we can compute the prices of the contract at an earlier time using the martingale property

$$V_X(n,\omega) = \mathbb{E}_n^X \left[V_X(n+1)\right](\omega) = p^X \cdot V_X(n+1,\omega H) + q^X \cdot V_X(n+1,\omega T)$$

for $n = 0, 1$. When $n = 1$, we get

$$\begin{aligned} V_X(1,H) &= \mathbb{E}_1^X \left[V_X(2)\right](H) \\ &= p^X \cdot V_X(2,HH) + q^X \cdot V_X(2,HT) = \tfrac{2}{3} \cdot \tfrac{33}{40} + \tfrac{1}{3} \cdot \tfrac{3}{10} = \tfrac{13}{20}, \end{aligned}$$

$$\begin{aligned} V_X(1,T) &= \mathbb{E}_1^X \left[V_X(2)\right](T) \\ &= p^X \cdot V_X(2,TH) + q^X \cdot V_X(2,TT) = \tfrac{2}{3} \cdot \tfrac{3}{10} + \tfrac{1}{3} \cdot 0 = \tfrac{1}{5}. \end{aligned}$$

At time $n = 0$, we have

$$\begin{aligned} V_X(0) &= \mathbb{E}_0^X \left[V_X(1)\right] \\ &= p^X \cdot V_X(1,H) + q^X \cdot V_X(1,T) = \tfrac{2}{3} \cdot \tfrac{13}{20} + \tfrac{1}{3} \cdot \tfrac{1}{5} = \tfrac{1}{2}. \end{aligned}$$

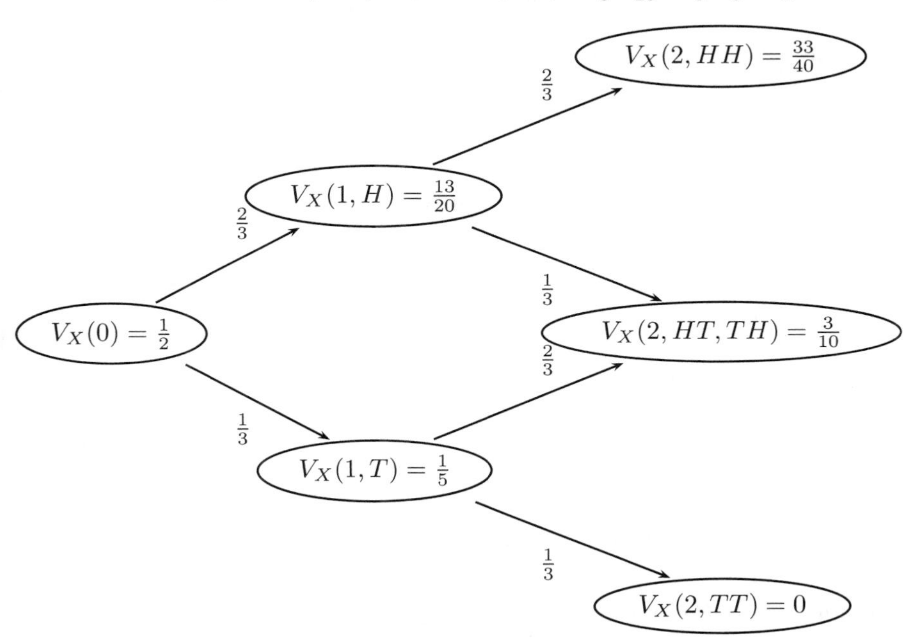

Note that the prices computed with respect to the reference asset Y and X are indeed consistent through the change of numeraire formula

$$\boxed{V_X(n) = V_Y(n) \cdot Y_X(n).} \tag{2.32}$$

Similarly, we have

$$\boxed{V_Y(n) = V_X(n) \cdot X_Y(n).} \tag{2.33}$$

For instance

$$V_X(0) = V_Y(0) \cdot Y_X(0) = 2 \cdot \tfrac{1}{4} = \tfrac{1}{2},$$

which agrees with the price computed from the martingale property of the price $V_X(n)$ under the $\mathbb{P}^X$ measure.

Note that we can also obtain the prices of this European call option from the Black–Scholes formula (1.77):

$$V = \left[\mathbb{P}_n^X(X_Y(2) \geq K)\right] \cdot X + \left[-K \cdot \mathbb{P}_n^Y(X_Y(2) \geq K)\right] \cdot Y.$$

At time $n = 0$,

$$\mathbb{P}^X(X_Y(2) \geq K) = \mathbb{P}^X(HH) + \mathbb{P}^X(HT) + \mathbb{P}^X(TH) = \tfrac{2}{3} \cdot \tfrac{2}{3} + \tfrac{2}{3} \cdot \tfrac{1}{3} + \tfrac{1}{3} \cdot \tfrac{2}{3} = \tfrac{8}{9},$$

$$\mathbb{P}^Y(X_Y(2) \geq K) = \mathbb{P}^Y(HH) + \mathbb{P}^Y(HT) + \mathbb{P}^Y(TH) = \tfrac{1}{3} \cdot \tfrac{1}{3} + \tfrac{1}{3} \cdot \tfrac{2}{3} + \tfrac{2}{3} \cdot \tfrac{1}{3} = \tfrac{5}{9},$$

resulting in

$$V_0 = \tfrac{8}{9} \cdot X_0 - \tfrac{14}{5} \cdot \tfrac{5}{9} \cdot Y_0 = \tfrac{8}{9} \cdot X_0 - \tfrac{14}{9} \cdot Y_0.$$

Thus we have

$$V_Y(0) = \tfrac{8}{9} \cdot X_Y(0) - \tfrac{14}{9} = \tfrac{8}{9} \cdot 4 - \tfrac{14}{9} = 2,$$

and

$$V_X(0) = \tfrac{8}{9} - \tfrac{14}{9} \cdot Y_X(0) = \tfrac{8}{9} - \tfrac{14}{9} \cdot \tfrac{1}{4} = \tfrac{1}{2}.$$

At time $n = 1$, we have

$$\mathbb{P}_1^X(X_Y(2) \geq K)(H) = \mathbb{P}^X(H) + \mathbb{P}^X(T) = 1,$$

$$\mathbb{P}_1^Y(X_Y(2) \geq K)(H) = \mathbb{P}^Y(H) + \mathbb{P}^Y(T) = 1,$$

which leads to

$$V_1(H) = X_1 - \tfrac{14}{5} \cdot Y_1.$$

Therefore

$$V_Y(1, H) = X_Y(1, H) - \tfrac{14}{5} = 8 - \tfrac{14}{5} = \tfrac{26}{5},$$

and

$$V_X(1, H) = 1 - \tfrac{14}{5} \cdot Y_X(1, H) = 1 - \tfrac{14}{5} \cdot \tfrac{1}{8} = \tfrac{13}{20}.$$

Similarly,

$$\mathbb{P}_1^X(X_Y(2) \geq K)(T) = \mathbb{P}^X(H) = \tfrac{2}{3},$$

$$\mathbb{P}_1^Y(X_Y(2) \geq K)(T) = \mathbb{P}^Y(H) = \tfrac{1}{3},$$

which leads to

$$V_1(T) = \tfrac{2}{3} \cdot X_1 - \tfrac{14}{5} \cdot \tfrac{1}{3} \cdot Y_1 = \tfrac{2}{3} \cdot X_1 - \tfrac{14}{15} \cdot Y_1.$$

Therefore

$$V_Y(1, T) = \tfrac{2}{3} \cdot X_Y(1, T) - \tfrac{14}{15} = \tfrac{2}{3} \cdot 2 - \tfrac{14}{15} = \tfrac{2}{5},$$

and

$$V_X(1, T) = \tfrac{2}{3} - \tfrac{14}{5} \cdot Y_X(1, T) = \tfrac{2}{3} - \tfrac{14}{5} \cdot \tfrac{1}{2} = \tfrac{1}{5}.$$

These results are indeed consistent with the prices we obtained from binomial pricing.

The hedging portfolio is given by

$$\begin{aligned}
P_0 &= \left(\frac{V_Y(1,H)-V_Y(1,T)}{X_Y(1,H)-X_Y(1,T)}\right)\cdot X + \left(\frac{V_X(1,H)-V_X(1,T)}{Y_X(1,H)-Y_X(1,T)}\right)\cdot Y \\
&= \left(\frac{\frac{26}{5}-\frac{2}{5}}{8-2}\right)\cdot X + \left(\frac{\frac{13}{20}-\frac{1}{5}}{\frac{1}{8}-\frac{1}{2}}\right)\cdot Y \\
&= \tfrac{4}{5}\cdot X - \tfrac{6}{5}\cdot Y,
\end{aligned}$$

$$\begin{aligned}
P_1(H) &= \left(\frac{V_Y(2,HH)-V_Y(2,HT)}{X_Y(2,HH)-X_Y(2,HT)}\right)\cdot X + \left(\frac{V_X(2,HH)-V_X(2,HT)}{Y_X(2,HH)-Y_X(2,HT)}\right)\cdot Y \\
&= \left(\frac{\frac{66}{5}-\frac{6}{5}}{16-4}\right)\cdot X + \left(\frac{\frac{33}{40}-\frac{3}{10}}{\frac{1}{16}-\frac{1}{4}}\right)\cdot Y \\
&= 1\cdot X - \tfrac{14}{5}\cdot Y,
\end{aligned}$$

and

$$\begin{aligned}
P_1(T) &= \left(\frac{V_Y(2,TH)-V_Y(2,TT)}{X_Y(2,TH)-X_Y(2,TT)}\right)\cdot X + \left(\frac{V_X(2,TH)-V_X(2,TT)}{Y_X(2,TH)-Y_X(2,TT)}\right)\cdot Y \\
&= \left(\frac{\frac{6}{5}-0}{4-1}\right)\cdot X + \left(\frac{\frac{3}{10}-0}{\frac{1}{4}-1}\right)\cdot Y \\
&= \tfrac{2}{5}\cdot X - \tfrac{2}{5}\cdot Y.
\end{aligned}$$

Note that the hedging positions in this case do not satisfy

$$\Delta^X(n) = \mathbb{P}_n^X(X_Y(2) \geq K),$$

or

$$\Delta^Y(n) = -K\cdot \mathbb{P}_n^Y(X_Y(2) \geq K)$$

in general, as one may expect from the Black–Scholes formula

$$V = \left[\mathbb{P}_n^X(X_Y(2) \geq K)\right]\cdot X + \left[-K\cdot \mathbb{P}_n^Y(X_Y(2) \geq K)\right]\cdot Y.$$

However, these formulas for hedging are valid in the geometric Brownian motion model, as we show in Chapter 3 Diffusion Models.

2.1.5 Probability Measures for Exotic No-Arbitrage Assets

Every no-arbitrage asset comes with its own martingale measure. So far we have identified the corresponding martingale measures associated with the assets Y and X. It is interesting to note that even self-financing portfolios created by trading in the assets X and Y generate their own martingale measures, as long as the value of the portfolio stays positive at all times. The resulting portfolio is itself a no-arbitrage asset.

Let us illustrate the martingale property on an asset A defined by a payoff

$$A_1 = \tfrac{1}{2}\left[X_Y(0) + X_Y(1)\right] \cdot Y_1. \tag{2.34}$$

The asset A is known as an **average asset** and it represents the average price of the asset X in terms of the asset Y. A contract whose payoff depends on the asset A defined above is known as an **Asian option.** It is obvious that A is a result of a self-financing strategy. One should start with a unit of an asset X, sell immediately half of the unit of X for $\frac{1}{2} \cdot X_Y(0)$ units of Y at time $n = 0$, and sell the remaining half of the unit of X for $\frac{1}{2} \cdot X_Y(1)$ units of Y at time $n = 1$. In terms of the formulas, we have

$$A_0 = X_0,$$

and

$$\Delta^X(0) = \tfrac{1}{2}.$$

The martingale measure associated with the asset A should assign probability to both possible outcomes ω, namely to $\omega = H$, and to $\omega = T$. The most straightforward way to determine $\mathbb{P}^A(H)$ and $\mathbb{P}^A(T)$ is from its relationships with the measure $\mathbb{P}^Y$ via the Radon–Nikodým derivative described in the formulas (1.69) and (1.72). Using these two relationships, we can write

$$\begin{aligned}
\mathbb{P}^A(\omega) &= Z(1,\omega) \cdot \mathbb{P}^Y(\omega) = \frac{A_Y(1,\omega)}{A_Y(0)} \cdot \mathbb{P}^Y(\omega) \\
&= \frac{\frac{1}{2}\left[X_Y(0) + X_Y(1)\right]}{X_Y(0)} \cdot \mathbb{P}^Y(\omega) \\
&= \tfrac{1}{2} \cdot \left(1 + \tfrac{X_Y(1,\omega)}{X_Y(0)}\right) \cdot \mathbb{P}^Y(\omega).
\end{aligned} \tag{2.35}$$

Thus we have

$$\mathbb{P}^A(H) = \tfrac{1}{2} \cdot (1+u) \cdot \frac{1-d}{u-d}, \tag{2.36}$$

and

$$\mathbb{P}^A(T) = \tfrac{1}{2} \cdot (1+d) \cdot \frac{u-1}{u-d}. \tag{2.37}$$

The pricing measure $\mathbb{P}^A$ is useful for pricing Asian options; see the following example, or Chapter 9 Asian Options.

Example 2.1
Consider a contract V that pays off

$$V_1 = \mathbb{I}(\omega = H) \cdot A_1.$$

Its price at time $n = 0$ is simply

$$V_A(0) = \mathbb{P}^A(H) = \tfrac{1}{2} \cdot (1+u) \cdot \frac{1-d}{u-d}.$$

The price with respect to the reference asset Y can be obtained from the change of numeraire formula

$$V_Y(0) = V_A(0) \cdot A_Y(0) = \tfrac{1}{2} \cdot (1+u) \cdot \frac{1-d}{u-d} \cdot X_Y(0).$$

□

2.2 Binomial Model with an Arbitrage Asset

Some contracts are written on one or several arbitrage assets and there is no way to replace them with a suitable no-arbitrage asset in the same way as for European options. This is, for instance, the case for American stock options with two underlying assets, a stock S, and the dollar \$. The American option pays off either $f^Y(X_Y(\tau))$ units of an asset Y, or $f^X(Y_X(\tau))$ units of an asset X at the exercise time $\tau \in [0, T]$. The exercise time τ is chosen by the holder of the option.

When the underlying assets are the stock S, and the dollar \$, the payoff is either $f^{\$}(S_{\$}(\tau))$ units of dollars \$, or $f^S(\$_S(\tau))$ units of the stock S at the exercise time $\tau \in [0, T]$. Since the payoff has to be delivered at an arbitrary time chosen by the holder of the option, it is not possible to substitute the dollar with the corresponding bond. When there is more than one possible delivery time, there is no corresponding no-arbitrage asset.

However, hedging of a contract whose underlying is one or more arbitrage assets must be done exclusively in no-arbitrage assets. For an American stock option, the pricing can be done with respect to both reference assets \$ and S, but the hedging is done in S and in a suitable no-arbitrage proxy of \$ such as a bond B^T. Thus we have three assets to consider: a stock S, a bond B^T, and the dollar \$.

Let us start with a one-step model at time n. The binomial model then assumes that the stock price in terms of dollars either increases by a factor u, or decreases by a factor d. Thus we have $S_{\$}(n+1, H) = u \cdot S_{\$}(n)$, and $S_{\$}(n+1, T) = d \cdot S_{\$}(n, T)$.

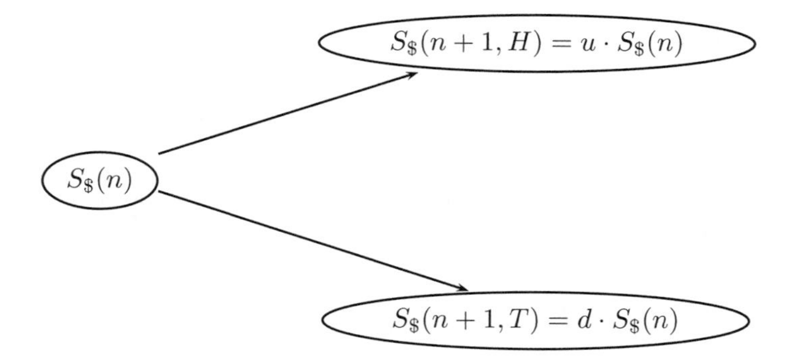

The bond has deterministic evolution given by $B^T_\$(n+1) = (1+r) \cdot B^T_\$(n)$.

$B^T_\$(n)$ → $B^T_\$(n+1) = (1+r) \cdot B^T_\$(n)$

Neither of these two evolutions are justified by the First Fundamental Theorem of Asset Pricing which considers only no-arbitrage assets. In particular, these prices are not martingales under the corresponding reference measures. The two no-arbitrage assets here are S and B^T. The First Fundamental Theorem of Asset Pricing says that the price evolution $S_{B^T}(n)$ is a $\mathbb{P}^T$ martingale (under the T-forward measure that corresponds to a B^T bond), and that the price evolution $B^T_S(n)$ is a $\mathbb{P}^S$ martingale (under the stock measure that corresponds to S). The measures $\mathbb{P}^T$ and $\mathbb{P}^S$ are the two measures that can be used for the pricing of an American option; there is no pricing measure that corresponds to the dollar.

From the price evolutions of $S_\$$ and $B^T_\$$, we can already determine the pricing measures $\mathbb{P}^T$ and $\mathbb{P}^S$. Note that

$$\begin{aligned} S_{B^T}(n+1,H) &= S_\$(n+1,H) \cdot \$_{B^T}(n+1) \\ &= u \cdot S_\$(n) \cdot \tfrac{1}{1+r} \cdot \$_{B^T}(n) = \tfrac{u}{1+r} \cdot S_{B^T}(n), \end{aligned}$$

and

$$\begin{aligned} S_{B^T}(n+1,T) &= S_\$(n+1,T) \cdot \$_{B^T}(n+1) \\ &= d \cdot S_\$(n) \cdot \tfrac{1}{1+r} \cdot \$_{B^T}(n) = \tfrac{d}{1+r} \cdot S_{B^T}(n). \end{aligned}$$

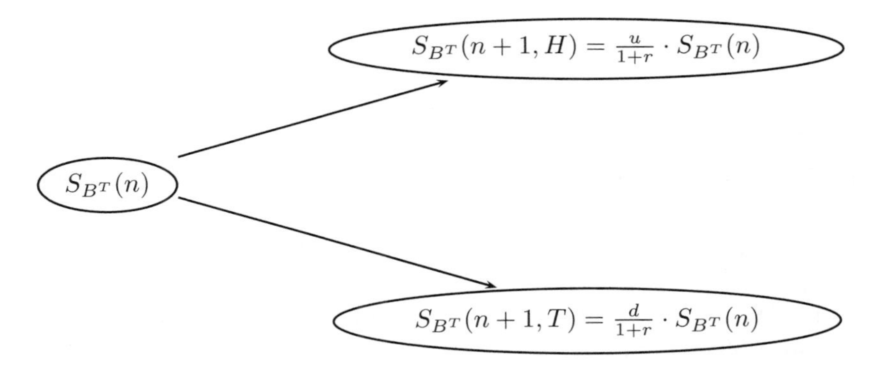

This is the same model as described in the previous section for two no-arbitrage assets with the exception that the scaling factors u and d are slightly modified. The u factor is now $\frac{u}{1+r}$, and the d factor is now $\frac{d}{1+r}$. The asset Y is now the bond B^T, and the asset X is now the stock S. Rewriting Equation (2.3) with the new scaling factors $\frac{u}{1+r}$ and $\frac{d}{1+r}$, we obtain that the probabilities that correspond to the T-forward measure are given by

$$\boxed{p^T(u,d) = \frac{1+r-d}{u-d}, \qquad q^T(u,d) = \frac{u-(1+r)}{u-d}.} \tag{2.38}$$

Similarly, by rewriting Equation (2.8) we obtain the stock measure

$$\boxed{p^S(u,d) = \frac{u}{1+r} \cdot \frac{1+r-d}{u-d}, \qquad q^S(u,d) = \frac{d}{1+r} \cdot \frac{u-(1+r)}{u-d}.} \tag{2.39}$$

Even when the contract depends on an arbitrage asset, we still have to compute its price with respect to a no-arbitrage asset and convert it back to the given arbitrage asset using the change of numeraire formula. When the contract is written on a stock and on dollars, we have two possible reference assets: a bond B^T, and a stock S. Recall that when V is a European option with a payoff V_N at time N, we can determine the price of this contract with respect to the reference asset B^N (maturity time $T = N$) using the martingale property of $V_{B^N}(n)$ under the $\mathbb{P}^T$ measure:

$$V_{B^N}(n) = \mathbb{E}_n^T[V_{B^N}(N)].$$

Using the tower property, it is possible to compute this expectation recursively from the relationship

$$V_{B^N}(n) = \mathbb{E}_n^T[V_{B^N}(n+1)] = p^T \cdot V_{B^N}(n+1,H) + q^T \cdot V_{B^N}(n+1,T), \tag{2.40}$$

for $n = N-1, N-2, \ldots, 0$. Since we assume

$$B_n^N = \frac{1}{(1+r)^{N-n}} \cdot \$_n,$$

Equation (2.40) can be rewritten in terms of the dollar as a reference asset as

$$\boxed{V_\$(n) = \tfrac{1}{1+r} \cdot \mathbb{E}_n^T\left[V_\$(n+1)\right] = \tfrac{1}{1+r} \cdot \left[p^T \cdot V_\$(n+1,H) + q^T \cdot V_\$(n+1,T)\right].} \tag{2.41}$$

This is a mainstream way of computing a price of a derivative contract in a binomial model, even for European-type options. Note that $V_\$(n)$ is not a $\mathbb{P}^T$ martingale, and thus the formula requires discounting the expectation by a factor $\frac{1}{1+r}$.

Similarly, the price of the contract with respect to the stock S is given by

$$V_S(n) = \mathbb{E}_n^S[V_S(N)]$$

using the martingale property of $V_S(n)$ under the $\mathbb{P}^S$ measure. Since V_S is a martingale, the corresponding recursive computation is given by

$$\boxed{V_S(n) = \mathbb{E}_n^S[V_S(n+1)] = p^S \cdot V_S(n+1,H) + q^S \cdot V_S(n+1,T),} \tag{2.42}$$

for $n = N-1, N-2, \ldots, 0$.

As we pointed out in the previous section, bringing an arbitrage asset to pricing European-type options only increases the dimensionality of the problem, a step that is not necessary. However, it is typical to use the approach presented in this section that needs three assets: a stock S, a bond B^N, and the dollar \$. The bond plays a role of a no-arbitrage proxy of the dollar, and is required for hedging. The hedging portfolio has no positions in arbitrage assets. A cleaner solution for European-type options is to consider only a stock S, and a bond B^N. The dollar value of a contract is easily determined from the dollar value of the stock

$$V_\$(n) = V_S(n) \cdot S_\$(n)$$

or from the dollar value of the bond

$$V_\$(n) = V_{B^N}(n) \cdot B_\$^N(n).$$

2.2.1 American Option Pricing in the Binomial Model

A general treatment of pricing American options is given in Chapter 7. In this section, we study pricing and hedging of an American stock option V in the binomial model. The holder of the contract has two possible actions at each time n: either exercise the option and collect the payoff, or keep the contract for the future. Both actions have value: immediate exercise has a value known as an **intrinsic value of the option;** keeping the option has a value known as a **continuation value of the option.** Thus the holder should evaluate the intrinsic and continuation value of the option, and choose the one with a higher price so as not to produce arbitrage opportunities for the seller of the option.

Recall that an American stock option pays off either $f^{\$}\left(S_{\$}(\tau)\right)$ units of the dollar \$ or $f^{S}\left(\$_{S}(\tau)\right)$ units of a stock S at the exercise time $\tau \in [0, T]$. The intrinsic value can be expressed in terms of both assets that enter this contract. In particular, the intrinsic value of the option at time n is $f^{\$}\left(S_{\$}(n)\right)$ when the dollar \$ is chosen as a reference asset, and it is $f^{S}\left(\$_{S}(n)\right)$ when a stock S is chosen as a reference asset. Similarly, the continuation value is given by $\frac{1}{1+r} \cdot \mathbb{E}_{n}^{T}\left[V_{\$}(n+1)\right]$ when the dollar \$ is chosen as a reference asset, and it is $\mathbb{E}_{n}^{S}\left[V_{S}(n+1)\right]$ when a stock S is chosen as a reference asset. Comparing the intrinsic and the continuation values in both cases, we find that

$$\boxed{V_{\$}(n) = \max\Big(f^{\$}\left(S_{\$}(n)\right), \tfrac{1}{1+r} \cdot \mathbb{E}_{n}^{T}\left[V_{\$}(n+1)\right]\Big)} \tag{2.43}$$

for the case when the dollar \$ is chosen as a reference asset, and

$$\boxed{V_{S}(n) = \max\Big(f^{S}\left(\$_{S}(n)\right), \mathbb{E}_{n}^{S}\left[V_{S}(n+1)\right]\Big)} \tag{2.44}$$

for the case when a stock S is chosen as a reference asset. Note that both expressions are equivalent and they are related by the change of numeraire formula

$$V_{\$}(n) = V_{S}(n) \cdot S_{\$}(n).$$

The option should be exercised the first time the option value coincides with its intrinsic value. In this case, the continuation value of the option is smaller or equal to the intrinsic value. Let τ^{*} be the optimal exercise policy. The exercise time is given by

$$\boxed{\tau^{*} = \min\Big\{n \geq 0 : V_{\$}(n) = f^{\$}\left(S_{\$}(n)\right)\Big\},} \tag{2.45}$$

which is equivalent to

$$\boxed{\tau^{*} = \min\Big\{n \geq 0 : V_{S}(n) = f^{S}\left(\$_{S}(n)\right)\Big\}.} \tag{2.46}$$

2.2.2 Hedging

The hedging of an American stock option must be done in no-arbitrage assets. An American stock option is written on two assets, a stock S, and the dollar \$. The hedging portfolio will take positions in a no-arbitrage asset S, and in a corresponding no-arbitrage proxy asset for \$ such as a zero coupon bond B^N. The hedging portfolio P_n takes the same form as in Equation (2.31) with X replaced by S and Y replaced by B^N:

$$P_n = \Delta^S(n) \cdot S + \Delta^T(n) \cdot B^T,$$

where

$$\Delta^S(n) = \frac{V_{B^N}(n+1,H) - V_{B^N}(n+1,T)}{S_{B^N}(n+1,H) - S_{B^N}(n+1,T)},$$

and

$$\begin{aligned}\Delta^T(n) &= \frac{V_S(n+1,H) - V_S(n+1,T)}{B^N_S(n+1,H) - B^N_S(n+1,T)} \\ &= \frac{V_{B^N}(n+1,T) \cdot S_{B^N}(n+1,H) - V_{B^N}(n+1,H) \cdot S_{B^N}(n+1,T)}{S_{B^N}(n+1,H) - S_{B^N}(n+1,T)}.\end{aligned}$$

We can express the hedging positions $\Delta^S(n)$ and $\Delta^T(n)$ in terms of dollar prices. Note that

$$\begin{aligned}\Delta^S(n) &= \frac{V_{B^N}(n+1,H) - V_{B^N}(n+1,T)}{S_{B^N}(n+1,H) - S_{B^N}(n+1,T)} \cdot \frac{B^N_\$(n+1)}{B^N_\$(n+1)} \\ &= \frac{V_\$(n+1,H) - V_\$(n+1,T)}{S_\$(n+1,H) - S_\$(n+1,T)}.\end{aligned}$$

The hedging position $\Delta^T(n)$ can be expressed either in terms of the prices with respect to a stock S as

$$\begin{aligned}\Delta^T(n) &= \frac{V_S(n+1,H) - V_S(n+1,T)}{B^N_S(n+1,H) - B^N_S(n+1,T)} \cdot \frac{\$_{B^N}(n+1)}{\$_{B^N}(n+1)} \\ &= \frac{V_S(n+1,H) - V_S(n+1,T)}{\$_S(n+1,H) - \$_S(n+1,T)} \cdot \$_{B^N}(n+1) \\ &= \frac{V_S(n+1,H) - V_S(n+1,T)}{\$_S(n+1,H) - \$_S(n+1,T)} \cdot (1+r)^{N-n-1},\end{aligned}$$

or in terms of the prices with respect to a dollar \$ as

$$\begin{aligned}&\Delta^T(n) = \\ &= \left[\frac{V_{B^N}(n+1,T) \cdot S_{B^N}(n+1,H) - V_{B^N}(n+1,H) \cdot S_{B^N}(n+1,T)}{S_{B^N}(n+1,H) - S_{B^N}(n+1,T)}\right] \times \\ &\qquad \times \left[\frac{B^N_\$(n+1)}{B^N_\$(n+1)}\right]^2 \\ &= \left[\frac{V_\$(n+1,T) \cdot S_\$(n+1,H) - V_\$(n+1,H) \cdot S_\$(n+1,T)}{S_\$(n+1,H) - S_\$(n+1,T)}\right] \times \$_{B^N}(n+1) \\ &= \left[\frac{V_\$(n+1,T) \cdot S_\$(n+1,H) - V_\$(n+1,H) \cdot S_\$(n+1,T)}{S_\$(n+1,H) - S_\$(n+1,T)}\right] \times (1+r)^{N-n-1}.\end{aligned}$$

Thus we can write

$$P_n = \left(\frac{V_\$(n+1,H) - V_\$(n+1,T)}{S_\$(n+1,H) - S_\$(n+1,T)}\right) \cdot S + \left(\frac{V_S(n+1,H) - V_S(n+1,T)}{\$_S(n+1,H) - \$_S(n+1,T)}\right) \cdot (1+r)^{N-n-1} \cdot B^N.$$

Since the interest rate r is assumed to be deterministic, the bond B^N is just a constant multiple of the money market account M. Thus we can alternatively invest in the money market instead of in the bond B^N. Assuming that time n is the reference time for the money market, we have that $M_n = 1\$_n$, and $M_{n+1} = (1+r)\$_{n+1}$. The relationship between M and B^N is given by $M_n = (1+r)^{N-n} \cdot B_n^N$. Let $\Delta^M(n)$ be the hedging position in the money market account M. Since $\Delta^M(n) \cdot M_n$ should have the same price as the bond position $\Delta^T(n) \cdot B_n^N$, we must have

$$\Delta^M(n) = (1+r)^{-(N-n)} \cdot \Delta^T(n).$$

This follows from the relationship

$$\Delta^M(n) \cdot M_n = \Delta^M(n) \cdot (1+r)^{N-n} \cdot B_n^N = \Delta^T(n) \cdot B_n^N.$$

In this case, we can also express the hedging portfolio as

$$P_n = \left(\frac{V_\$(n+1,H) - V_\$(n+1,T)}{S_\$(n+1,H) - S_\$(n+1,T)}\right) \cdot S + \left(\frac{V_S(n+1,H) - V_S(n+1,T)}{\$_S(n+1,H) - \$_S(n+1,T)}\right) \cdot \frac{1}{1+r} \cdot M. \tag{2.47}$$

2.2.3 Numerical Example

Let us consider a binomial model for S, B^N and \$ with the following parameters: $u = 2$, $d = \frac{1}{2}$, $r = \frac{1}{4}$, $S_\$(0) = 4$. Let us consider an American put option with a payoff $(5 \cdot \$_\tau - S_\tau)^+$. The dollar price of the stock has the following evolution

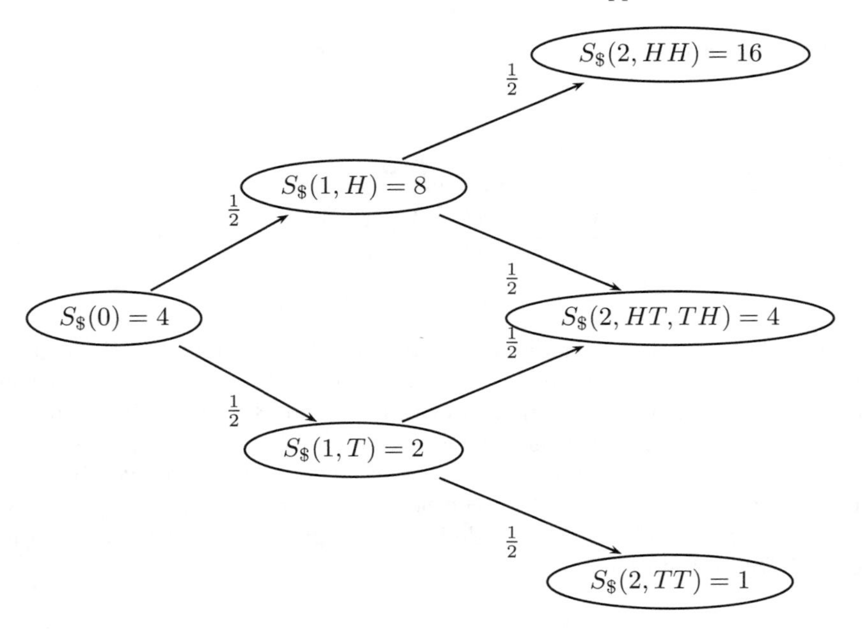

The probabilities $p^T = \mathbb{P}^T(H)$ and $q^T = \mathbb{P}^T(T)$ are given by

$$p^T = \frac{1+r-d}{u-d} = \frac{1+\frac{1}{4}-\frac{1}{2}}{2-\frac{1}{2}} = \frac{1}{2},$$

and

$$q^T = \frac{u-(1+r)}{u-d} = \frac{2-(1+\frac{1}{4})}{2-\frac{1}{2}} = \frac{1}{2}.$$

The price process $S_\$(n)$ satisfies

$$\begin{aligned} S_\$(n,\omega) &= \tfrac{1}{1+r} \cdot \mathbb{E}_n^T \left[S_\$(n+1)\right](\omega) \\ &= \tfrac{1}{1+r} \cdot \left[p^T \cdot S_\$(n+1,\omega H) + q^T \cdot S_\$(n+1,\omega T)\right] \end{aligned}$$

for $n = 0, 1$.

Let V denote the American option with a payoff $(5 \cdot \$_\tau - S_\tau)^+$ at exercise time τ, where $\tau \leq 2$. Let us determine the price of this contract $V_\$$ with respect to the dollar \$. The intrinsic value function $f^\$$ for the dollar as a reference asset is

$$f^\$(x) = (5-x)^+,$$

which means that the option pays off

$$(5 - S_\$(\tau))^+$$

units of a dollar \$ at the exercise time τ. The price of the American option is computed by comparing the intrinsic value and the continuation value of the option:

$$V_{\$}(n) = \max\Big(f^{\$}\left(S_{\$}(n)\right), \tfrac{1}{1+r} \cdot \mathbb{E}_n^T\left[V_{\$}(n+1)\right]\Big). \tag{2.48}$$

The computation of the price is done recursively, starting from the maturity of the option. At $N = 2$, there is no continuation possible, and thus the value of the option is equal to its intrinsic value $f^{\$}\left(S_{\$}(2)\right)$:

$$V_{\$}(2, HH) = 0, \quad V_{\$}(2, HT) = V_{\$}(2, TH) = 1, \quad V_{\$}(2, TT) = 4.$$

We can compute the price $V_{\$}(n)$ for $n = 1, 0$ using formula (2.48). We first compute $V_{\$}(1)$ by comparing the intrinsic value and the continuation value of the option

$$\begin{aligned}
V_{\$}(1, H) &= \max\Big(f^{\$}\left(S_{\$}(1, H)\right), \tfrac{1}{1+r} \cdot \mathbb{E}_1^T\left[V_{\$}(2)\right](H)\Big) \\
&= \max\Big(f^{\$}\left(S_{\$}(1, H)\right), \tfrac{1}{1+r} \cdot \left[p^T \cdot V_{\$}(2, HH) + q^T \cdot V_{\$}(2, HT)\right]\Big) \\
&= \max\left((5-8)^+, \tfrac{1}{1+\frac{1}{4}} \cdot \left[\tfrac{1}{2} \cdot 0 + \tfrac{1}{2} \cdot 1\right]\right) = \max(0, \tfrac{2}{5}) = \tfrac{2}{5}.
\end{aligned}$$

The intrinsic value of the option is zero, and thus it is optimal not to exercise the option in this scenario. On the other hand

$$\begin{aligned}
V_{\$}(1, T) &= \max\Big(f^{\$}\left(S_{\$}(1, T)\right), \tfrac{1}{1+r} \cdot \mathbb{E}_1^T\left[V_{\$}(2)\right](T)\Big) \\
&= \max\Big(f^{\$}\left(S_{\$}(1, T)\right), \tfrac{1}{1+r} \cdot \left[p^T \cdot V_{\$}(2, TH) + q^T \cdot V_{\$}(2, TT)\right]\Big) \\
&= \max\left((5-2)^+, \tfrac{1}{1+\frac{1}{4}} \cdot \left[\tfrac{1}{2} \cdot 1 + \tfrac{1}{2} \cdot 4\right]\right) = \max(3, 2) = 3,
\end{aligned}$$

and thus it is better to exercise the option since the intrinsic value is larger than the continuation value. At time $n = 0$, we have

$$\begin{aligned}
V_{\$}(0) &= \max\Big(f^{\$}\left(S_{\$}(0)\right), \tfrac{1}{1+r} \cdot \mathbb{E}_0^T\left[V_{\$}(1)\right]\Big) \\
&= \max\Big(f^{\$}\left(S_{\$}(0)\right), \tfrac{1}{1+r} \cdot \left[p^T \cdot V_{\$}(1, H) + q^T \cdot V_{\$}(1, T)\right]\Big) \\
&= \max\left((5-4)^+, \tfrac{1}{1+\frac{1}{4}} \cdot \left[\tfrac{1}{2} \cdot \tfrac{2}{5} + \tfrac{1}{2} \cdot 3\right]\right) = \max(1, \tfrac{34}{25}) = \tfrac{34}{25}
\end{aligned}$$

and thus it is better to hold the option. Thus the price of the option is given by the following binomial tree. Note that in the scenario when $\omega = T$, it is optimal to exercise the option, and continuation of the option is suboptimal. Suboptimal continuation is depicted with a dashed line.

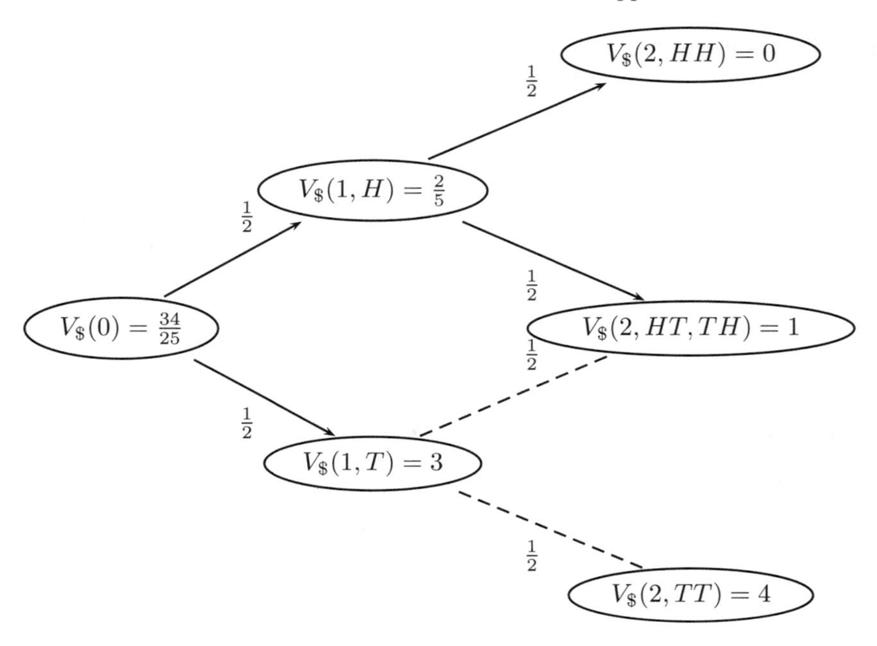

Let us consider the reference asset S. The evolution of the $\$_S$ price is given by

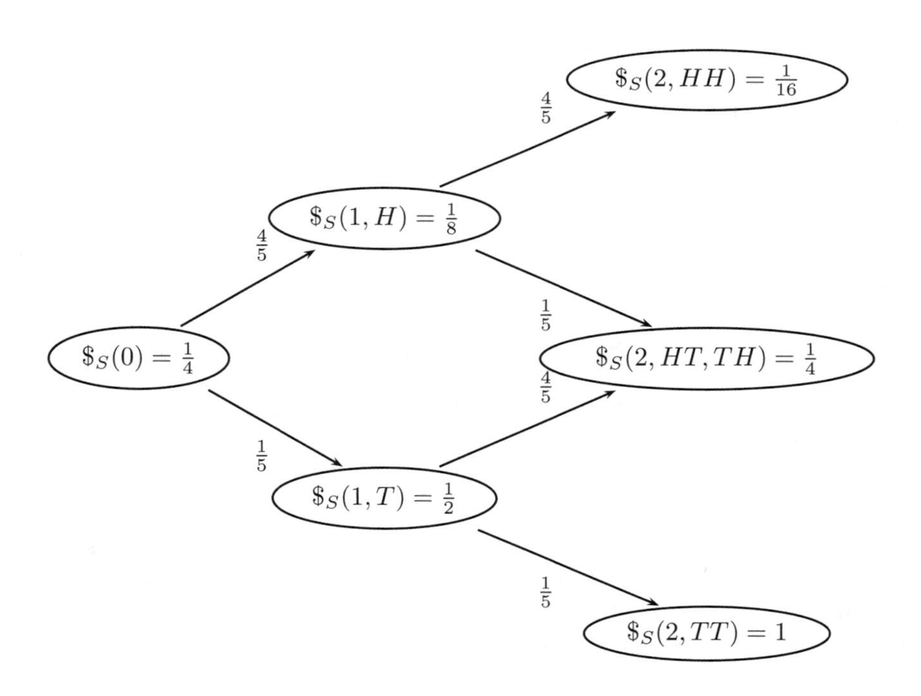

The probabilities $p^S = \mathbb{P}^S(H)$ and $q^S = \mathbb{P}^S(T)$ are given by

$$p^S = \frac{u}{1+r} \cdot \frac{1+r-d}{u-d} = \frac{2}{1+\frac{1}{4}} \cdot \frac{1+\frac{1}{4}-\frac{1}{2}}{2-\frac{1}{2}} = \frac{4}{5},$$

and

$$q^S = \frac{d}{1+r} \cdot \frac{u-(1+r)}{u-d} = \frac{\frac{1}{2}}{1+\frac{1}{4}} \cdot \frac{2-(1+\frac{1}{4})}{2-\frac{1}{2}} = \frac{1}{5}.$$

The price process $\$_S(n)$ satisfies

$$\begin{aligned} \$_S(n,\omega) &= (1+r) \cdot \mathbb{E}_n^S \left[\$_S(n+1)\right](\omega) \\ &= (1+r) \cdot \left[p^S \cdot \$_S(n+1,\omega H) + q^S \cdot \$_S(n+1,\omega T)\right] \end{aligned}$$

for $n = 0, 1$.

The intrinsic value of the option in terms of the stock is given by $f^S(x) = (5 \cdot x - 1)^+$. Similar to the case when the dollar is chosen as a reference asset, the price of the American option is computed by comparing the intrinsic value and the continuation value of the option:

$$V_S(n) = \max\Big(f^S\left(\$_S(n)\right), \mathbb{E}_n^S\left[V_S(n+1)\right]\Big). \tag{2.49}$$

The computation of the price is done recursively, starting from the maturity of the option. At $N = 2$, there is no continuation possible, and thus the value of the option is equal to its intrinsic value $f^S\left(\$_S(2)\right)$:

$$V_S(2,HH) = 0, \quad V_S(2,HT) = V_S(2,TH) = \tfrac{1}{4}, \quad V_S(2,TT) = 4.$$

We can compute the price $V_\$(n)$ for $n = 1, 0$ using the formula (2.49). We first compute $V_S(1)$ by comparing the intrinsic value and the continuation value of the option

$$\begin{aligned} V_S(1,H) &= \max\Big(f^S\left(\$_S(1,H)\right), \mathbb{E}_1^S\left[V_S(2)\right](H)\Big) \\ &= \max\Big(f^S\left(\$_S(1,H)\right), p^S \cdot V_S(2,HH) + q^S \cdot V_S(2,HT)\Big) \\ &= \max\left((\tfrac{5}{8}-1)^+, [\tfrac{4}{5} \cdot 0 + \tfrac{1}{5} \cdot \tfrac{1}{4}]\right) = \max(0, \tfrac{1}{20}) = \tfrac{1}{20}. \end{aligned}$$

The intrinsic value of the option is zero, and thus it is optimal not to exercise the option in this scenario. On the other hand

$$\begin{aligned} V_S(1,T) &= \max\Big(f^S\left(\$_S(1,T)\right), \mathbb{E}_1^S\left[V_S(2)\right](T)\Big) \\ &= \max\Big(f^S\left(\$_S(1,T)\right), p^S \cdot V_S(2,TH) + q^S \cdot V_S(2,TT)\Big) \\ &= \max\left((\tfrac{5}{2}-1)^+, [\tfrac{4}{5} \cdot \tfrac{1}{4} + \tfrac{1}{5} \cdot 4]\right) = \max(\tfrac{3}{2}, 1) = \tfrac{3}{2}, \end{aligned}$$

and thus it is better to exercise the option since the intrinsic value is larger than the continuation value. At time $n = 0$, we have

$$\begin{aligned}V_S(0) &= \max\Big(f^S\left(\$_S(0)\right), \mathbb{E}_0^S\left[V_S(1)\right]\Big)\\ &= \max\Big(f^S\left(\$_S(0)\right), p^S \cdot V_S(1,H) + q^S \cdot V_S(1,T)\Big)\\ &= \max\left(\left(\tfrac{5}{4}-1\right)^+, \left[\tfrac{4}{5}\cdot\tfrac{1}{20}+\tfrac{1}{5}\cdot\tfrac{3}{2}\right]\right) = \max\left(\tfrac{1}{4}, \tfrac{17}{50}\right) = \tfrac{17}{50}\end{aligned}$$

and thus it is better to continue. Thus the price of the option is given by the following binomial tree. Note that in the scenario when $\omega = T$, it is optimal to exercise the option, and continuation of the option is suboptimal. Suboptimal continuation is depicted with a dashed line.

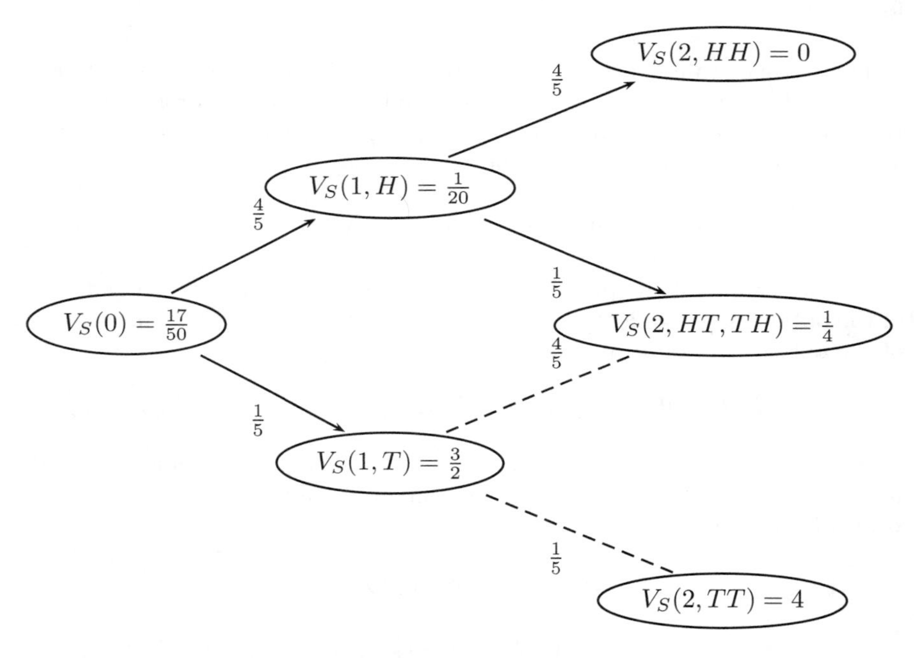

The prices of the American option V with respect to the dollar \$ and the stock S are related by the change of numeraire formula

$$\boxed{V_\$(n) = V_S(n) \cdot S_\$(n).} \tag{2.50}$$

Similarly, we have

$$\boxed{V_S(n) = V_\$(n) \cdot \$_S(n).} \tag{2.51}$$

The reader can check that this is indeed the case.

The hedging portfolio is given by

$$\begin{aligned}
P_0 &= \left(\frac{V_\$(1,H) - V_\$(1,T)}{S_\$(1,H) - S_\$(1,T)}\right) \cdot S \\
&\qquad + \left(\frac{V_S(1,H) - V_S(1,T)}{\$_S(1,H) - \$_S(1,T)}\right) \cdot (1+r)^1 \cdot B^2 \\
&= \left(\frac{\frac{2}{5} - 3}{8-2}\right) \cdot S + \left(\frac{\frac{1}{20} - \frac{3}{2}}{\frac{1}{8} - \frac{1}{2}}\right) \cdot \left(\frac{5}{4}\right) \cdot B^2 \\
&= -\tfrac{13}{30} \cdot S + \tfrac{29}{6} \cdot B^2 = -\tfrac{13}{30} \cdot S + \tfrac{232}{75} \cdot M
\end{aligned}$$

at time $n = 0$, and by

$$\begin{aligned}
P_1(H) &= \left(\frac{V_\$(2,HH) - V_\$(2,HT)}{S_\$(2,HH) - S_\$(2,HT)}\right) \cdot S \\
&\qquad + \left(\frac{V_S(2,HH) - V_S(2,HT)}{\$_S(2,HH) - \$_S(2,HT)}\right) \cdot B^2 \\
&= \left(\frac{0-1}{16-4}\right) \cdot S + \left(\frac{0 - \frac{1}{4}}{\frac{1}{16} - \frac{1}{4}}\right) \cdot B^2 \\
&= -\tfrac{1}{12} \cdot S + \tfrac{4}{3} \cdot B^2 = -\tfrac{1}{12} \cdot S + \tfrac{16}{15} \cdot M
\end{aligned}$$

at time $n = 1$. The positions in the money market M correspond to the dollar positions at each time, but the positions in the money market gain interest.

Note that the holder of the option should exercise the contract in the scenario when $\omega = T$. However, if the holder fails to exercise the option in that scenario, the seller of the option can keep the difference between the intrinsic and continuation values. The seller can create a hedging portfolio for the suboptimal continuation that costs only the continuation value of the option. The difference of the intrinsic and continuation value is given by

$$\begin{aligned}
&f^\$\left(S_\$(1,T)\right) - \tfrac{1}{1+r} \cdot \mathbb{E}_1^T\left[V_\$(2)\right](T) = \\
&\quad = f^\$\left(S_\$(1,T)\right) - \tfrac{1}{1+r} \cdot \left[p^T \cdot V_\$(2,TH) + q^T \cdot V_\$(2,TT)\right] = \\
&\qquad = (5-2)^+ - \tfrac{1}{1+\frac{1}{4}} \cdot \left[\tfrac{1}{2} \cdot 1 + \tfrac{1}{2} \cdot 4\right] = 3 - 2 = 1
\end{aligned}$$

when the reference asset is the dollar \$, or

$$\begin{aligned}
&f^S\left(\$_S(1,T)\right) - \mathbb{E}_1^S\left[V_S(2)\right](T) = \\
&\quad = f^S\left(\$_S(1,T)\right) - \left[p^S \cdot V_S(2,TH) + q^S \cdot V_S(2,TT)\right] = \\
&\qquad = \left(\tfrac{5}{2} - 1\right)^+ - \left[\tfrac{4}{5} \cdot \tfrac{1}{4} + \tfrac{1}{5} \cdot 4\right] = \tfrac{3}{2} - 1 = \tfrac{1}{2}
\end{aligned}$$

when the reference asset is a stock S. Thus the seller of the option can keep this difference

$$1\$ = \tfrac{1}{2} \cdot S,$$

and use the remaining funds

$$2\$ = 1 \cdot S$$

to hedge the option. We get

$$\begin{aligned} P_1(T) &= \left(\frac{V_\$(2,TH) - V_\$(2,TT)}{S_\$(2,TH) - S_\$(2,TT)}\right) \cdot S \\ &\qquad + \left(\frac{V_S(2,TH) - V_S(2,TT)}{\$_S(2,TH) - \$_S(2,TT)}\right) \cdot B^2 \\ &= \left(\frac{1-4}{4-1}\right) \cdot S + \left(\frac{\frac{1}{4}-4}{\frac{1}{4}-1}\right) \cdot B^2 \\ &= -1 \cdot S + 5 \cdot B^2 = -1 \cdot S + 4 \cdot M. \end{aligned}$$

References and Further Reading

The binomial model was introduced by Cox et al. (1979). An interested reader may also refer to standard textbooks of Cox and Rubinstein (1985), or Shreve (2004a).

Exercises

2.1 Consider a one-step binomial model for two no-arbitrage assets X and Y with parameters $0 < d < 1 < u$, and initial price $X_Y(0)$, meaning that $X_Y(1,H) = u \cdot X_Y(0)$, and $X_Y(1,T) = d \cdot X_Y(0)$.

(a) Compute the price of a contract that pays off $V_1 = \mathbb{I}(\omega = H) \cdot Y_1$ (in terms of a reference asset Y, and the probability measure $\mathbb{P}^Y$).

(b) Find the hedging portfolio P_0 for this contract, i.e., determine $\Delta^X(0)$ and $\Delta^Y(0)$ such that

$$P_0 = \Delta^X(0) \cdot X + \Delta^Y(0) \cdot Y,$$

with

$$P_Y(0) = V_Y(0), \quad P_Y(1) = V_Y(1).$$

(c) Compute the price of a contract that pays off $U_1 = \mathbb{I}(\omega = H) \cdot X_1$ (in terms of a reference asset X, and the probability measure $\mathbb{P}^X$).

(d) Find the hedging portfolio P_0 for this contract, i.e., determine $\Delta^X(0)$ and $\Delta^Y(0)$ such that

$$P_0 = \Delta^X(0) \cdot X + \Delta^Y(0) \cdot Y,$$

with

$$P_X(0) = U_X(0), \quad P_X(1) = U_X(1).$$

2.2 Consider a one-step binomial model for two no-arbitrage assets X and Y with parameters $0 < d < 1 < u$, and initial price $X_Y(0) = 1$. Find the price and the hedge of a contract V that pays off

$$V_1 = \max(X_1, Y_1).$$

Compute the price of the contract V using both martingale measures $\mathbb{P}^Y$ and $\mathbb{P}^X$.

2.3 Consider a two-step binomial model with general $u > 1 > d > 0$.

(a) Find the price and the hedging portfolio for a contract V that pays off

$$V_2 = \left[\tfrac{1}{2}X_Y(1) + \tfrac{1}{2}X_Y(2)\right] \cdot Y_2.$$

Note that the price and the hedge do not depend on the choice of parameters u and d.

(b) How can one lock an arbitrage opportunity if somebody is offering to buy or to sell V_0 for $1.05 \times X_0$?

(c) Determine the price and the hedge of a contract A that pays off

$$A_2 = (V_Y(2) - 2)^+ \cdot Y_2,$$

using the parameters $u = 2$, $d = \frac{1}{2}$, and $X_Y(0) = 4$.

2.4 Consider an American contract V that pays off $\max(5\$_\tau, S_\tau) = \max(5, S_\$(\tau)) \cdot \$_\tau$ in a two-step binomial model with parameters $u = 2$, $d = \frac{1}{2}$, $r = \frac{1}{4}$, and $S_\$(0) = 4$.

(a) Find the price $V_\$(n)$ for $n = 0, 1, 2$. Determine the optimal stopping strategy τ^*.

(b) Find the hedging portfolio.

(c) What is the price and an optimal stopping strategy for an American contract that pays off $\min(5\$_\tau, S_\tau)$?

2.5 Consider a two-step binomial model with two no-arbitrage assets X and Y and parameters $u > 1 > d > 0$. Let A be a contract with a payoff

$$A_2 = \tfrac{1}{3}\left(X_Y(0) + X_Y(1) + X_Y(2)\right) \cdot Y_2.$$

This is an average asset. Assume that $u = 2$, $d = \frac{1}{2}$, $X_Y(0) = 4$. Consider a contract that pays off $V_2 = (A_2 - X_2)^+$ (Asian floating strike option).

(a) Compute the prices $V_Y(n)$ using the measure $\mathbb{P}^Y$. Note that

$$V_Y(2) = (A_Y(2) - X_Y(2))^+.$$

(b) Compute $V_X(n)$ using the measure $\mathbb{P}^X$. Note that

$$V_X(2) = (A_X(2) - 1)^+.$$

(c) Determine the hedging strategy for V in the assets X and Y.

(d) The asset A admits a model independent hedge ($\Delta^X(0) = \frac{2}{3}, \Delta^X(1) = \frac{1}{3}$), and it is a no-arbitrage asset. Therefore there is a probability measure $\mathbb{P}^A$ under which the prices of no-arbitrage assets with respect to the reference asset A are martingales. Determine $\mathbb{P}^A$ by giving $\mathbb{P}^A(HH)$, $\mathbb{P}^A(HT)$, $\mathbb{P}^A(TH)$, $\mathbb{P}^A(TT)$. These are functions of parameters u and d. The easiest way is to employ the Radon–Nikodým derivative with respect to a martingale measure $\mathbb{P}^Y$. Recall that

$$\mathbb{P}^A(\omega) = \frac{A_Y(2,\omega)}{A_Y(0)} \cdot \mathbb{P}^Y(\omega).$$

Determine also $\mathbb{P}^A(H)$ and $\mathbb{P}^A(T)$ (probabilities after one time step) using a similar formula, and $\mathbb{P}^A(HH|H)$, $\mathbb{P}^A(HT|H)$, $\mathbb{P}^A(TH|T)$, $\mathbb{P}^A(TT|T)$ (conditional probabilities from time 1 to time 2).

(e) Compute $V_A(n)$ using the measure $\mathbb{P}^A$. Note that

$$V_A(2) = (1 - X_A(2))^+.$$

Chapter 3

Diffusion Models

This chapter introduces diffusion models. Under very broad conditions, all no-arbitrage models of a continuous price evolution are diffusion models. In other words, every continuous evolution of the price can be expressed as an Ito's integral. This result is known as a Martingale Representation Theorem.

Diffusion models of price use Brownian motion to represent market noise. Since the market noise itself can take negative values, it does not serve as a good model for the prices. However, we can take the corresponding stochastic exponential which is a positive martingale, and thus is perfectly suitable for a no-arbitrage model of a price process. The simplest model assumes a constant volatility that leads to a geometric Brownian motion. While most of the real price processes do not have constant volatility, this assumption still results in reasonable models for prices and hedges of complex financial instruments. Moreover, the prices of many financial products in the geometric Brownian motion model admit closed form solutions, and thus they are easy to use.

In order to compute the prices of financial derivatives, we need to determine the martingale measures that correspond to all the assets relevant to the given contract. For instance, the Black–Scholes formula for the price of the European call option uses both probability measures $\mathbb{P}^X$ and $\mathbb{P}^Y$ that are associated with the assets X and Y. The probability measure $\mathbb{P}^X$ can be determined from the evolution of the inverse price $Y_X(t)$, and this price has to be a $\mathbb{P}^X$ martingale. It turns out that the evolution of the inverse price $Y_X(t)$ is also a geometric Brownian motion, but the market noise W^X is associated with the reference asset X.

Diffusion models have one important property: every no-arbitrage asset comes with its own market noise. An asset Y has a market noise W^Y, and an asset X has a market noise W^X. Although W^Y and W^X are perfectly correlated in the geometric Brownian motion model, we can always identify the market noise that comes with each individual asset. Even more complicated assets, such as a power option, or an average asset, come with its own market noise. This fact will be used for pricing barrier, lookback, and Asian options in the subsequent chapters.

The first section introduces the geometric Brownian motion model, and studies the evolution of the prices X_Y and Y_X under the corresponding martingale measures $\mathbb{P}^Y$ and $\mathbb{P}^X$. We also show that the measures $\mathbb{P}^Y$ and $\mathbb{P}^X$ have the interpretation of how costly a given event is if settled in terms of the asset Y, or in terms of the asset X, respectively. The second section introduces general European contracts. European contracts are contracts on two assets that are defined by the payoff function, which can be expressed in terms of each reference asset Y or X. The two payoff functions are related by a formula known as a perspective mapping. Some contracts remain the same if the roles of the assets Y and X is switched in the payoff function; for instance the best of the two assets defined as $\max(X_T, Y_T)$ is the same as $\max(Y_T, X_T)$. The best of the two assets naturally leads to European call and put options with the payoff

$$(X_T - K \cdot Y_T)^+ = \max(X_T, K \cdot Y_T) - K \cdot Y_T.$$

We give examples of European call and put options that appear in different markets: a stock option, a currency option, an exchange option, or a caplet. Their prices and the hedging portfolios are given by the Black–Scholes formula. We compute all prices in terms of the no-arbitrage assets so that we can employ the First Fundamental Theorem of Asset Pricing directly. In order to get the prices in terms of a dollar, an arbitrage asset, we can trivially apply the change of numeraire formula to the prices computed with respect to no-arbitrage assets.

The price of a contingent claim can be computed by two alternative methods: by computing the conditional expectation, or by solving the associated partial differential equation. The case of European options is usually simple enough to obtain closed form formulas, but both approaches also work for more complicated products when no close formula is known. The conditional expectation can be approximated by Monte Carlo methods, and the partial differential equation can be solved numerically by applying finite difference techniques.

The primary goal of contingent pricing is to find the dollar price of a given contract. **Our text suggests to compute the price of a contingent claim with respect to a no-arbitrage asset first, such as a corresponding bond, and then convert it to the dollar price using the change of numeraire. This approach is valid in general, and it has clear computational advantages when the contingent claim is more complex, such as in the case of exotic options.** However, the dollar prices of European claims also satisfy a certain and more complicated partial differential equation that is obtained by discounting to the dollar prices of the underlying assets. But this partial differential equation does not hold in general, it assumes a deterministic evolution of the interest rate. We mention it in our text since the partial differential equation in terms of dollars is the most widely used in practice. For simple contracts, such as for European options, it does not make a difference to compute the prices under different reference assets (arbitrage or no-arbitrage) since the price of the contract is simple to determine. The only loss when computing the dollar prices directly from the corresponding partial differential equation approach is that the approach does not apply to stochastic interest rates. In that case one should compute the prices in terms of the bond, and convert it to dollar prices by changing the numeraire.

For more complex products, such as for barrier, lookback, or Asian options, using the no-arbitrage asset as a numeraire leads to significant computational advantages. On the other hand, American options have to use dollar values in order to compare the intrinsic and the continuation values, and the partial differential equation in terms of dollars has to be used. In the case of the American option, it is the setup of the contract that forces us to use the partial differential equation in terms of a dollar.

We also discuss how to construct the hedging portfolios for European contracts. The hedging must always be done in the two underlying no-arbitrage assets. We determine the hedging positions in both assets. We can also get a similar expression for the hedging positions in terms of the dollar price functions. The hedging positions for European call options are bounded in both assets; the position in the asset X is always between zero and one, and the hedging position in the asset Y is always between minus the strike K and zero.

We also briefly introduce stochastic volatility models. The price of the contract is still considered to be Markov, but it depends on two parameters: the price $X_Y(t)$ of the asset X and stochastic volatility $\xi(t)$. The resulting partial differential equation for the price of the derivative security becomes two dimensional in space. The chapter is concluded with an example of a European option contract in the foreign exchange market which is just a special case of the general approach presented in the previous text.

3.1 Geometric Brownian Motion

Assume that the two assets X and Y are no-arbitrage assets. We have seen that the price $X_Y(t)$ must be a $\mathbb{P}^Y$ martingale in order to prevent any arbitrage opportunity. In continuous time, a general martingale can be written as a sum of a martingale with continuous paths and a purely discontinuous martingale:

$$\mathcal{M}(t) = \mathcal{M}^c(t) + \mathcal{M}^d(t). \tag{3.1}$$

Continuous martingales adapted to a filtration $\mathcal{F}_t^W$ generated by a Brownian motion W are in fact diffusions; they can be represented as stochastic integrals with respect to Brownian motion. Thus

$$\mathcal{M}^c(t) = \mathcal{M}^c(0) + \int_0^t \phi(s)dW(s), \tag{3.2}$$

where $\phi(t)$ is adapted to $\mathcal{F}_t^W$. This result is known as a **Martingale Representation Theorem** (Theorem A.3).

This chapter focuses on price models with continuous paths. The process $X_Y(t)$ must have the form

$$dX_Y(t) = \phi(t)dW(t).$$

Let us start with the simple but very popular model when

$$\phi(t) = \sigma X_Y(t).$$

The price process $X_Y(t)$ follows

$$dX_Y(t) = \sigma X_Y(t)dW^Y(t), \tag{3.3}$$

which is known as a **geometric Brownian motion.** The parameter σ is referred to as **volatility**. Volatility is inherent to diffusion models. Similar to price, volatility is a pairwise relationship between two assets X and Y. The price X_Y of the asset X with respect to a reference asset Y may have very different volatility than the price X_Z with respect to a different reference asset Z. For instance, a typical dollar stock price $S_\$$ is more volatile than a stock price S_I taken with respect to a market index I. Sometimes we will denote by σ_{xy} the volatility that corresponds to the assets X and Y.

A natural question is how the measure $\mathbb{P}^Y$ is determined. Under $\mathbb{P}^Y$, the driving process $W^Y(t)$ is a Brownian motion. Also the above stochastic differential equation has the following solution:

$$\boxed{X_Y(t) = X_Y(0) \cdot \exp\left(\sigma W^Y(t) - \tfrac{1}{2}\sigma^2 t\right).} \tag{3.4}$$

Note that $X_Y(t)$ is a $\mathbb{P}^Y$ martingale.

In order to compute the prices of European options and other derivative securities, we also need to determine the probability measure $\mathbb{P}^X$. The role of X and Y should be exchangeable in models that preserve the symmetry between both assets. Mathematically, this requirement translates to

$$\mathcal{L}_t^Y\left(\frac{X_Y(T)}{X_Y(t)}\right) = \mathcal{L}_t^X\left(\frac{Y_X(T)}{Y_X(t)}\right), \tag{3.5}$$

meaning that the price increment $\frac{X_Y(T)}{X_Y(t)}$ under the probability measure $\mathbb{P}^Y$ should have the same distribution as the price increment $\frac{Y_X(T)}{Y_X(t)}$ under the probability measure $\mathbb{P}^X$. Therefore we need to have a description of the dynamics of the inverse price, $Y_X(t)$, that would be analogous to the dynamics of the original price $X_Y(t)$. Ideally, the evolution of this price should have the same form as (3.3), but the dynamics are already determined by Ito's formula (see Appendix):

$$\begin{aligned} dY_X(t) = dX_Y(t)^{-1} &= -X_Y(t)^{-2}dX_Y(t) + \tfrac{1}{2}\cdot 2X_Y(t)^{-3}d^2X_Y(t) \\ &= -\sigma Y_X(t)dW^Y(t) + \sigma^2 Y_X(t)dt \\ &= \sigma Y_X(t)\cdot\left(-dW^Y(t) + \sigma dt\right). \end{aligned} \tag{3.6}$$

Given the exchangeability argument of X and Y, we should also have

$$dY_X(t) = \sigma Y_X(t)dW_t^X, \tag{3.7}$$

which is the same stochastic differential equation as (3.3) with X and Y flipped, and with a different Brownian motion $W^X(t)$ under the measure $\mathbb{P}^X$. The solution of the above stochastic differential equation is given by

$$\boxed{Y_X(t) = Y_X(0)\cdot\exp\left(\sigma W^X(t) - \tfrac{1}{2}\sigma^2 t\right).} \tag{3.8}$$

In diffusion models, each reference asset Y has its own market noise that is represented by one or several Brownian motions $W^{i,Y}(t)$. Other reference assets, such as an asset X, have different market noise that is represented by Brownian motions $W^{i,X}(t)$. Obviously, the Brownian motions W^Y and W^X are related. In the above case, we just have one Brownian motion for each asset, and the relationship between $W^X(t)$ and $W^Y(t)$ follows from the equation

$$dY_X(t) = \sigma Y_X(t)\cdot\left(-dW^Y(t) + \sigma dt\right) = \sigma Y_X(t)dW^X(t). \tag{3.9}$$

Thus we must have

$$dW^X(t) = -dW^Y(t) + \sigma dt,$$

or in other words,

$$W^X(t) = -W^Y(t) + \sigma t. \tag{3.10}$$

Note that a symmetric relationship holds as well

$$W^Y(t) = -W^X(t) + \sigma t. \tag{3.11}$$

REMARK 3.1 Some authors define $dW^X(t)$ as $dW^Y(t) + \sigma dt$ which is an equivalent definition since the Brownian motion is symmetric and thus $dW^Y(t)$ has the same distribution as $-dW^Y(t)$. However, such a definition would break the symmetry of the price formulas for X and Y, and thus it is more appropriate to use $dW^X(t) = -dW^Y(t) + \sigma dt$. □

The two Brownian motions $W^Y(t)$ and $W^X(t)$ are perfectly correlated with a correlation coefficient of -1:

$$dW^Y(t) \cdot dW^X(t) = -1 \cdot dt.$$

This makes sense since when $X_Y(t)$ goes up, the inverse price $Y_X(t)$ goes down, and vice versa.

From the financial representation of the Radon–Nikodým derivative we have

$$\mathbb{Z}(t) = \frac{d\mathbb{P}^X_t}{d\mathbb{P}^Y_t} = \frac{X_Y(t)}{X_Y(0)} = \exp\left(\sigma W^Y(t) - \tfrac{1}{2}\sigma^2 t\right) = \exp\left(-\sigma W^X(t) + \tfrac{1}{2}\sigma^2 t\right). \tag{3.12}$$

The concept of equivalent treatment of both X and Y is also supported by the following theorem.

***THEOREM 3.1* (Girsanov).**
Let $W^Y(t)$ be a $\mathbb{P}^Y$ Brownian motion. Then $W^X(t) = -W^Y(t) + \sigma t$ is a $\mathbb{P}^X$ Brownian motion, where $\mathbb{Z}(t) = \frac{d\mathbb{P}^X_t}{d\mathbb{P}^Y_t} = \frac{X_Y(t)}{X_Y(0)} = \exp\left(\sigma W^Y(t) - \tfrac{1}{2}\sigma^2 t\right)$.

REMARK 3.2 The two measures $\mathbb{P}^Y$ and $\mathbb{P}^X$ may disagree on the drift of the Brownian motion. More specifically,

$$\mathbb{E}^Y[W^X(t)] = \mathbb{E}^Y[-W^Y(t) + \sigma t] = \sigma t,$$

but

$$\mathbb{E}^X[W^X(t)] = 0.$$

The last statement can be proved by a change of the measure argument (1.66)

$$\begin{aligned}
\mathbb{E}^X[W^X(t)] &= \mathbb{E}^X[-W^Y(t) + \sigma t] = \mathbb{E}^Y[(-W^Y(t))Z(t)] + \sigma t \\
&= \mathbb{E}^Y[-W^Y(t) \cdot \exp\left(\sigma W^Y(t) - \tfrac{1}{2}\sigma^2 t\right)] + \sigma t \\
&= -\exp(-\tfrac{1}{2}\sigma^2 t) \cdot \mathbb{E}^Y[W^Y(t) \cdot \exp\left(\sigma W^Y(t)\right)] + \sigma t \\
&= -\exp(-\tfrac{1}{2}\sigma^2 t) \cdot \tfrac{d}{d\sigma}\mathbb{E}^Y[\exp\left(\sigma W^Y(t)\right)] + \sigma t \\
&= -\exp(-\tfrac{1}{2}\sigma^2 t) \cdot \tfrac{d}{d\sigma}[\exp(\tfrac{1}{2}\sigma^2 t)] + \sigma t \\
&= 0.
\end{aligned}$$

□

Note that we have

$$\frac{d^2X_Y(t)}{X_Y(t)^2} = \sigma^2 dt, \tag{3.13}$$

as well as

$$\frac{d^2Y_X(t)}{Y_X(t)^2} = \sigma^2 dt, \tag{3.14}$$

and thus the volatility of $X_Y(t)$ is the same as the volatility of $Y_X(t)$. It does not matter which of the two assets, X or Y, is chosen as a numeraire. For instance the volatility of the dollar/euro exchange rate is the same as the volatility of the euro/dollar exchange rate. This is true even when the volatility is stochastic.

Having the closed form expressions for the price $X_Y(T)$ from Equation (3.3) and for the price $Y_X(T)$ from Equation (3.8), we can determine the prices of Arrow–Debreu securities that pay off either $\mathbb{I}_A(\omega)$ units of an asset Y at time T, or $\mathbb{I}_A(\omega)$ units of an asset X at the same time. Let us consider events A of the form

$$A = \{\omega \in \Omega : X_Y(T,\omega) \geq K\}$$

for a given constant K. A is a set of scenarios where the terminal price of $X_Y(T)$ exceeds a level K. Let us determine the price of a contract U that pays off

$$U_T = \mathbb{I}_A(\omega) \cdot Y_T.$$

Since the price of this contract is a martingale under the $\mathbb{P}^Y$ measure, we have

$$U_Y(t) = \mathbb{E}_t^Y [\mathbb{I}_A(\omega)] = \mathbb{P}_t^Y(A).$$

The event

$$A = \{X_Y(T) \geq K\}$$

is equivalent to

$$X_Y(t) \cdot \exp\left(\sigma(W^Y(T) - W^Y(t)) - \tfrac{1}{2}\sigma^2(T-t)\right) \geq K,$$

or in other words

$$-\frac{W^Y(T) - W^Y(t)}{\sqrt{T-t}} \leq \frac{1}{\sigma\sqrt{T-t}} \cdot \log\left(\tfrac{X_Y(t)}{K}\right) - \tfrac{1}{2}\sigma\sqrt{T-t}.$$

Since $-\frac{W^Y(T)-W^Y(t)}{\sqrt{T-t}}$ has a normal distribution with zero mean and a unit variance $N(0,1)$ under the probability measure $\mathbb{P}^Y$, the probability of the event A is given by

$$\mathbb{P}_t^Y(A) = \mathbb{P}_t^Y(X_Y(T) \geq K) = N\left(\tfrac{1}{\sigma\sqrt{T-t}} \cdot \log\left(\tfrac{X_Y(t)}{K}\right) - \tfrac{1}{2}\sigma\sqrt{T-t}\right), \tag{3.15}$$

where $N(\cdot)$ is a cumulative distribution function of a standard normal variable

$$N(x) = \int_{-\infty}^{x} \frac{1}{\sqrt{2\pi}} \cdot e^{-\frac{y^2}{2}} dy.$$

We can determine the price of the Arrow–Debreu security V that pays off $\mathbb{I}_A(\omega)$ units of X at time T in a similar fashion. At time T we have

$$V_T = \mathbb{I}_A(\omega) \cdot X_T.$$

The price $V_X(t)$ is a $\mathbb{P}^X$ martingale, and thus

$$V_X(t) = \mathbb{E}_t^X[\mathbb{I}_A(\omega)] = \mathbb{P}_t^X(A).$$

The event

$$A = \{X_Y(T) \geq K\}$$

is equivalent to

$$X_Y(t) \cdot \exp\left(-\sigma(W^X(T) - W^X(t)) + \tfrac{1}{2}\sigma^2(T-t)\right) \geq K.$$

Here we used the fact that

$$\begin{aligned} X_Y(T) = \frac{1}{Y_X(T)} &= \frac{1}{Y_X(t) \cdot \exp\left(\sigma(W^X(T) - W^X(t)) - \frac{1}{2}\sigma^2(T-t)\right)} \\ &= X_Y(t) \cdot \exp\left(-\sigma(W^X(T) - W^X(t)) + \tfrac{1}{2}\sigma^2(T-t)\right). \end{aligned}$$

We need to express the price of $X_Y(T)$ in terms of the Brownian motion $W^X(t)$ in order to determine the probability of the event A using the $\mathbb{P}^X$ measure. The event A is equivalent to

$$\frac{W^X(T) - W^X(t)}{\sqrt{T-t}} \leq \frac{1}{\sigma\sqrt{T-t}} \cdot \log\left(\frac{X_Y(t)}{K}\right) + \tfrac{1}{2}\sigma\sqrt{T-t}.$$

Therefore

$$\mathbb{P}_t^X(A) = \mathbb{P}_t^X(X_Y(T) \geq K) = N\left(\tfrac{1}{\sigma\sqrt{T-t}} \cdot \log\left(\tfrac{X_Y(t)}{K}\right) + \tfrac{1}{2}\sigma\sqrt{T-t}\right). \quad (3.16)$$

The Arrow–Debreu securities $U_T = \mathbb{I}_A(\omega) \cdot X_T$ and $V_T = \mathbb{I}_A(\omega) \cdot Y_T$ are also known as **digital options.** Determination of their hedging portfolios is the subject of Exercise 3.4.

REMARK 3.3

It is interesting to note that when $K = X_Y(0)$, we have

$$\mathbb{P}^Y(X_Y(T) \geq X_Y(0)) = N\left(-\tfrac{1}{2}\sigma\sqrt{T}\right) < \tfrac{1}{2},$$

and

$$\mathbb{P}^X(X_Y(T) \geq X_Y(0)) = N\left(\tfrac{1}{2}\sigma\sqrt{T}\right) > \tfrac{1}{2}.$$

A delivery of a unit of Y when the price X_Y of the asset X moves up requires less than a $\frac{1}{2}$ unit of Y to start with. On the other hand, a delivery of a unit of X on the same event requires more than a $\frac{1}{2}$ unit of an asset X. In this sense, the asset Y is "cheaper" (it requires a smaller fraction of the underlying asset) to deliver than the asset X on the up movement of the price X_Y. □

3.2 General European Contracts

A general European-type contract pays off either $f^Y(X_Y(T))$ units of an asset Y, or $f^X(Y_X(T))$ units of an asset X at time T. In order that these two payoffs correspond to the same contract, we must have

$$f^Y(X_Y(T)) \cdot Y = f^X(Y_X(T)) \cdot X$$

or in other words,

$$f^Y(X_Y(T)) \cdot Y = f^X\left(\frac{1}{X_Y(T)}\right) \cdot X_Y(T) \cdot Y.$$

Therefore the two payoff functions f^Y and f^X are linked by the following symmetric relationship

$$\boxed{f^Y(x) = f^X\left(\tfrac{1}{x}\right) \cdot x,} \quad \text{or} \quad \boxed{f^X(x) = f^Y\left(\tfrac{1}{x}\right) \cdot x,} \tag{3.17}$$

which is valid for $0 < x < \infty$, meaning that neither the asset X nor the asset Y is worthless. Note that the payoff function depends on a choice of the reference asset. The formulas that link functions f^Y and f^X are known as a **perspective mapping.** A financial contract with a non-negative payoff function $f^Y(x)$ is known as an **option.** Note that $f^Y(x) \geq 0$ is equivalent to $f^X(x) \geq 0$, so the definition of the option does not depend on the choice of the reference asset. An option of this type is also known as a **plain vanilla option.** The perspective mapping also preserves convexity; $f^Y(x)$ is convex if and only if $f^X(x)$ is convex (see Exercise 3.1).

Example 3.1 **The best asset and the worst asset**
The simplest contract on two assets one can think of is the best of the two assets, or the worst of the two assets. The best of the two assets contract pays off $\max(X_T, Y_T)$ at time T; the worst of the two assets contract pays off $\min(X_T, Y_T)$ at time T. These contracts are completely symmetric since

$$\max(X_T, Y_T) = \max(Y_T, X_T),$$

and

$$\min(X_T, Y_T) = \min(Y_T, X_T).$$

When the best of the two assets contract is settled in the asset Y, the contract pays off $\max(X_Y(T), 1)$ units of Y. Similarly, when the best of the two assets contract is settled in the asset X, the contract pays off $\max(Y_X(T), 1)$ units of X. The payoff functions for the best of the two assets are thus given by

$$f^Y(x) = \max(x, 1),$$

and

$$f^X(x) = f^Y(\tfrac{1}{x}) \cdot x = \max(\tfrac{1}{x}, 1) \cdot x = \max(1, x).$$

Note that we have $f^X(x) = f^Y(x)$. Analogously, the payoff functions for the worst of the two assets are given by

$$f^Y(x) = \min(x, 1),$$

$$f^X(x) = f^Y(\tfrac{1}{x}) \cdot x = \min(x, 1).$$

Note that the payoff of the best asset contract can be re-expressed in the following form

$$\max(X_T, Y_T) = (X_T - Y_T)^+ + Y_T = (Y_T - X_T)^+ + X_T,$$

where $x^+ = \max(x, 0)$, leading us to contracts known as the call and the put options. □

The most typical traded contract that has the feature of paying the best asset is a **convertible bond.** One of the payments of the convertible bond is $\max(S_T, K \cdot B_T^T)$, so the holder of this contract can choose between the equity position in the asset S, and K units of the bond B^T at the expiration time T.

However, the logic of the financial markets is to allow for maximal leverage, and in this respect, the contract that delivers the best asset is not ideal as it ties down a portion of the capital of the investor that can be used otherwise. Instead, one can trade just the differences between the best asset and the asset itself, which requires significantly less capital. The contract on the difference of the best asset and the asset itself is known as a call option. Formally, a **European call option** $V^{EC}(X, K \cdot Y, T)$ is a contract that pays off

$$\boxed{(X_T - K \cdot Y_T)^+,} \tag{3.18}$$

where X and Y are two assets. The constant K is known as the **strike.** The relationship between the European call option and the contract that delivers the best asset is given by

$$\max(X_T, K \cdot Y_T) = (X_T - K \cdot Y_T)^+ + K \cdot Y_T.$$

In the contract that delivers we may rescale one of the assets by a factor of K to achieve a better proportionality of the assets X and Y. Note that the European call option is a combination of two Arrow–Debreu securities

$$\boxed{(X_T - K \cdot Y_T)^+ = \mathbb{I}(X_Y(T) \geq K) \cdot X - K \cdot \mathbb{I}(X_Y(T) \geq K) \cdot Y.} \tag{3.19}$$

The first Arrow–Debreu security pays off $\mathbb{I}(X_Y(T) \geq K)$ units of the asset X, the second Arrow–Debreu security pays off $\mathbb{I}(X_Y(T) \geq K)$ units of the asset Y.

A closely related contract to a European call option is a **European put option** $V^{EP}(K \cdot Y, X, T)$ with a payoff

$$\boxed{(K \cdot Y_T - X_T)^+.} \tag{3.20}$$

The put option is also related to the contract that delivers the best asset by

$$\max(X_T, K \cdot Y_T) = (K \cdot Y_T - X_T)^+ + X_T.$$

The only difference between the call and the put option is which of the two available assets is chosen to be subtracted from the payoff of the contract on the best asset. Since this choice is arbitrary, the call option on assets X and $K \cdot Y$ is the same contract as a put option on assets $K \cdot Y$ and X. This relationship is known as the **put-call duality:**

$$\boxed{V^{EC}(X, K \cdot Y, T) = V^{EP}(K \cdot Y, X, T) = K \cdot V^{EP}(Y, \tfrac{X}{K}, T).} \tag{3.21}$$

Another simple relationship between European call and European put options is a **put-call parity.** Note that

$$X_T - K \cdot Y_T = (X_T - K \cdot Y_T)^+ - (K \cdot Y_T - X_T)^+, \tag{3.22}$$

where $X_T - K \cdot Y_T$ is a payoff of a forward contract $F(X, K \cdot Y, T)$. The relationship between the forward contract and the corresponding call and put options holds at all times $t \leq T$:

$$\boxed{F(X, K \cdot Y, T) = V^{EC}(X, K \cdot Y, T) - V^{EP}(X, K \cdot Y, T).} \tag{3.23}$$

A European call option payoff can be written in the following equivalent ways

$$(X_T - K \cdot Y_T)^+ = (X_Y(T) - K)^+ \cdot Y_T = (1 - K \cdot Y_X(T))^+ \cdot X_T. \tag{3.24}$$

When the European call option is settled in the asset Y, the payoff is given by

$$\boxed{(X_Y(T) - K)^+ \cdot Y} \tag{3.25}$$

which corresponds to a payoff function $f^Y(x) = (x - K)^+$. The holder receives $(X_Y(T) - K)^+$ units of Y at time T. Similarly, the European call option settled in the asset X has the payoff

$$\boxed{(1 - K \cdot Y_X(T))^+ \cdot X} \tag{3.26}$$

which corresponds to a payoff function $f^X(x) = (1 - K \cdot x)^+$. Note that $f^X(\frac{1}{x}) \cdot x = (1 - K \cdot \frac{1}{x})^+ \cdot x = (x - K)^+ = f^Y(x)$. The holder receives $(1 - K \cdot Y_X(T))^+$ units of X at time T.

European-type contracts can always be expressed in terms of two no-arbitrage assets.

REMARK 3.4 European option as a contract on two no-arbitrage assets
A European option can always be expressed as a contract on two no-arbitrage assets. The payoff of a European option is defined as $f^Y(X_Y(T))$ units of an asset Y, or $f^X(Y_X(T))$ units of an asset X at time T for general assets with positive price X and Y. When X or Y is an arbitrage asset, such as the dollar \$, we can substitute an arbitrage asset X (or Y) with a corresponding no-arbitrage asset U or V that delivers a unit of an asset X or an asset Y at time T. In particular, we have

$$U_T = X_T, \quad V_T = Y_T.$$

Thus the European option payoff can be re-expressed as $f^Y(U_V(T)) = f^V(U_V(T))$ units of an asset V, or $f^X(V_U(T)) = f^U(V_U(T))$ units of an asset U at time T for two no-arbitrage assets U and V. This substitution is not possible when there is no fixed delivery of the option payoff such as in the case of American options. □

***Example 3.2* European call option in different markets**

Stock option When the asset X is a stock S, and the asset Y is a dollar **\$**, we have a European stock option

$$(S_T - K \cdot \$_T)^+. \tag{3.27}$$

Note that the existing literature typically omits the fact that the strike is in fact multiplied by the dollar **\$**. This notation means that the holder of the option has the right to increase his position in the stock S by one unit, and decrease his position in the dollar **\$** by K units at time T.

Should the contract be settled in dollars, one can write the payoff as

$$(S_{\$}(T) - K)^+ \cdot \$_T. \tag{3.28}$$

The holder receives $(S_{\$}(T) - K)^+$ units of the dollar **\$** at time T. As noted earlier, the European option on a stock may also be settled in terms of a bond B^T, a contract that delivers 1 **\$** at time T, so that $B^T_T = 1\$_T$. In this case, the payoff of the option may be written as

$$(S_T - K \cdot B^T_T)^+. \tag{3.29}$$

This fact is useful in the pricing of this option. In contrast to the dollar, the bond does not create arbitrage opportunities in time. Therefore it can be used as a natural reference asset for pricing this option. The option can be settled entirely in the bond

$$(S_{B^T}(T) - K)^+ \cdot B^T_T, \tag{3.30}$$

or in the stock

$$\left(1 - K \cdot B^T_S(T)\right)^+ \cdot S_T. \tag{3.31}$$

Exchange option When the asset X is a stock S^1, and the asset Y is another stock S^2, the corresponding European call option

$$(S^1_T - K \cdot S^2_T)^+ \tag{3.32}$$

is known as an **exchange option.** The natural reference asset for pricing this option is either the stock S^1, or the stock S^2. Adding another reference asset, such as a dollar **\$**, for pricing this option would only increase the dimensionality of the problem.

Currency option When the asset X is a euro **€**, and the asset Y is a dollar **\$** (or any other currencies), we have a European currency option

$$(€_T - K \cdot \$_T)^+. \tag{3.33}$$

A European currency option can be settled in the dollar or in the euro only

$$(€_T - K \cdot \$_T)^+ = (€_{\$}(T) - K)^+ \cdot \$_T = (1 - K \cdot \$_{€}(T))^+ \cdot €_T.$$

In order to express the payoff in terms of no-arbitrage assets only, we can take a foreign bond $B^{€,T}$ that delivers a unit of a foreign currency **€** at time T, and a domestic bond B^T that delivers a unit of a domestic currency **\$** at time T. The payoff of the currency option is equivalent to

$$(€_T - K \cdot \$_T)^+ = (B^{€,T}_T - K \cdot B^T_T)^+.$$

Caplet A caplet is an option on a LIBOR that pays off

$$(L(T,T) - K)^+ \cdot \$_{T+\delta}. \tag{3.34}$$

The LIBOR $L(T,T)$ is observed at time T, but the contract is settled at a later time $T+\delta$ in a dollar **\$**. Here it is not entirely obvious what the corresponding assets X and Y should be. But from the definition of the LIBOR

$$L(T,T) = \frac{B_\$^T(T) - B_\$^{T+\delta}(T)}{\delta B_\$^{T+\delta}(T)} = [B^T - B^{T+\delta}]_{\delta B^{T+\delta}}(T),$$

and using the fact that $B_{T+\delta}^{T+\delta} = \$_{T+\delta}$, we can rewrite the payoff as

$$\begin{aligned}(L(T,T) - K)^+ \cdot \$_{T+\delta} \\ &= \left([B^T - B^{T+\delta}]_{\delta B^{T+\delta}}(T) - K\right)^+ \cdot B_{T+\delta}^{T+\delta} \\ &= \tfrac{1}{\delta} \cdot \left([B^T - B^{T+\delta}]_T - K \cdot \delta B_{T+\delta}^{T+\delta}\right)^+.\end{aligned}$$

Thus the asset X is a combination of two bonds $[B^T - B^{T+\delta}]$, and the asset Y is $\delta B^{T+\delta}$.

□

REMARK 3.5 European call option price
We have already seen that a European call option is just a combination of two Arrow–Debreu securities

$$(X_T - K \cdot Y_T)^+ = \mathbb{I}(X_Y(T) \geq K) \cdot X_T - K \cdot \mathbb{I}(X_Y(T) \geq K) \cdot Y_T. \tag{3.35}$$

The first Arrow–Debreu security costs $\mathbb{P}_t^X(X_Y(T) \geq K)$ units of the asset X, the second Arrow–Debreu security costs $\mathbb{P}^Y(X_Y(T) \geq K)$ units of the asset Y. Therefore we have the following result:

THEOREM 3.2 Black–Scholes formula
The price of a European option contract $V^{EC}(X, K \cdot Y, T)$ with the payoff $(X_T - K \cdot Y_T)^+$ is given by

$$\boxed{V^{EC}(X, K \cdot Y, T) = \mathbb{P}_t^X(X_Y(T) \geq K) \cdot X - K\mathbb{P}_t^Y(X_Y(T) \geq K) \cdot Y.} \tag{3.36}$$

Recall from the previous section (Equations (3.16) and (3.15)) that for the geometric Brownian motion model, we have

$$\mathbb{P}_t^X(X_Y(T) \geq K) = N\left(\tfrac{1}{\sigma\sqrt{T-t}} \cdot \log\left(\tfrac{X_Y(t)}{K}\right) + \tfrac{1}{2}\sigma\sqrt{T-t}\right),$$

and

$$\mathbb{P}_t^Y(X_Y(T) \geq K) = N\left(\frac{1}{\sigma\sqrt{T-t}} \cdot \log\left(\frac{X_Y(t)}{K}\right) - \frac{1}{2}\sigma\sqrt{T-t}\right).$$

Thus in the geometric Brownian motion model, the Black–Scholes formula simplifies to

$$\boxed{V^{EC}(X, K \cdot Y, T) = [N(d_+)] \cdot X_t + [-K \cdot N(d_-)] \cdot Y_t,} \tag{3.37}$$

where

$$\boxed{d_\pm = \frac{1}{\sigma\sqrt{T-t}} \cdot \log\left(\frac{1}{K} \cdot X_Y(t)\right) \pm \frac{1}{2}\sigma\sqrt{T-t}.} \tag{3.38}$$

□

REMARK 3.6 Option on a dividend-paying asset
When one of the underlying assets is a stock S that pays dividends, which is an arbitrage asset, we can transform the problem using a no-arbitrage asset $\widetilde{S}$ that represents the stock S plus the dividends. Assume for instance a constant continuous dividend yield $\delta = a(t)$, in which case the relationship between S and $\widetilde{S}$ in (1.31) simplifies to

$$\widetilde{S}_T = e^{\delta T} S_T. \tag{3.39}$$

Consider a European call option V with a payoff

$$V_T = (S_T - K \cdot \$_T)^+,$$

which is written in terms of two arbitrage assets: the dividend paying stock S and the dollar \$. Given the relationship between S and $\widetilde{S}$ in (3.39), we can rewrite the option payoff in terms of two no-arbitrage assets as

$$V_T = (e^{-\delta T}\widetilde{S}_T - K \cdot B_T^T)^+.$$

The Black–Scholes formula now applies, and the price of the call option at time $t = 0$ is given

$$\begin{aligned}
&\mathbb{P}^{\tilde{S}}(\widetilde{S}_{B^T}(T) \geq e^{\delta T}K) \cdot e^{-\delta T}\widetilde{S}_0 - K\mathbb{P}_t^T(\widetilde{S}_{B^T}(T) \geq e^{\delta T}K) \cdot B_0^T \\
&\quad = \mathbb{P}^{\tilde{S}}(\widetilde{S}_{B^T}(T) \geq e^{\delta T}K) \cdot e^{-\delta T}S_0 - K\mathbb{P}_t^T(\widetilde{S}_{B^T}(T) \geq e^{\delta T}K) \cdot B_0^T.
\end{aligned}$$

The only difference is that the dividend yield modifies the strike from K to $e^{\delta T}K$, and the stock S is discounted by the factor $e^{-\delta T}$. In the geometric Brownian motion model we get at time t

$$V_t = [N(d_+)] \cdot e^{\delta(T-t)}S_t + [-K \cdot N(d_-)] \cdot B_t^T, \tag{3.40}$$

where

$$d_\pm = \frac{1}{\sigma\sqrt{T-t}} \cdot \log\left(\frac{1}{K} \cdot X_Y(t)\right) + (-\delta \pm \tfrac{1}{2}\sigma)\sqrt{T-t}. \tag{3.41}$$

□

REMARK 3.7 Money as a reference asset
We have seen that European-type contracts can be expressed in terms of two no-arbitrage assets X and Y which also serve as natural reference assets for pricing a given European option V. Thus for pricing a general European contract, one first determines the price $V_Y(t)$ or $V_X(t)$ in terms of the no-arbitrage assets Y or X. The dollar price $V_\$(t)$ follows immediately from the change of numeraire formula

$$V_\$(t) = V_Y(t) \cdot Y_\$(t) = V_X(t) \cdot X_\$(t).$$

Let us illustrate how to compute the dollar price of a European call option on a stock and a dollar with a payoff

$$V_T^{EC} = (S_T - K \cdot \$_T)^+.$$

Since a dollar does not have a martingale measure $\mathbb{P}^\$$, we have to compute the price of the European call option using the First Fundamental Theorem of Asset Pricing either in terms of a stock S, or a bond B^T. This leads to the Black–Scholes formula (3.37), which takes the following form:

$$V_t^{EC} = [N(d_+)] \cdot S_t + [-K \cdot N(d_-)] \cdot B_t^T,$$

where

$$d_\pm = \tfrac{1}{\sigma\sqrt{T-t}} \cdot \log\left(\tfrac{1}{K} \cdot S_{B^T}(t)\right) \pm \tfrac{1}{2}\sigma\sqrt{T-t}.$$

We can rewrite the Black–Scholes formula in terms of prices with respect to a bond B^T as

$$V_{B^T}^{EC}(t) = N(d_+) \cdot S_{B^T}(t) - K \cdot N(d_-).$$

Multiplying the above equation by the dollar price of the bond $B_\$^T(t)$ and using the change of numeraire formula, we obtain the dollar price of the European call option

$$\begin{aligned} V_\$^{EC}(t) &= V_{B^T}^{EC}(t) \cdot B_\$^T(t) \\ &= N(d_+) \cdot S_{B^T}(t) \cdot B_\$^T(t) - K \cdot N(d_-) \cdot B_\$^T(t) \\ &= N(d_+) \cdot S_\$(t) - K \cdot N(d_-) \cdot B_\$^T(t). \end{aligned}$$

The formula for $d_\pm$ can also be expressed in terms of dollar prices as

$$d_\pm = \tfrac{1}{\sigma\sqrt{T-t}} \cdot \log\left(\tfrac{1}{K} \cdot S_\$(t) \cdot \$_{B^T}(t)\right) \pm \tfrac{1}{2}\sigma\sqrt{T-t}.$$

If we further assume a deterministic term structure evolution with a constant interest rate r,

$$B_t^T = e^{-r(T-t)} \cdot \$_t,$$

the above relationships simplify to

$$V_{\$}^{EC}(t) = S_{\$}(t) \cdot N(d_+) - K \cdot e^{-r(T-t)} \cdot N(d_-), \tag{3.42}$$

with

$$d_\pm = \frac{1}{\sigma\sqrt{T-t}} \cdot \left[\log\left(\frac{1}{K} \cdot S_{\$}(t)\right) + \left(r \pm \tfrac{1}{2}\sigma^2\right)(T-t)\right]. \tag{3.43}$$

This is the Black–Scholes formula expressed in terms of the dollar prices. Note that we had to assume a deterministic interest rate r in order to simplify the Black–Scholes formula (3.37) that applies also to stochastic interest rates. □

Table 3.1 summarizes payoffs of various contracts. Options with the power and the logarithmic payoff do not appear directly on the market, but they are related to barrier and lookback options as we will see in the following text. Note that the payoff function $f^Y(x)$ that corresponds to Y being chosen as a reference asset may have a different form than the payoff function $f^X(x)$ that corresponds to X being chosen as a reference asset. But the two payoff functions f^Y and f^X represent the same contract. One can think of switching roles of the assets X and Y, in which case we would get a new contract with a payoff $f^X(X_Y(T))$ units of Y. This is a **dual contract** to the original contract that pays off $f^Y(X_Y(T))$ units of Y. When we know the price of an original contract, we also know the price of the dual contract by switching the roles of X and Y.

We have already seen that the call option with a payoff $f^Y(x) = (x-K)^+$ is a dual contract to the put option with a payoff $f^X(x) = (1 - K \cdot x)^+$. The contract that pays off the best asset $\max(X_T, Y_T)$ is dual to itself as $f^Y(x) = f^X(x) = \max(x, 1)$. The role of X and Y can be switched and it does not change the contract as $\max(X_T, Y_T) = \max(Y_T, X_T)$. Similarly, the worst asset $\min(X_T, Y_T)$ is also dual to itself. The following example illustrates the concept of the dual contracts of the power options.

TABLE 3.1: Contracts on two assets.

Contract	Payoff	$f^Y(x)$	$f^X(x)$
Digital	$\mathbb{I}_A(X_Y(T)) \cdot Y_T$	$\mathbb{I}_A(x)$	$\mathbb{I}_A(\frac{1}{x}) \cdot x$
Best Asset	$\max(X_Y, K \cdot Y_T)$	$\max(x, K)$	$\max(K \cdot x, 1)$
Worst Asset	$\min(X_Y, K \cdot Y_T)$	$\min(x, K)$	$\min(K \cdot x, 1)$
Call	$(X_T - K \cdot Y_T)^+$	$(x-K)^+$	$(1 - K \cdot x)^+$
Put	$(K \cdot Y_T - X_T)^+$	$(K-x)^+$	$(K \cdot x - 1)^+$
Forward	$X_T - K \cdot Y_T$	$x - K$	$1 - K \cdot x$
Power	$[X_Y(T)]^\alpha \cdot Y_T$	x^α	$x^{1-\alpha}$
Logarithm	$\log(X_Y(T)) \cdot Y_T$	$\log(x)$	$-x \cdot \log(x)$

***Example 3.3* Dual contracts of power options**

A power option R^α pays off

$$R_T^\alpha = [X_Y(T)]^\alpha \cdot Y_T.$$

Power options are useful in pricing barrier and lookback options. The dual contract switches the roles of the assets X and Y; it pays off

$$[Y_X(T)]^\alpha \cdot X_T.$$

This can be rewritten as

$$[Y_X(T)]^\alpha \cdot X_T = [X_Y(T)]^{-\alpha} \cdot X_Y(T) \cdot Y_T = [X_Y(T)]^{1-\alpha} \cdot Y_T.$$

Thus the payoff function x^α has a dual payoff function $x^{1-\alpha}$.

Note that when $\alpha = 0$, the corresponding power option R^0 coincides with the asset Y. When $\alpha = 1$, the corresponding power option R^1 is the asset X. This suggests that for $\alpha \in (0,1)$, the resulting power option R^α creates an asset that is a combination of the assets X and Y. When $\alpha > 1$, the power option R^α leverages the position in the asset X. Similarly, when $\alpha < 0$, the power option R^α leverages the position in the asset Y. This is supported by the following argument. When X is comparable to Y in terms of price, meaning $X_Y(T) \approx 1$, we have

$$[X_Y(T)]^\alpha \approx 1 + \alpha(X_Y(T) - 1) = (1-\alpha) + \alpha X_Y(T)$$

according to the first order Taylor expansion around 1. Rewriting this relationship in terms of the assets, we have

$$R_T^\alpha \approx (1-\alpha) \cdot Y_T + \alpha \cdot X_T. \tag{3.44}$$

Clearly, when $\alpha \in (0,1)$, the power option is approximately a linear combination of the assets X and Y with positive weights. In particular, the power option $R^{1/2}$ corresponding to a **square root asset** is approximately just an average of the two assets X and Y. The square root asset is the only power option that is dual to itself, meaning that one can swap the roles of the assets X and Y without changing the contract.

When $\alpha > 1$, the power option corresponds to having a long position in the asset X, and a short position in the asset Y. When $\alpha < 0$, the situation is reversed, and the power option represents a long position in the asset Y, and a short position in the asset X. The hedging position $\Delta^X(t)$ of the power option indeed has the same sign as α, and the hedging position $\Delta^Y(t)$ has the same sign as $1-\alpha$. This confirms that the approximation from (3.44) is reasonable. The reader should determine the price and the hedging portfolio of the power options in Exercise 5.1. □

3.3 Price as an Expectation

For pricing a general European claim V, we can use either reference asset Y or X in order to determine the price of V:

$$V = V_Y(t) \cdot Y = V_X(t) \cdot X.$$

In Markovian models, which include geometric Brownian motion, we can express these prices in terms of the price functions u^Y and u^X defined as

$$V_Y(t) = u^Y(t, X_Y(t)), \qquad V_X(t) = u^X(t, Y_X(t)).$$

The functions u^Y and u^X are linked by

$$u^Y(t, X_Y(t)) = u^X(t, Y_X(t)) \cdot X_Y(t),$$

or by

$$u^X(t, Y_X(t)) = u^Y(t, X_Y(t)) \cdot Y_X(t).$$

Therefore we have the following symmetric relationship

$$\boxed{u^Y(t,x) = u^X(t, \tfrac{1}{x}) \cdot x,} \quad \text{or} \quad \boxed{u^X(t,x) = u^Y(t, \tfrac{1}{x}) \cdot x} \tag{3.45}$$

for $0 < x < \infty$, which is known as a **perspective mapping.** Note that we have $u^Y(T,x) = f^Y(x)$, and $u^X(T,x) = f^X(x)$, so the terminal price of the contract agrees with the payoff function. We have already seen that $f^Y(x) = f^X(\frac{1}{x}) \cdot x$, and $f^X(x) = f^Y(\frac{1}{x}) \cdot x$, which is just a special case of the relationship between the prices $u^Y(t,x)$, and $u^X(t,x)$.

Recall that the payoff of European options can always be written in terms of two no-arbitrage assets: U that agrees to deliver an asset X at time T, and V that agrees to deliver an asset Y at time T. It is easy to see that the contract to deliver a no-arbitrage asset is the asset itself, so the substitution of the underlying for a no-arbitrage asset makes sense only when one of the underlying assets is an arbitrage asset, such as in the case of the dollar or other currencies. Therefore without loss of generality, we may assume that the European option is settled in terms of two no-arbitrage assets.

Given that European options can be expressed in terms of two no-arbitrage assets, the First Fundamental Theorem of Asset Pricing states that the price of V in terms of the reference asset Y is a $\mathbb{P}^Y$ martingale, and the price of V in terms of the reference asset X is a $\mathbb{P}^X$ martingale. This gives us a stochastic representation of the contingent claim price

$$V_Y(t) = \mathbb{E}_t^Y\left[V_Y(T)\right] = \mathbb{E}_t^Y\left[f^Y(X_Y(T))\right], \tag{3.46}$$

when the asset Y is used as a numeraire, and

$$V_X(t) = \mathbb{E}_t^X\left[V_X(T)\right] = \mathbb{E}_t^X\left[f^X\left(Y_X(T)\right)\right], \tag{3.47}$$

when the asset X is used as a numeraire. The number of units $\mathbb{E}_t^Y\left[f^Y\left(X_Y(T)\right)\right]$ of Y that is needed in order to acquire the contract V is the price of the contract in terms of the reference asset Y. Similarly, the number of units $\mathbb{E}_t^X\left[f^X\left(Y_X(T)\right)\right]$ of X that is needed in order to acquire the contract V is the price of the contract in terms of the reference asset X.

When the prices $V_Y(t)$ and $V_X(t)$ are Markovian in the prices $X_Y(t)$ and $Y_X(t)$, the price functions u^Y and u^X have the following representations

$$u^Y(t,x) = \mathbb{E}^Y\left[f^Y\left(X_Y(t)\right)|X_Y(t) = x\right], \tag{3.48}$$

and

$$u^X(t,x) = \mathbb{E}^X\left[f^X\left(Y_X(t)\right)|Y_X(t) = x\right]. \tag{3.49}$$

The price processes V_Y and V_X are indeed Markovian in the geometric Brownian motion model.

When the price processes $X_Y(t)$ and $Y_X(t)$ are geometric Brownian motions, we can compute the price functions u^Y and u^X directly by computing the conditional expected value. For the function u we have

$$\begin{aligned} u^Y(t,x) &= \mathbb{E}^Y\left[f^Y\left(X_Y(T)\right)|X_Y(t) = x\right] \\ &= \mathbb{E}^Y\left[f^Y\left(X_Y(t)\cdot\exp\left(\sigma W^Y(T-t) - \tfrac{1}{2}\sigma^2(T-t)\right)\right)|X_Y(t) = x\right] \\ &= \mathbb{E}^Y\left[f^Y\left(x\cdot\exp\left(\sigma W^Y(T-t) - \tfrac{1}{2}\sigma^2(T-t)\right)\right)|X_Y(t) = x\right] \\ &= \int_{-\infty}^{\infty} f^Y\left(x\cdot\exp\left(\sigma y\sqrt{T-t} - \tfrac{1}{2}\sigma^2(T-t)\right)\right)\cdot\tfrac{1}{\sqrt{2\pi}}\exp\left(-\tfrac{y^2}{2}\right)dy. \end{aligned} \tag{3.50}$$

We have used the fact that

$$X_Y(T) = X_Y(t)\cdot\exp\left(\sigma W^Y(T-t) - \tfrac{1}{2}\sigma^2(T-t)\right),$$

and that $\frac{W^Y(T-t)}{\sqrt{T-t}}$ has a normal distribution $N(0,1)$.

Similarly, the function u^X can be determined from the following formula:

$$\begin{aligned} u^X(t,x) &= \mathbb{E}^X\left[f^X\left(Y_X(T)\right)|Y_X(t) = x\right] \\ &= \mathbb{E}^X\left[f^X\left(Y_X(t)\cdot\exp\left(\sigma W^X(T-t) - \tfrac{1}{2}\sigma^2(T-t)\right)\right)|Y_X(t) = x\right] \\ &= \mathbb{E}^X\left[f^X\left(x\cdot\exp\left(\sigma W^X(T-t) - \tfrac{1}{2}\sigma^2(T-t)\right)\right)|Y_X(t) = x\right] \\ &= \int_{-\infty}^{\infty} f^X\left(x\cdot\exp\left(\sigma y\sqrt{T-t} - \tfrac{1}{2}\sigma^2(T-t)\right)\right)\cdot\tfrac{1}{\sqrt{2\pi}}\exp\left(-\tfrac{y^2}{2}\right)dy. \end{aligned} \tag{3.51}$$

Example 3.4

Consider a European call option with a payoff $(X_T - K \cdot Y_T)^+$. When Y is chosen as a reference asset, the payoff function is given by $f^Y(x) = (x - K)^+$. Thus we have

$$\begin{aligned} u^Y(t,x) &= \int_{-\infty}^{\infty} f^Y\left(x \cdot \exp\left(\sigma y\sqrt{T-t} - \tfrac{1}{2}\sigma^2(T-t)\right)\right) \cdot \tfrac{1}{\sqrt{2\pi}} \exp\left(-\tfrac{y^2}{2}\right) dy \\ &= \int_{-\infty}^{\infty} \left(x \cdot \exp\left(\sigma y\sqrt{T-t} - \tfrac{1}{2}\sigma^2(T-t)\right) - K\right)^+ \\ &\qquad\qquad \times \tfrac{1}{\sqrt{2\pi}} \exp\left(-\tfrac{y^2}{2}\right) dy \\ &= x \cdot N\left(\tfrac{1}{\sigma\sqrt{T-t}} \cdot \left[\log\left(\tfrac{x}{K}\right) + \tfrac{1}{2}\sigma^2(T-t)\right]\right) \\ &\qquad -K \cdot N\left(\tfrac{1}{\sigma\sqrt{T-t}} \cdot \left[\log\left(\tfrac{x}{K}\right) - \tfrac{1}{2}\sigma^2(T-t)\right]\right). \end{aligned}$$

When X is chosen as a reference asset, the payoff function is given by $f^X(x) = f^Y(\frac{1}{x}) \cdot x = (1 - K \cdot x)^+$, and thus we have

$$\begin{aligned} u^X(t,x) &= \int_{-\infty}^{\infty} f^X\left(x \cdot \exp\left(\sigma y\sqrt{T-t} - \tfrac{1}{2}\sigma^2(T-t)\right)\right) \cdot \tfrac{1}{\sqrt{2\pi}} \cdot \exp\left(-\tfrac{y^2}{2}\right) dy \\ &= \int_{-\infty}^{\infty} \left(1 - K \cdot x \cdot \exp\left(\sigma y\sqrt{T-t} - \tfrac{1}{2}\sigma^2(T-t)\right)\right)^+ \\ &\qquad\qquad \times \tfrac{1}{\sqrt{2\pi}} \exp\left(-\tfrac{y^2}{2}\right) dy \\ &= N\left(\tfrac{1}{\sigma\sqrt{T-t}} \cdot \left[\log\left(\tfrac{1}{K\cdot x}\right) + \tfrac{1}{2}\sigma^2(T-t)\right]\right) \\ &\qquad -K \cdot x \cdot N\left(\tfrac{1}{\sigma\sqrt{T-t}} \cdot \left[\log\left(\tfrac{1}{K\cdot x}\right) - \tfrac{1}{2}\sigma^2(T-t)\right]\right). \end{aligned}$$

The reader may check that the price functions u^Y and u^X indeed satisfy $u^X(t,x) = u^Y(t, \frac{1}{x}) \cdot x$. □

3.4 Connections with Partial Differential Equations

Let us assume that the price $X_Y(t)$ follows the geometric Brownian motion model

$$dX_Y(t) = \sigma X_Y(t) dW^Y(t).$$

We point out in this section that the price functions u^Y and u^X satisfy a certain partial differential equation.

THEOREM 3.3
The price function $u^Y(t,x) = \mathbb{E}^Y\left[f^Y(X_Y(T))\,|X_Y(t)=x\right]$ *satisfies the partial differential equation*

$$\boxed{u_t^Y(t,x) + \tfrac{1}{2}\sigma^2 x^2 u_{xx}^Y(t,x) = 0} \tag{3.52}$$

with the terminal condition

$$u^Y(T,x) = f^Y(x), \tag{3.53}$$

and the boundary condition

$$u^Y(t,0) = f^Y(0). \tag{3.54}$$

The price function $u^X(t,x) = \mathbb{E}^X\left[f^X(Y_X(T))\,|Y_X(t)=x\right]$ *satisfies the partial differential equation*

$$\boxed{u_t^X(t,x) + \tfrac{1}{2}\sigma^2 x^2 u_{xx}^X(t,x) = 0} \tag{3.55}$$

with the terminal condition

$$u^X(T,x) = f^X(x), \tag{3.56}$$

and the boundary condition

$$u^X(t,0) = f^X(0). \tag{3.57}$$

REMARK 3.8
The partial differential equations (3.52) and (3.55) are also known as the **Black–Scholes partial differential equations.** □

PROOF Let

$$u^Y(t,x) = \mathbb{E}^Y\left[f^Y(X_Y(T))\,|X_Y(t)=x\right]$$

be the price of the contract with respect to the reference asset Y. According to Ito's formula, the option price has the following dynamics:

$$\begin{aligned}
du^Y(t,X_Y(t)) &= u_t^Y(t,X_Y(t))\,dt + u_x^Y(t,X_Y(t))\,dX_Y(t) \\
&\quad + \tfrac{1}{2}u_{xx}^Y(t,X_Y(t))\,d^2X_Y(t) \\
&= \left[u_t^Y(t,X_Y(t)) + \tfrac{1}{2}\sigma^2 X_Y(t)^2 u_{xx}^Y(t,X_Y(t))\right]dt \\
&\quad + u_x^Y(t,X_Y(t))\,dX_Y(t).
\end{aligned}$$

Since $u^Y(t,X_Y(t))$ is a $\mathbb{P}^Y$ martingale, the dt term of this equation must vanish for all values of $X_Y(t)$, and thus the following partial differential equation for the price of the option must hold:

$$u_t^Y(t,x) + \tfrac{1}{2}\sigma^2 x^2 u_{xx}^Y(t,x) = 0,$$

with the terminal condition

$$u^Y(T, x) = f^Y(x).$$

The case when $x = 0$ represents the situation when $X_Y(t) = 0$ (the asset X becomes worthless), and thus the value of $X_Y(T)$ will also be zero. Thus the payoff of the option will be $f^Y(0)$ units of an asset Y at time T. Thus the value of the contract at time t is $u^Y(t, 0) = f^Y(0)$.

We can apply the same technique using the no-arbitrage asset X as a numeraire when the payoff of the contract is $f^X(Y_X(T))$ units of an asset X, leading to the partial differential equation (3.55). ▯

REMARK 3.9 The prices of X and Y satisfy the Black–Scholes partial differential equation
Partial differential equation (3.52)

$$u_t^Y(t, x) + \tfrac{1}{2}\sigma^2 x^2 u_{xx}^Y(t, x) = 0$$

has two trivial solutions that correspond to the payoff functions $f^Y(x) = 1$ and $f^Y(x) = x$. When the payoff function is $f^Y(x) = 1$, the price function $u^Y(t, x)$ is also identically equal to one, and the partial differential equation (3.52) is satisfied. In financial terms, the payoff function $f^Y(x) = 1$ corresponds to the delivery of a unit of an asset Y at time T. This is a contract to deliver an asset Y, and its price at any given time $t \leq T$ is one unit of an asset Y. Thus we have $u^Y(t, x) = 1$ as a solution. When the payoff function is $f^Y(x) = x$, the price function $u^Y(t, x)$ is also equal to x, and the partial differential equation (3.52) is satisfied. In financial terms, the payoff function $f^Y(x) = x$ corresponds to the delivery of a unit of an asset X at time T (it is $X_Y(t)$ units of an asset Y). This is a contract to deliver an asset X at time T and its price at any given time $t \leq T$ is one unit of an asset X. Thus we have $u^Y(t, x) = x$ as a solution.

Similarly, the partial differential equation (3.55)

$$u_t^X(t, x) + \tfrac{1}{2}\sigma^2 x^2 u_{xx}^X(t, x) = 0$$

also has two trivial solutions that correspond to the payoff functions $f^X(x) = 1$ and $f^X(x) = x$. In financial terms, the payoff function $f^X(x) = 1$ corresponds to the delivery of an asset X, the payoff function $f^X(x) = x$ corresponds to the delivery of an asset Y. ▯

Example 3.5
The European option V with a payoff $V_T = (X_T - K \cdot Y_T)^+$ has an associated payoff function $f^Y(x) = (x - K)^+$, or $f^X(x) = (1 - K \cdot x)^+$. The $V_Y(t) =$

$u^Y(t, X_Y(t))$ price satisfies the partial differential equation (3.52) and the $V_X(t) = u^X(t, Y_X(t))$ price satisfies the partial differential equation (3.55). When the asset X becomes worthless, or in other words when $X_Y(t) = 0$, the option will also be worthless as $f^Y(0) = 0$, giving us the boundary condition $u^Y(t, 0) = 0$. The asset X will not serve as a reference asset in this case, but the price of the contract can still be expressed in terms of the asset Y. On the other hand, when the asset Y becomes worthless, $Y_X(t) = 0$, the option will pay off a unit of the asset X, which corresponds to $f^X(0) = 1$. This gives the boundary condition $u^X(t, 0) = 1$, the asset X can still be used as a numeraire. Note that the boundary conditions when one of the prices is zero do not have a perspective mapping counterpart as the perspective mapping applies only to cases when the prices are positive. When one of the assets becomes worthless, it still makes sense to use the remaining asset with a positive price as a numeraire, but the pricing problem cannot be solved using the worthless asset. □

3.5 Money as a Reference Asset

It is also possible to write the Black–Scholes partial differential equation in terms of the dollar \$ as a reference asset. Let X be a stock S, and Y be a bond B^T. A contract V that pays off $f^T(S_{B^T}(T))$ units of a bond B^T at time T can equivalently be expressed as

$$V_T = f^T(S_{B^T}(T)) \cdot B_T^T = f^\$(S_\$(T)) \cdot \$_T,$$

a contract that pays off $f^\$(S_\$(T))$ units of a dollar \$ at time T. The payoff functions in terms of a bond B^T and a dollar \$ agree: $f^T(x) = f^\$(x)$. The contract V at time t can be also expressed in the following equivalent ways:

$$V_t = V_{B^T}(t) \cdot B_t^T = V_\$(t) \cdot \$_t = V_S(t) \cdot S_t.$$

Let $u^T(t, S_{B^T}(t)) = V_{B^T}(t)$ be the price of the contract V in terms of a bond B^T, and let

$$v^\$(t, S_\$(t)) = V_\$(t) \tag{3.58}$$

be the price of the contract V in terms of a dollar \$. We are using a different letter v for the dollar price in order to distinguish it from the prices u that use only no-arbitrage assets. Let us also assume $B_\$^T(t) = e^{-r(T-t)}$. Since

$$V_t = u^T(t, S_{B^T}(t)) \cdot B_t^T = v^\$(t, S_\$(t)) \cdot \$_t,$$

we get the following relationship between u^T and $v^\$$:

$$\boxed{v^\$(t, x) = e^{-r(T-t)} \cdot u^T(t, e^{r(T-t)} x),} \tag{3.59}$$

and

$$\boxed{u^T(t,x) = e^{r(T-t)} \cdot v^\$(t, e^{-r(T-t)}x).} \tag{3.60}$$

We have seen that the price function u^T satisfies the partial differential equation

$$u_t^T(t,x) + \tfrac{1}{2}\sigma^2 x^2 u_{xx}^T(t,x) = 0.$$

Using the relationship between the functions u^T and $v^\$$, we find that

$$u_t^T(t,x) = e^{r(T-t)} \cdot \Big(-rv^\$(t, e^{-r(T-t)}x) + v_t^\$(t, e^{-r(T-t)}x) + r\left(e^{-r(T-t)}x\right) v_x^\$(t, e^{-r(T-t)}x) \Big),$$

and

$$u_{xx}^T(t,x) = e^{r(T-t)}\left(e^{-r(T-t)}\right)^2 v_{xx}^\$(t, e^{-r(T-t)}x).$$

After substitution of x for $e^{-r(T-t)}x$, we conclude that $v^\$$ satisfies the following partial differential equation

$$\boxed{-rv^\$(t,x) + v_t^\$(t,x) + rxv_x^\$(t,x) + \tfrac{1}{2}\sigma^2 x^2 v_{xx}^\$(t,x) = 0.} \tag{3.61}$$

The terminal condition is given by

$$v^\$(T,x) = f^T(x) = f^\$(x), \tag{3.62}$$

and the boundary condition is

$$v^\$(t,0) = e^{-r(T-t)} \cdot u^T(t,0) = e^{-r(T-t)} \cdot f^T(0). \tag{3.63}$$

The Black–Scholes partial differential equation in the form of (3.61) is widely used since it directly determines the price of a contract in terms of a dollar. **However, the partial differential equation (3.61) has two limitations. First, it applies only when the interest rate r is deterministic. Second, its form is more complicated than the Black–Scholes partial differential equation (3.52) obtained for two no-arbitrage assets S and B^T.** The pricing of European options is still relatively straightforward, so the advantage of using no-arbitrage assets for pricing is small. Therefore using no-arbitrage assets in pricing is more important for complex financial products, such as exotic options.

Example 3.6
We have seen that the price of the European call option with a payoff $(S_T - K \cdot B_T^T)^+$ is given by

$$u^T(t,x) = x \cdot N\left(\tfrac{1}{\sigma\sqrt{T-t}} \cdot \left[\log\left(\tfrac{x}{K}\right) + \tfrac{1}{2}\sigma^2(T-t)\right]\right) - K \cdot N\left(\tfrac{1}{\sigma\sqrt{T-t}} \cdot \left[\log\left(\tfrac{x}{K}\right) - \tfrac{1}{2}\sigma^2(T-t)\right]\right). \quad (3.64)$$

Using the relationship $v^\$(t,x) = e^{-r(T-t)} \cdot u^T(t, e^{r(T-t)}x)$, we can express the dollar price of the option as

$$v^\$(t,x) = x \cdot N\left(\tfrac{1}{\sigma\sqrt{T-t}} \cdot \left[\log\left(\tfrac{x}{K}\right) + (r + \tfrac{1}{2}\sigma^2)(T-t)\right]\right) - Ke^{-r(T-t)} \cdot N\left(\tfrac{1}{\sigma\sqrt{T-t}} \cdot \left[\log\left(\tfrac{x}{K}\right) + (r - \tfrac{1}{2}\sigma^2)(T-t)\right]\right). \quad (3.65)$$

This is the best-known form of the Black–Scholes formula. One can verify that $v^\$(t,x)$ from (3.65) satisfies the Black–Scholes partial differential equation (3.61). □

Similarly, we can define the price function v^S in terms of \$ and S as a reference asset by

$$V_t = v^S(t, \$_S(t)) \cdot S_t. \quad (3.66)$$

The relationship between V^S and the price function u^S defined as

$$V_t = u^S(t, B_S^T(t)) \cdot S_t \quad (3.67)$$

is given by

$$\boxed{v^S(t,x) = u^S(t, e^{-r(T-t)}x),} \qquad \boxed{u^S(t,x) = v^S(t, e^{r(T-t)}x).} \quad (3.68)$$

Using the relationship between the price functions v^S and u^S, we can obtain a partial differential equation for v^S. Since u^S satisfies the partial differential equation

$$u_t^S(t,x) + \tfrac{1}{2}\sigma^2 x^2 u_{xx}^S(t,x) = 0,$$

the function v^S satisfies the partial differential equation

$$\boxed{v_t^S(t,x) - rxv_x^S(t,x) + \tfrac{1}{2}\sigma^2 x^2 v_{xx}^S(t,x) = 0.} \quad (3.69)$$

The terminal condition is

$$v^S(T,x) = u^S(T,x) = f^S(x), \quad (3.70)$$

and the boundary condition is

$$v^S(t,0) = u^S(t,0) = f^S(0). \quad (3.71)$$

3.6 Hedging

Let us determine the hedging portfolio for a general European option contract V.

THEOREM 3.4
The hedging portfolio P_t of the European option is given by

$$\boxed{P_t = \left[u_x^Y(t, X_Y(t))\right] \cdot X + \left[u^Y(t, X_Y(t)) - u_x^Y(t, X_Y(t)) \cdot X_Y(t)\right] \cdot Y,} \tag{3.72}$$

or equivalently by

$$\boxed{P_t = \left[u^X(t, Y_X(t)) - u_x^X(t, Y_X(t)) \cdot Y_X(t)\right] \cdot X + \left[u_x^X(t, Y_X(t))\right] \cdot Y.} \tag{3.73}$$

PROOF The hedging portfolio is in the form

$$P_t = \Delta^X(t) \cdot X + \Delta^Y(t) \cdot Y,$$

and has dynamics of the form

$$dP_Y(t) = \Delta^X(t, X_Y(t))\, dX_Y(t).$$

We also have

$$dV_Y(t) = du^Y(t, X_Y(t)) = u_x^Y(t, X_Y(t))\, dX_Y(t).$$

In order to have

$$P_t = V_t,$$

at all times, the hedge of this contract must satisfy

$$\boxed{\Delta^X(t, X_Y(t)) = u_x^Y(t, X_Y(t)) = \frac{\partial V_Y(t)}{\partial X_Y(t)}.} \tag{3.74}$$

The hedging position Δ^X in the asset X is the sensitivity of the price of the contract $V_Y(t)$ to the changes of the underlying price $X_Y(t)$. The hedge position Δ^Y in the asset Y follows from

$$\Delta^Y(t) = P_Y(t) - \Delta^X(t) \cdot X_Y(t) = u^Y(t, X_Y(t)) - u_x^Y(t, X_Y(t)) \cdot X_Y(t).$$

When X is chosen as a reference asset, the price dynamics of the hedging portfolio P are given by

$$dP_X(t) = \Delta^Y(t, Y_X(t))\, dY_X(t).$$

We also have

$$dV_X(t) = du^X\left(t, Y_X(t)\right) = u_x^X\left(t, Y_X(t)\right) dY_X(t),$$

and thus in order to have

$$P_t = V_t,$$

the hedging position Δ^Y must satisfy

$$\boxed{\Delta^Y\left(t, Y_X(t)\right) = u_x^X\left(t, Y_X(t)\right) = \frac{\partial V_X(t)}{\partial Y_X(t)}.} \tag{3.75}$$

The hedging position $\Delta^X(t)$ in the asset X follows from

$$\Delta^X(t) = P_X(t) - \Delta^Y(t) \cdot Y_X(t) = u^X(t, Y_X(t)) - u_x^X(t, Y_X(t)) \cdot Y_X(t).$$

□

Recall that the prices in terms of the functions u^Y and u^X are related by the following symmetric relationship known as a perspective mapping:

$$\boxed{u^Y(t,x) = u^X(t, \tfrac{1}{x}) \cdot x,} \quad \text{or} \quad \boxed{u^X(t,x) = u^Y(t, \tfrac{1}{x}) \cdot x.}$$

We can connect the pricing partial differential equations for u^Y and u^X through the above relationship. The function u^Y solves Equation (3.52):

$$u_t^Y(t,x) + \tfrac{1}{2}\sigma^2 x^2 u_{xx}^Y(t,x) = 0.$$

We can rewrite this partial differential equation in terms of u^X using the following identities:

$$\begin{aligned}
u_t^Y(t,x) &= u_t^X(t, \tfrac{1}{x}) \cdot x, \\
u_x^Y(t,x) &= u^X(t, \tfrac{1}{x}) - \tfrac{1}{x} \cdot u_x^X(t, \tfrac{1}{x}), \\
u_{xx}^Y(t,x) &= -\tfrac{1}{x^2} \cdot u_x^X(t, \tfrac{1}{x}) + \tfrac{1}{x^2} \cdot u_x^X(t, \tfrac{1}{x}) + \tfrac{1}{x^3} \cdot u_{xx}^X(t, \tfrac{1}{x}) \\
&= \tfrac{1}{x^3} \cdot u_{xx}^X(t, \tfrac{1}{x}).
\end{aligned}$$

Substituting for $u_t^Y(t,x)$ and $u_{xx}^Y(t,x)$ in (3.52) we get

$$u_t^Y(t,x) + \tfrac{1}{2}\sigma^2 x^2 u_{xx}^Y(t,x) = u_t^X(t, \tfrac{1}{x}) \cdot x + \tfrac{1}{2}\sigma^2 x^2 \tfrac{1}{x^3} \cdot u_{xx}^X(t, \tfrac{1}{x}) = 0,$$

which leads to

$$u_t^X(t, \tfrac{1}{x}) + \tfrac{1}{2}\sigma^2 \tfrac{1}{x^2} \cdot u_{xx}^X(t, \tfrac{1}{x}) = 0.$$

After making the substitution $\frac{1}{x} \to x$, we can rewrite the above partial differential equation as

$$u_t^X(t,x) + \tfrac{1}{2}\sigma^2 x^2 u_{xx}^X(t,x) = 0,$$

which is Equation (3.55). This is an independent derivation of this partial differential equation using the relationship between u^Y and u^X. Note that the partial differential equation for u^Y and u^X takes the same form, so it is completely symmetric with respect to the choice of the reference asset. This is not the case for more complex products, such as for Asian options.

We have previously seen that the hedging portfolio is given by

$$P_t = \Delta^X(t) \cdot X + \Delta^Y(t) \cdot Y = \left[u_x^Y\left(t, X_Y(t)\right)\right] \cdot X + \left[u_x^X\left(t, Y_X(t)\right)\right] \cdot Y,$$

when using both price functions u^Y and u^X, or in other words,

$$\boxed{P_t = \left[\frac{\partial V_Y(t)}{\partial X_Y(t)}\right] \cdot X + \left[\frac{\partial V_X(t)}{\partial Y_X(t)}\right] \cdot Y.} \tag{3.76}$$

Using the relationship between u^Y and u^X:

$$u_x^Y(t,x) = u^X(t, \tfrac{1}{x}) - \tfrac{1}{x} \cdot u_x^X(t, \tfrac{1}{x}),$$

or

$$u_x^X(t,x) = u^Y(t, \tfrac{1}{x}) - \tfrac{1}{x} \cdot u_x^Y(t, \tfrac{1}{x}),$$

we can also write

$$P_t = \Big[u_x^Y\left(t, X_Y(t)\right)\Big] \cdot X + \Big[u^Y(t, X_Y(t)) - u_x^Y(t, X_Y(t)) \cdot X_Y(t)\Big] \cdot Y,$$

or equivalently

$$P_t = \Big[u^X(t, Y_X(t)) - u_x^X(t, Y_X(t)) \cdot Y_X(t)\Big] \cdot X + \Big[u_x^X\left(t, Y_X(t)\right)\Big] \cdot Y.$$

This confirms Theorem 3.4.

Example 3.7 Hedging of the forward contract

The forward contract pays off $X_T - K \cdot Y_T$, which corresponds to the payoff functions $f^Y(x) = x - K$, and $f^X(x) = 1 - K \cdot x$. The price of the forward contract is trivially given by $u^Y(t,x) = x - K$, and $u^X(t,x) = 1 - K \cdot x$. Therefore the hedging portfolio is given by

$$P_t = \left[u_x^Y(t,x)\right] \cdot X_t + \left[u_x^X(t,x)\right] \cdot Y_t = X_t - K \cdot Y_t.$$

The hedge is static; one buys one unit of the asset X and sells K units of Y. The forward contract can be thought of as a combination of two contracts to deliver: one that delivers a unit of an asset X and one that delivers $-K$ units (or in other words shorts K units) of an asset Y. A contract to deliver an asset X at time T is trivial: it is the asset X itself. One simply buys the asset and holds it until expiration. A similar argument applies to the asset Y. Note that the hedge of the forward contract is model independent; it does not depend on the evolution of the price $X_Y(t)$. □

Example 3.8 Hedging of the European call option

We have seen that the hedging position of a general European option in the asset X is given by

$$\Delta_t^X(t, X_Y(t)) = u_x^Y(t, X_Y(t)).$$

This further simplifies when the payoff function is given by $f^Y(x) = (x-K)^+$. We have that

$$\begin{aligned} u_x^Y(t,x) &= \tfrac{d}{dx}\mathbb{E}_t^Y\left(X_Y(T) - K\right)^+ \\ &= \tfrac{d}{dx}\mathbb{E}^Y\left[\left(x \cdot \tfrac{X_Y(T)}{X_Y(t)} - K\right)^+ |X_Y(t) = x\right] \\ &= \tfrac{d}{dx}\mathbb{E}^X\left[\left(x - K \cdot \tfrac{Y_X(T)}{Y_X(t)}\right)^+ |X_Y(t) = x\right] \\ &= \mathbb{P}_t^X(X_Y(T) \geq K). \end{aligned}$$

Thus we have

$$\Delta^X(t) = \mathbb{P}_t^X(X_Y(T) \geq K) = N\left(\tfrac{1}{\sigma\sqrt{T-t}} \cdot \log\left(\tfrac{1}{K} \cdot X_Y(t)\right) + \tfrac{1}{2}\sigma\sqrt{T-t}\right). \tag{3.77}$$

Similarly we get

$$\begin{aligned} \Delta^Y(t) &= -K \cdot \mathbb{P}_t^Y(X_Y(T) \geq K) \\ &= -K \cdot N\left(\tfrac{1}{\sigma\sqrt{T-t}} \cdot \log\left(\tfrac{1}{K} \cdot X_Y(t)\right) - \tfrac{1}{2}\sigma\sqrt{T-t}\right). \end{aligned} \tag{3.78}$$

□

Hedging of an option that has a dollar as an underlying asset has to be done in a stock S and in the bond B^T (or equivalently in the money market M). Thus the hedging portfolio P_t is in the form

$$P_t = \Delta^S(t) \cdot S + \Delta^T(t) \cdot B^T.$$

We have already seen that

$$\Delta^S(t) = u_x^T(t, S_{B^T}(t)), \qquad \text{and} \qquad \Delta^T(t) = u_x^S\left(t, B_S^T(t)\right).$$

We can also express the hedging positions in terms of the price functions $v^\$$ and v^S. Since $u^T(t,x) = e^{r(T-t)} \cdot v^\$(t, e^{-r(T-t)}x)$, we have

$$u_x^T(t,x) = v_x^\$(t, e^{-r(T-t)}x),$$

and thus

$$\Delta^S(t) = u_x^T(t, S_{B^T}(t)) = u_x^T(t, S_\$(t) \cdot \$_{B^T}(t)) \\ = u_x^T\left(t, e^{r(T-t)} \cdot S_\$(t)\right) = v_x^\$(t, S_\$(t)).$$

The hedging position in the bond B^T can be obtained from the dollar price of the hedging portfolio

$$P_\$(t) = \Delta^S(t) \cdot S_\$(t) + \Delta^T(t) \cdot B_\$^T(t),$$

or in other words,

$$v^\$(t, S_\$(t)) = v_x^\$(t, S_\$(t)) \cdot S_\$(t) + \Delta^T(t) \cdot e^{-r(T-t)}.$$

Thus we have

$$\Delta^T(t) = e^{r(T-t)} \cdot \left[v^\$(t, S_\$(t)) - v_x^\$(t, S_\$(t)) \cdot S_\$(t)\right].$$

Similarly, we can express the hedging portfolio in terms of the price function v^S. Since $u^S(t,x) = v^S(t, e^{r(T-t)}x)$, we have

$$u_x^S(t,x) = v_x^S(t, e^{r(T-t)}x) \cdot e^{r(T-t)},$$

and thus

$$\Delta^T(t) = u_x^S\left(t, B_S^T(t)\right) = u_x^S\left(t, \$_S(t) \cdot B_\$^T(t)\right) \\ = u_x^S\left(t, e^{-r(T-t)} \cdot \$_S(t)\right) = e^{r(T-t)} \cdot v_x^S(t, \$_S(t)).$$

The hedging position $\Delta^S(t)$ can be obtained from

$$\begin{aligned}\Delta^S(t) &= P_S(t) - \Delta^T(t) \cdot B_S^T(t) \\ &= v^S(t, \$_S(t)) - e^{r(T-t)} \cdot v_x^S(t, \$_S(t)) \cdot \$_S(t) \cdot B_\$^T(t) \\ &= v^S(t, \$_S(t)) - v_x^S(t, \$_S(t)) \cdot \$_S(t).\end{aligned}$$

COROLLARY 3.1

The hedging portfolio is given by

$$P_t = \left[v_x^\$(t, S_\$(t))\right] \cdot S \\ + \left[e^{r(T-t)} \cdot \left[v^\$(t, S_\$(t)) - v_x^\$(t, S_\$(t)) \cdot S_\$(t)\right]\right] \cdot B^T, \quad (3.79)$$

or

$$P_t = \left[\left[v^S(t, \$_S(t)) - v_x^S(t, \$_S(t)) \cdot \$_S(t)\right]\right] \cdot S \\ + \left[e^{r(T-t)} \cdot v_x^S(t, \$_S(t))\right] \cdot B^T. \quad (3.80)$$

Assuming that $M_{\$}(t) = 1$, or equivalently stated, $M_t = e^{r(T-t)} \cdot B_t^T$, we can also express the hedging portfolio in term of the stock S and the money market M as

$$P_t = \left[v_x^{\$}\left(t, S_{\$}(t)\right)\right] \cdot S + \left[v^{\$}\left(t, S_{\$}(t)\right) - v_x^{\$}\left(t, S_{\$}(t)\right) \cdot S_{\$}(t)\right] \cdot M, \qquad (3.81)$$

or

$$P_t = \left[\left[v^S\left(t, \$_S(t)\right) - v_x^S\left(t, \$_S(t)\right) \cdot \$_S(t)\right]\right] \cdot S + \left[v_x^S\left(t, \$_S(t)\right)\right] \cdot M. \qquad (3.82)$$

3.7 Properties of European Call and Put Options

An option is **in the money** at time t if $f^Y(X_Y(t)) > 0$. If the option were to expire immediately at time t, its holder would collect a positive payoff. An option is **deep in the money** if it is in the money and $f^Y(X_Y(T)) > 0$ with high probability, meaning that the option is likely to expire with a positive payoff. An option is **out of the money** at time t if $f^Y(X_Y(t)) = 0$. An option is **deep out of the money** if it is out of the money, and $f^Y(X_Y(T)) = 0$ with high probability, meaning that the option is likely to expire worthless. An option is **at the money** if $f^Y(X_Y(t) + \epsilon) > 0$ and $f^Y(X_Y(t) - \epsilon) = 0$ for $\epsilon > 0$. An at the money option is a boundary case between in the money and out of the money option.

Given the hedge representation for a European call option

$$\Delta^X(t) = \mathbb{P}_t^X(X_Y(T) \geq K),$$

and

$$\Delta^Y(t) = -K \cdot \mathbb{P}_t^Y(X_Y(T) \geq K),$$

we can see that

$$0 \leq \Delta^X(t) \leq 1, \qquad \text{and} \qquad -K \leq \Delta^Y(t) \leq 0.$$

Moreover, if the option is deep out of the money, the option is almost worthless, and the corresponding hedge is $\Delta^X(t) \approx 0$, and $\Delta^Y(t) \approx 0$. On the other hand, if the option is deep in the money, $\Delta^X(t) \approx 1$, $\Delta^Y(t) \approx -K$, and the European option contract is close to a forward $X_t - K \cdot Y_t$.

Another interesting observation is to see what happens when the maturity of the option approaches infinity, or equivalently, when the volatility approaches infinity. Recall that the price of a European call option is given by

$$V^{EC}(X, K \cdot Y, T) = N(d_+) \cdot X - K \cdot N(d_-) \cdot Y,$$

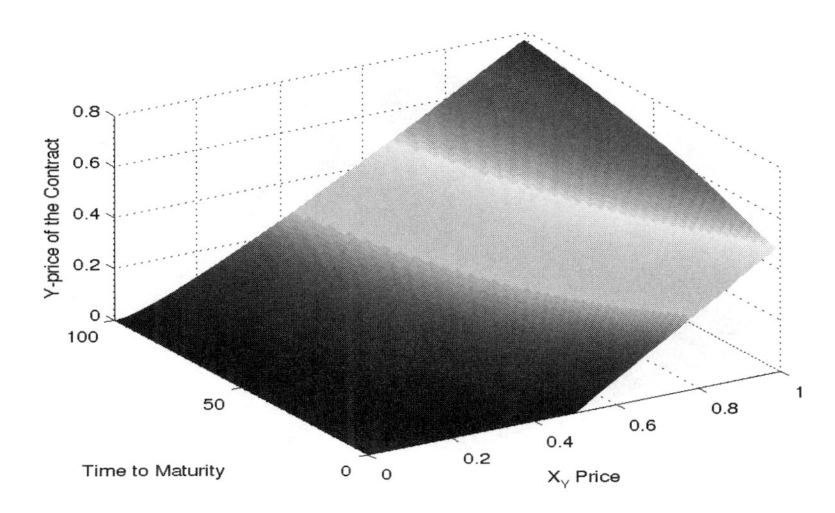

FIGURE 3.1: The price $V_Y(t)$ of a European option contract with a payoff $(X_T - K \cdot Y_T)^+$ with parameters $K = \frac{1}{2}$, $\sigma = 0.2$, as a function of price $X_Y(t)$, and time to maturity T. We have considered unrealistically large maturities in order to show the limiting behavior of the option price.

where

$$d_{\pm} = \frac{1}{\sigma\sqrt{T}} \cdot \log\left(\frac{1}{K} \cdot X_Y(0)\right) \pm \frac{1}{2}\sigma\sqrt{T},$$

and so the price is a function of a factor $\sigma\sqrt{T}$. For instance, doubling the volatility has the same effect on the option price as quadrupling time. When $T \to \infty$, we simply have

$$\lim_{T\to\infty} V_Y^{EC} = X_Y$$

since $d_+ \to \infty$, and $d_- \to -\infty$. Therefore for large T, $V_Y(0) \approx X_Y(0)$, and the hedge is to hold a unit of an asset X and have no position in the asset Y. Figure 3.1 shows the price V_Y of a European call option with a payoff $(X_T - \frac{1}{2}Y_T)^+$ as a function of the price $X_Y(t)$ of the underlying asset X, and time to maturity $T - t$. Note that when $t = T$, the price of the contract is simply the payoff $(x - \frac{1}{2})^+$. On the other hand, for large maturities the price of the contract is approximately X_Y, so the price of V_Y becomes approximately linear in X_Y.

Figures 3.2 and 3.3 show the corresponding hedging positions in the underlying assets X and Y as a function of the price $X_Y(t)$ and time to maturity $T - t$. Note that the hedging position in the asset X is between 0 and 1, and the hedging position in the asset Y is between $-\frac{1}{2}$ and 0. For short maturities, the hedging position in the asset X should be close to 1 when the option is in

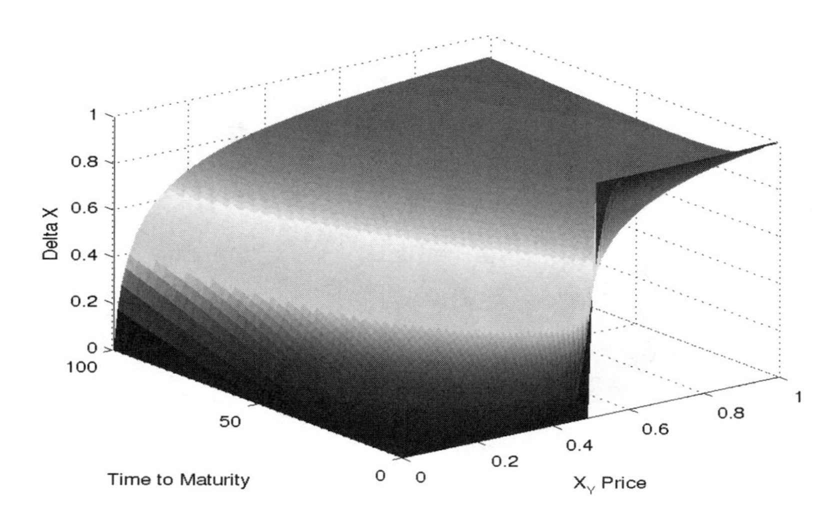

FIGURE 3.2: The hedging position in the asset X for the European option contract $(X_T - K \cdot Y_T)^+$ with parameters $K = \frac{1}{2}$, $\sigma = 0.2$, as a function of the price $X_Y(t)$ and time to maturity $T - t$. Note that the hedging position in the asset X is between 0 and 1.

the money, but it should be close to 0 when the option is out of the money. There is a jump in the hedging position at the strike price at the time of maturity. For large maturities, the hedging position in the asset X should be close to 1.

Similarly, for short maturities, the hedging position in the asset Y should be close to $-\frac{1}{2}$ when the option is in the money, and it should be close to 0 when the option is out of the money. For long maturities, the hedging position in the asset Y should be close to 0.

Figure 3.4 shows a sample path of X_Y in a geometric Brownian motion model, and the corresponding price of the European option V_Y. Figure 3.5 shows the corresponding hedging position in the underlying assets X and Y. Note that the hedging positions start to change dramatically when the time is close to maturity. The reason is that the price of the underlying asset happens to be near the strike price when the option is close to maturity, and the corresponding hedging position in the asset X takes the values close to 0 or 1 depending whether the option is out of the money or in the money. We observe a similar behavior for the hedging position in the asset Y.

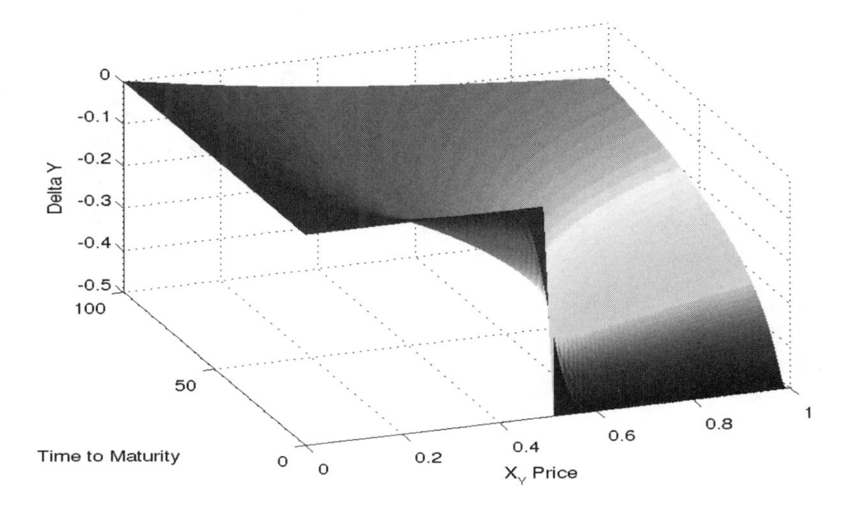

FIGURE 3.3: The hedging position in the asset Y for the European option contract $(X_T - K \cdot Y_T)^+$ with parameters $K = \frac{1}{2}$, $\sigma = 0.2$, as a function of the price $X_Y(t)$ and time to maturity $T - t$. Note that the hedging position in the asset Y is between $-\frac{1}{2}$ and 0.

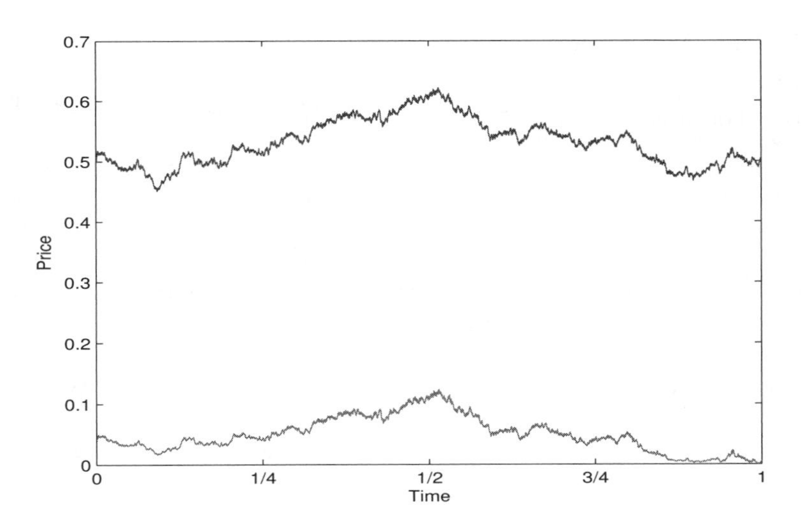

FIGURE 3.4: The price $X_Y(t)$ of an asset X in terms of the reference asset Y (top), and the price $V_Y(t)$ of a European option contract with a payoff $(X_T - K \cdot Y_T)^+$ with parameters $X_Y(0) = \frac{1}{2}$, $K = \frac{1}{2}$, $\sigma = 0.2$, $T = 1$ (bottom).

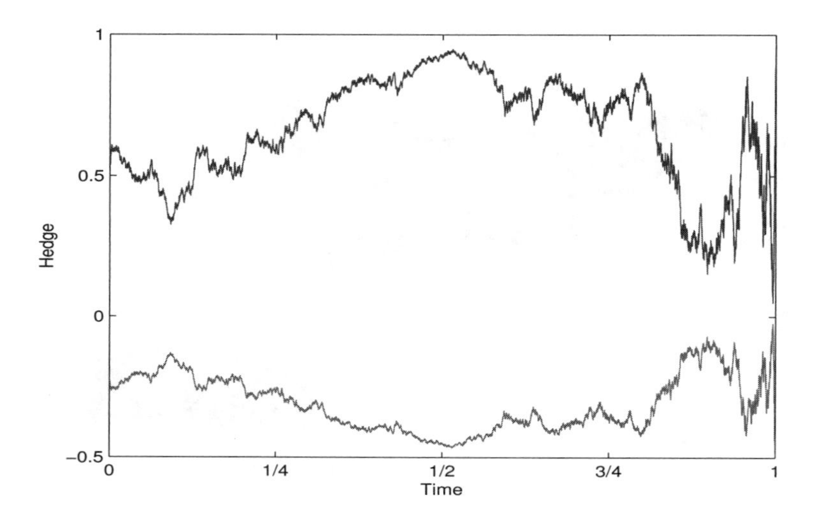

FIGURE 3.5: The hedging position in the asset X (top) and Y (bottom) for the European option contract $(X_T - K\cdot Y_T)^+$ with parameters $X_Y(0) = \frac{1}{2}$, $K = \frac{1}{2}$, $\sigma = 0.2$, $T = 1$. Note that the hedging position in the asset X is between 0 and 1, and the hedging position in the asset Y is between $-\frac{1}{2}$ and 0.

REMARK 3.10 Greeks

Greeks measure sensitivities of the prices of the portfolio (or in particular a single financial contract) to the changes of the parameters of the model. They describe how the price of the portfolio would change if the parameters change. Note that the price of the portfolio is given relative to the reference asset, so one can define portfolio sensitivities for any price function. The traditional definition of greeks applies to the price function $v^\$$, but it would make even better sense to apply it to the price function u^Y, or u^X. The assets Y and X have no time value (in contrast to a dollar \$), and thus the corresponding greeks would not be influenced by the time decay of the reference asset.

Delta is the sensitivity of the price u^Y with respect to the price of the underlying X_Y:

$$\Delta(t) = u^Y_x(t, X_Y(t)) = \frac{\partial u^Y(t, X_Y(t))}{\partial X_Y(t)}. \tag{3.83}$$

Gamma is the sensitivity of Δ with respect to the price of the underlying X_Y, which is the same as the second derivative of u^Y with respect to X_Y:

$$\Gamma(t) = u^Y_{xx}(t, X_Y(t)) = \frac{\partial^2 u^Y(t, X_Y(t))}{\partial X^2_Y(t)}. \tag{3.84}$$

Theta is the sensitivity of the price u^Y with respect to time t:

$$\Theta(t) = u_t^Y(t, X_Y(t)) = \frac{\partial u^Y(t, X_Y(t))}{\partial t}. \tag{3.85}$$

Vega is the sensitivity of the price u^Y with respect to the volatility σ:

$$\nu(t) = u_\sigma^Y(t, X_Y(t)) = \frac{\partial u^Y(t, X_Y(t))}{\partial \sigma}. \tag{3.86}$$

Rho is the sensitivity of the price u^Y with respect to the interest rate r:

$$\rho(t) = u_r^Y(t, X_Y(t)) = \frac{\partial u^Y(t, X_Y(t))}{\partial r}. \tag{3.87}$$

□

Example 3.9
Consider an option with a payoff $(X_T - K \cdot Y_T)^+$. Its price is given by the Black–Scholes formula

$$\begin{aligned} u^Y(t, X_Y(t)) = xN\left(\tfrac{1}{\sigma\sqrt{T-t}} \cdot \log\left(\tfrac{X_Y(t)}{K}\right) + \tfrac{1}{2}\sigma\sqrt{T-t}\right) \\ - KN\left(\tfrac{1}{\sigma\sqrt{T-t}} \cdot \log\left(\tfrac{X_Y(t)}{K}\right) - \tfrac{1}{2}\sigma\sqrt{T-t}\right). \end{aligned}$$

The corresponding greeks are given by

$$\begin{aligned}
\Delta(t) &= N\left(\tfrac{1}{\sigma\sqrt{T-t}} \cdot \log\left(\tfrac{X_Y(t)}{K}\right) + \tfrac{1}{2}\sigma\sqrt{T-t}\right), \\
\Gamma(t) &= \frac{1}{X_Y(t)\sigma\sqrt{T-t}} \cdot \phi\left(\tfrac{1}{\sigma\sqrt{T-t}} \cdot \log\left(\tfrac{X_Y(t)}{K}\right) + \tfrac{1}{2}\sigma\sqrt{T-t}\right), \\
\theta(t) &= -\tfrac{1}{2} \cdot \frac{\sigma X_Y(t)}{\sqrt{T-t}} \cdot \phi\left(\tfrac{1}{\sigma\sqrt{T-t}} \cdot \log\left(\tfrac{X_Y(t)}{K}\right) + \tfrac{1}{2}\sigma\sqrt{T-t}\right), \\
\nu(t) &= X_Y(t)\sqrt{T-t} \cdot \phi\left(\tfrac{1}{\sigma\sqrt{T-t}} \cdot \log\left(\tfrac{X_Y(t)}{K}\right) + \tfrac{1}{2}\sigma\sqrt{T-t}\right), \\
\rho(t) &= 0.
\end{aligned}$$

The sensitivity ρ turns out to be zero since the price evolution X_Y is not influenced by the changes of the interest rate (assets X and Y have no time value). The changes of the interest rate would influence contracts that depend on the assets with time value, such as a dollar \$. □

3.8 Stochastic Volatility Models

When we have a contingent claim V whose payoff depends on the assets X and Y, its price $V_Y(t)$ can depend on the entire price evolution $X_Y(s)$ up to time t. It can also depend on several additional external processes $\xi^i(s)$, such as a random process that represents stochastic volatility. In this case we can write

$$V_Y(t) = u^Y(t, \{X_Y(s)\}_{s=0}^t, \{\xi^i(s)\}_{s=0}^t).$$

While this expression would explain the price process $V_Y(t)$ in full, it would be prohibitively complicated to model the price of $V_Y(t)$ using infinitely many possible values from $\{X_Y(s)\}_{s=0}^t$ and $\{\xi^i(s)\}_{s=0}^t$. Thus it is desirable to express such dependence using only a small number of factors that would explain the price evolution $V_Y(t)$ sufficiently well.

A common approach to price modeling is to use the Markov property:

$$V_Y(t) = u^Y(t, \{X_Y(s)\}_{s=0}^t, \{\xi^i(s)\}_{s=0}^t) = u^Y(t, X_Y(t), \xi^i(t)),$$

which says that the only relevant information about the future evolution of the process $V_Y(t)$ is given by the present values of the underlying processes $X_Y(t)$ and $\xi^i(t)$.

The simplest models that we considered in the previous text assume no external processes $\xi^i(t)$, and the price of the contract V can be written as

$$V(t) = u^Y(t, X_Y(t)) \cdot Y = u^X(t, Y_X(t)) \cdot X.$$

More general models of the asset prices consider a stochastic evolution of volatility. The price of a contract V depends on the price of the underlying asset $X_Y(t)$, and on a process $\xi(t)$ that represents the volatility

$$V(t) = u^Y(t, X_Y(t), \xi(t)) \cdot Y = u^X(t, Y_X(t), \xi(t)) \cdot X.$$

This model has two sources of uncertainty, and it is not possible in general to hedge such contracts perfectly with only two assets X and Y. A general rule for a complete market is to have $n+1$ assets for n sources of noise, which is not the case here. Thus stochastic volatility models are not complete in general and a perfect replication of an arbitrary contingent claim may no longer be possible. As mentioned earlier, the volatility is the same for both $X_Y(t)$ and $Y_X(t)$.

Let us assume that the price process follows

$$dX_Y(t) = g(t, \xi(t))X_Y(t)dW^Y(t), \tag{3.88}$$

where $\xi(t)$ is a stochastic process in the form

$$d\xi(t) = \alpha(t, \xi(t))dt + \beta(t, \xi(t))dW^\xi(t). \tag{3.89}$$

We assume that the two Brownian motions W^Y and W^ξ are correlated:

$$dW^Y(t) \cdot dW^\xi(t) = \rho dt.$$

Note that the price process $X_Y(t)$ is a $\mathbb{P}^Y$ martingale. The process $\xi(t)$ is a parameter of the model, and as such it can have an arbitrary evolution. In particular, it does not need to be a martingale.

Example 3.10

A popular stochastic volatility model is the **Heston model**, which is given by the following choice of the functions g, α, and β:

$$g(t,\xi) = \sqrt{\xi}, \quad \alpha(t,\xi) = a - b \cdot \xi, \quad \beta(t,\xi) = \sigma\sqrt{\xi}.$$

In this case we can write

$$dX_Y(t) = \sqrt{\xi(t)} \cdot X_Y(t) dW^Y(t),$$

and

$$d\xi(t) = (a - b \cdot \xi(t))dt + \sigma\sqrt{\xi(t)} dW^\xi(t).$$

□

Let V be a contingent claim whose price $V_Y(t)$ depends only on $X_Y(t)$ and on $\xi(t)$. We can write

$$V_Y(t) = u^Y(t, X_Y(t), \xi(t)).$$

Since $V_Y(t)$ is a $\mathbb{P}^Y$ martingale, we can obtain a partial differential equation for the price function u^Y. We have

$$\begin{aligned} du^Y(t, X_Y(t), \xi(t)) &= u_t^Y dt + u_x^Y dX_Y(t) + u_\xi^Y d\xi(t) \\ &\quad + \tfrac{1}{2} u_{xx}^Y d^2 X_Y(t) + u_{x\xi}^Y dX_Y(t) d\xi(t) + \tfrac{1}{2} u_{\xi\xi}^Y d^2\xi(t) \\ &= \Big[u_t^Y + \alpha(x,\xi) u_\xi^Y + \tfrac{1}{2} g^2 X_Y(t)^2 u_{xx}^Y \\ &\quad + \rho\beta g X_Y(t) u_{x\xi}^Y + \tfrac{1}{2}\beta^2 u_{\xi\xi}^Y\Big] dt \\ &\quad + g X_Y(t) u_x^Y + \beta u_\xi^Y dW^\xi(t). \end{aligned}$$

Since the dt term must be zero, we get a partial differential equation for u^Y:

$$\begin{aligned} u_t^Y(t,x,\xi) &+ \alpha(t,\xi) u_\xi^Y(t,x,\xi) + \tfrac{1}{2} g(t,\xi)^2 x^2 u_{xx}^Y(t,x,\xi) \\ &+ \rho\beta(t,\xi) g(t,\xi) x u_{x\xi}^Y(t,x,\xi) + \tfrac{1}{2}\beta(t,\xi)^2 u_{\xi\xi}^Y(t,x,\xi) = 0. \end{aligned} \tag{3.90}$$

Similarly, we can study the evolution of the inverse price that takes the same form

$$dY_X(t) = g(t, \xi(t)) \cdot Y_X(t) dW^X(t),$$

where

$$dW^X(t) = -dW^Y(t) + g(t, \xi(t)) dt.$$

This follows from Ito's formula

$$\begin{aligned} dY_X(t) = dX_Y(t)^{-1} &= -Y_X(t)^2 dX_Y(t) + \tfrac{1}{2} \cdot 2Y_X(t)^3 d^2 X_Y(t) \\ &= -g(t, \xi(t)) \cdot Y_X(t) dW^Y(t) + g(t, \xi(t))^2 \cdot Y_X(t) dt \\ &= g(t, \xi(t)) \cdot Y_X(t) dW^X(t). \end{aligned}$$

The correlation between $W^X(t)$ and $W^\xi(t)$ is given by

$$dW^X(t) \cdot dW^\xi(t) = (-dW^Y(t) + g(t, \xi(t))dt) \cdot dW^\xi(t) = -\rho dt.$$

The only difference is that the correlation coefficient takes an opposite sign. Thus if we have

$$V_X(t) = u^X(t, Y_X(t), \xi(t)),$$

the partial differential equation for u^X differs only in the sign that corresponds to the correlation coefficient. Therefore u^X satisfies

$$\begin{aligned} u_t^X(t,x,\xi) + \alpha(x,\xi) u_\xi^X(t,x,\xi) + \tfrac{1}{2} g(t,\xi)^2 x^2 u_{xx}^X(t,x,\xi) \\ - \rho\beta(t,\xi) g(t,\xi) x u_{x\xi}^X(t,x,\xi) + \tfrac{1}{2}\beta(t,\xi)^2 u_{\xi\xi}^X(t,x,\xi) = 0. \quad (3.91) \end{aligned}$$

3.9 Foreign Exchange Market

This section studies contracts traded on foreign exchange markets. Let an asset X be the domestic currency \$, and an asset Y be the foreign currency €. Let

$$€_\$(t)$$

denote the amount of a domestic currency that is needed to acquire a unit of a foreign currency at time t. The quantity $€_\$(t)$ is known as an **exchange rate,** but in fact this is just a special case of the price $X_Y(t)$, where $X = €$, and $Y = \$$. Thus we can apply the results we have obtained in the previous sections for the case of the foreign exchange market.

The foreign exchange market is an excellent example to illustrate the relative concept of prices since both the domestic and the foreign currencies are legitimate choices for the reference asset. Whether a currency is domestic or foreign depends on which country one lives in. For some people \$ is the domestic currency and € is the foreign currency, but for other people € is the domestic currency and \$ is the foreign currency. Thus it makes perfect sense to study the inverse exchange rate

$$\frac{1}{€_\$(t)} = \$_€(t).$$

Note that $€_\$(t)$ and $\$_€(t)$ are prices, not assets that could be bought or sold. Moreover, the currencies themselves are arbitrage assets, and thus one needs to immediately acquire a suitable no-arbitrage asset for it in order not to lose value, such as a bond that is denominated in the corresponding currency.

3.9.1 Forwards

Let us consider first a **forward** contract on the foreign exchange with a payoff

$$€_T - K \cdot \$_T = (€_\$(T) - K) \cdot \$_T = (1 - \$_€(T)) \cdot €_T.$$

at time T. Let us write this contract in terms of no-arbitrage assets. There is a corresponding foreign bond that delivers one € at time T. We will denote this no-arbitrage asset by $B^{€,T}$. Similarly, there is a domestic bond B^T that delivers one \$ at time T. Therefore the forward contract is equivalent to the contract with a payoff

$$B_T^{€,T} - K \cdot B_T^T.$$

Let us denote the forward contract by $F(B^{€,T}, K \cdot B^T, T)$, and let us compute its price. The contract depends on two no-arbitrage assets, namely on $B^{€,T}$, and B^T. A possible numeraire for pricing is B^T, a domestic bond maturing at time T. We have

$$F_{B^T}(t) = \mathbb{E}_t^T\left[\left(B^{€,T} - K \cdot B^T\right)_{B^T}(T)\right] = B_{B^T}^{€,T}(t) - K. \tag{3.92}$$

The last identity follows from the fact that the price of the bond $B^{€,T}$ in terms of the bond B^T is a martingale under the T-forward measure that corresponds to B^T as a reference asset. Thus we conclude that the forward contract is equal to

$$F_t(B^{€,T}, K \cdot B^T, T) = B_t^{€,T} - K \cdot B_t^T. \tag{3.93}$$

Note that this is a model-independent formula (we have not assumed any particular dynamics). A **forward exchange rate** is a choice of $\bar{K}$ that makes the forward contract equal to zero:

$$F_0(B^{€,T}, \bar{K} \cdot B^T, T) = 0. \tag{3.94}$$

Solving this equation, we get

$$\boxed{\bar{K} = B_{B^T}^{€,T}(0).} \tag{3.95}$$

If we assume constant interest rates for both the domestic and the foreign zero coupon bond, we can express the above relationship in terms of the exchange rate $€_\$(0)$. The domestic bond price in terms of the domestic currency is

$$B_t^T = e^{-r(T-t)} \cdot \$_t,$$

and the foreign bond price in terms of the foreign currency is

$$B_t^{€,T} = e^{-r^F(T-t)} \cdot €_t.$$

Thus we can write

$$B_{B^T}^{€,T}(t) = B_{€}^{€,T}(t) \cdot €_{\$}(t) \cdot \$_{B^T}(t)$$
$$= e^{-r^F(T-t)} \cdot €_{\$}(t) \cdot e^{r(T-t)} = e^{(r-r^F)(T-t)} \cdot €_{\$}(t).$$

In other words,

$$\boxed{\bar{K} = B_{B^T}^{€,T}(0) = e^{(r-r^F)T} \cdot €_{\$}(0).} \tag{3.96}$$

We can consider a similar contract on the inverse exchange rate $\$_{€}(T)$

$$\$_T - K^f \cdot €_T = \left(\$_{€}(T) - K^f\right) \cdot €_T = (1 - K^f \cdot €_{\$}(T)) \cdot \$_T.$$

The corresponding no-arbitrage assets are B^T and $B^{€,T}$. We can rewrite the payoff of the contract as

$$B_T^T - K^f \cdot B_T^{€,T}.$$

If we denote this contract by $F(B^T, K^f \cdot B^{€,T}, T)$, and choose the corresponding foreign bond $B^{€,T}$ as a numeraire, we get

$$F_{B^{€,T}}(t) = \mathbb{E}_t^{€,T}\left[\left(B^T - K^f \cdot B^{€,T}\right)_{B^{€,T}}(T)\right] = B_{B^{€,T}}^T(t) - K^f. \tag{3.97}$$

The last equation follows from the fact that the price of B^T in terms of $B^{€,T}$ is a martingale under the measure that corresponds to $B^{€,T}$ as a numeraire. The forward contract is equal to

$$F_t(B^T, K^f \cdot B^{€,T}, T) = B_t^T - K^f \cdot B_t^{€,T}. \tag{3.98}$$

The corresponding forward exchange rate from the point of view of the foreign currency is a choice of $\bar{K}^f$ that makes the value of the forward contract zero:

$$F_0(B^T, \bar{K}^f \cdot B^{€,T}, T) = 0. \tag{3.99}$$

Solving for $\bar{K}^f$, we get

$$\boxed{\bar{K}^f = B_{B^{€,T}}^T(0) = e^{(r^F-r)T} \cdot \$_{€}(0) = \frac{1}{\bar{K}}.} \tag{3.100}$$

Note that the forward exchange rates as seen from the domestic currency and from the foreign currency point of view are linked through $\bar{K} = \frac{1}{\bar{K}^f}$.

3.9.2 Options

European-type contracts on foreign exchange are special cases of general European contracts where the roles of the no-arbitrage assets X and Y are played by no-arbitrage assets $B^{€,T}$ and B^T. For instance, a call option with payoff

$$(€_T - K \cdot \$_T)^+$$

can be rewritten in terms of the no-arbitrage assets as

$$(B_T^{€,T} - K \cdot B_T^T)^+.$$

We have a special case of the Black–Scholes formula that is also known as Garman–Kohlhagen formula.

REMARK 3.11 Garman–Kohlhagen formula

The value $V_t^{EC}(B^{€,T}, KB^T, T)$ of a European option contract with a payoff $\left(B^{€,T} - K \cdot B_T^T\right)^+$ is given by

$$\begin{aligned} V^{EC}(B^{€,T}, KB^T, T) &= \mathbb{P}_t^{€,T}(B_{B^T}^{€,T}(T) \geq K) \cdot B_t^{€,T} \\ &\quad - K \cdot \mathbb{P}_t^T(B_{B^T}^{€,T}(T) \geq K) \cdot B_t^T \qquad (3.101) \\ &= \mathbb{P}_t^{€,T}(€_\$(T) \geq K) \cdot e^{-r^f(T-t)} \cdot €_T \\ &\quad - K \cdot \mathbb{P}_t^T(€_\$(T) \geq K) \cdot e^{-r(T-t)} \cdot \$_T. \end{aligned}$$

Moreover, the corresponding deltas are given in the geometric Brownian motion model by

$$\Delta^{€,T}(t) = \mathbb{P}_t^{€,T}(B_{B^T}^{€,T}(T) \geq K) = \mathbb{P}_t^{€,T}(€_\$(T) \geq K), \qquad (3.102)$$

and

$$\Delta^T(t) = -K \cdot \mathbb{P}_t^T(B_{B^T}^{€,T}(T) \geq K) = -K \cdot \mathbb{P}_t^T(€_\$(T) \geq K). \qquad (3.103)$$

□

References and Further Reading

The first introduction of Brownian motion to finance is by Bachelier (1900). He used it as a model for stock prices, although Brownian motion can take negative values. The general theory of stochastic calculus was developed by Ito (1944). Merton (1969) was the first person who used it in finance. Samuelson (1965, 1973) argued that geometric Brownian motion is a good model for stock prices. The first derivation of the Black–Scholes formula appears in Black and Scholes (1973). A similar result appears in the independent work of Merton (1973). The Black–Scholes formula is quite robust to model misspecifications as shown in El Karoui et al. (1998). Garman and Kohlhagen (1983) found the analogous formula to the Black–Scholes for currency options.

Girsanov's theorem is due to Girsanov (1960), although the result for the constant σ appeared already in Cameron and Martin (1944). The principle of exchangeability of the reference assets appears implicitly already in Carr and Bowie (1994), and more recently the symmetries between the pricing martingale measures are explored in detail in Carr and Lee (2009). Hoogland and Neumann (2001a,b) explored the symmetries in pricing with respect to different reference assets using the partial differential equation approach. They noticed the advantages of using no-arbitrage assets for the numeraire. Wystup (2008) applied symmetry analysis in the foreign exchange market and showed various relationships of greeks for specific options. Preservation of convexity by the perspective mapping is shown for instance in Hiriart-Urruty and Lemarechal (1993).

Books that explain the numerical implementation of partial differential equations in detail are, for instance, Tavella and Randall (2000) or Duffy (2006). Monte Carlo methods for pricing financial derivatives are employed for instance in the papers of Boyle (1977), Boyle et al. (1997), Broadie and Glasserman (1996), or in the monographs Glasserman (2003), Jaeckel (2002), or Korn et al. (2010). Books on pricing derivative contracts under stochastic volatility include Lewis (2000), Gatheral (2006), Fouque et al. (2000), or Rebonato (2004). The Heston stochastic volatility model was introduced in Heston (1993).

Exercises

3.1 Prove that the perspective mapping preserves convexity: If $f^Y(x)$ is a convex function, then the function $f^X(x) = f^Y(\frac{1}{x}) \cdot x$, $x > 0$ is also convex.

(a) Show the result for the twice differentiable function f^Y.
Hint: Twice differentiable convex functions satisfy $[f^Y(x)]'' \geq 0$. Show that this implies $[f^X(x)]'' \geq 0$.

(b) Show the result for a general convex function $f^Y(x)$.

3.2 Show that

$$X_Y(t) \cdot \phi(d_+) - K \cdot \phi(d_-) = 0,$$

where $\phi(x) = \frac{1}{\sqrt{2\pi}} \cdot e^{-\frac{x^2}{2}}$, and

$$d_{\pm} = \frac{1}{\sigma\sqrt{T-t}} \cdot \log\left(\frac{X_Y(t)}{K}\right) \pm \frac{1}{2}\sigma\sqrt{T-t}.$$

3.3 Show that the price

$$u^Y(t,x) = N\left(\frac{1}{\sigma\sqrt{T-t}} \cdot \log\left(\frac{x}{K}\right) - \frac{1}{2}\sigma\sqrt{T-t}\right)$$

of the Arrow–Debreu security V that pays off $V_T = \mathbb{I}(X_Y(T) \geq K) \cdot Y_T$ in the geometric Brownian motion model satisfies the partial differential equation

$$u_t^Y(t,x) + \frac{1}{2}\sigma^2 x^2 u_{xx}^Y(t,x) = 0,$$

with the terminal condition $u^Y(T,x) = \mathbb{I}(x \geq K)$, and the boundary condition $u^Y(t,0) = 0$.

3.4 **(a)** Find the hedging portfolio for an Arrow–Debreu security V that pays off $V_T = \mathbb{I}(X_Y(T) \geq K) \cdot Y_T$ in a geometric Brownian motion model.

(b) Find the hedging portfolio for an Arrow–Debreu security V that pays off $U_T = \mathbb{I}(X_Y(T) \geq K) \cdot X_T$ in the same model.

(c) Combining the results from (a) and (b), find the hedging portfolio for a contract W that pays off $W_T = U_T - K \cdot V_T$. Note that W is a European call option.

3.5 Assume that $X_Y(0) = 1$, and consider a contract that pays off the more valuable asset at time T; i.e., the payoff is in the form

$$\max(X_T, Y_T) = X\mathbb{I}\left(X_Y(T) \geq 1\right) + Y\mathbb{I}\left(X_Y(T) < 1\right).$$

The price $X_Y(t)$ follows Geometric Brownian Motion:

$$dX_Y(t) = \sigma X_Y(t) dW^Y(t).$$

(a) Compute the price of this contract.

(b) Determine the hedge of the contract: $\Delta^X(t)$ and $\Delta^Y(t)$.

3.6 Let the price process $X_Y(t)$ follow the geometric Brownian motion

$$dX_Y(t) = \sigma X_Y(t) dW^Y(t),$$

and let K be a general positive constant.

(a) What is the price of a contract that pays a unit of Y when $X_Y(T) \geq K$ when $T \to \infty$? How would you hedge such a contract?

(b) What is the price of a contract that pays a unit of X when $X_Y(T) \geq K$ when $T \to \infty$? How would you hedge such a contract?

(c) Determine the price and the hedge of a European call option with the payoff $(X_T - K \cdot Y_T)^+$ when $T \to \infty$.

Chapter 4

Interest Rate Contracts

This chapter covers financial contracts that depend on interest rates. The market that trades interest rate contracts is known as a fixed income market. The main message of this chapter is that the interest rate traded on the exchanges or on the over-the-counter markets is just a special case of the concept of price, and thus the pricing techniques developed in the previous chapters immediately apply to the interest rate market as well. The interest rate is distinct in the sense that the corresponding reference asset is not a currency, but rather a bond, or a portfolio of bonds. Thus the numeraire approach to use a bond as a natural reference asset is especially helpful in the determination of the price of the interest rate contracts.

The most traded interest rate contracts depend on the London Interbank Offered Rate, or LIBOR for short. Spot LIBOR $L(T,T)$ represents a simple annualized interest rate that corresponds to borrowing money for a period between times T and $T+\delta$, where δ is typically 3 months. The market also trades contracts that depend on the value of LIBOR in the future. The simplest contract, the backset LIBOR, agrees to deliver $L(T,T)$ units of a dollar at time $T+\delta$. The price of this contract is a forward LIBOR $L(t,T)$, a simple interest rate that corresponds to borrowing money between times T and $T+\delta$ as seen at an earlier time t. The forward LIBOR is in fact a price, where the reference asset is a bond with maturity $T+\delta$.

Exchanges trade on the LIBORs as using futures contracts rather than forward contracts. The connection between futures and forwards was discussed in detail in Section 1.10, and we will not revisit this analysis in this chapter. The futures market on the LIBOR has one specific convention as the settlement price is quoted as $100 \cdot (1 - L(T,T))$ rather than simply $L(T,T)$. For example, the interest rate 1.98% corresponds to the quote of 98.02 on the futures market for LIBOR. An option on the LIBOR rate is known as either a caplet (call option on LIBOR) or a floorlet (put option on LIBOR). Since LIBOR is a price, the price of the caplet and the floorlet is given by the Black–Scholes formula, but it is known as the Black caplet formula in the fixed income market.

Another widely traded contract is a swap. A holder of a swap agrees to receive floating interest rate payments for an exchange of fixed interest rate payments. A swap rate is the level of the fixed interest rate payments that makes the value of the swap equal to zero. Similar to the case of the forward LIBOR, forward swap rates are particular instances of a price, where the reference asset is a portfolio of bonds. The right to enter a swap contract is known as a swaption, and the price of the swaption is also given by the Black–Scholes formula.

Models of the interest rate consider instantaneous rates corresponding to a situation when the borrowed money is returned in an infinitesimal instant. This is the case of a forward rate $f(t,T)$, an instantaneous rate for borrowing money at time T as seen at time t. The forward rates must follow a no-arbitrage evolution given by the Heath–Jarrow–Morton model (HJM for short). It turns out that the spot rate models are just special cases of the HJM model.

4.1 Forward LIBOR

The **forward LIBOR** is defined as a simple interest rate that corresponds to borrowing money over the time interval between T and $T+\delta$ as seen at time $t \leq T$. We denote the forward LIBOR by $L(t,T)$. Suppose that \$ 1 is borrowed at time T, and assume that $L(t,T)$ is the simple interest rate for the period between T and $T+\delta$. Then the agent should return $1+\delta L(t,T)$ dollars at time $T+\delta$. Thus $L(t,T)$ can be defined by the following relationship:

$$(1+\delta L(t,T)) \cdot B_t^{T+\delta} = B_t^T. \tag{4.1}$$

The right hand side of the above relationship indicates that one dollar will be delivered at time T. The left hand side indicates that $(1+\delta L(t,T))$ dollars will be returned at time $T+\delta$. Therefore

$$\delta B_t^{T+\delta} \cdot L(t,T) = B_t^T - B_t^{T+\delta}, \tag{4.2}$$

or in other words

$$\boxed{L(t,T) = \left[B^T - B^{T+\delta}\right]_{\delta B^{T+\delta}}(t).} \tag{4.3}$$

Notice that the forward LIBOR is in fact a price, where the asset X is a portfolio $[B^T - B^{T+\delta}]$ (long the B^T bond, and short the $B^{T+\delta}$ bond), and the reference asset Y is $\delta B^{T+\delta}$. Note that the forward LIBOR is an example of a price process where the reference asset is not the currency itself. The fact that we can treat the forward LIBOR as a price means that we can apply the general theory of pricing contingent claims.

It should be also noted that $L(t,T)$ as a price continues to evolve up to time T. At time T, the B^T bond expires, so the last meaningful observation of the forward LIBOR is $L(T,T)$. Therefore after time T, the LIBOR $L(T,T)$ is known, and it is a constant. Typical contracts that depend on the LIBOR $L(T,T)$ are settled at time $T+\delta$, a delay of δ units of time after observing its value at time T. The most traded contracts have δ equal to 3 months $= \frac{1}{4}$ year.

Since the forward LIBOR is a price, it is a martingale under the forward measure $\mathbb{P}^{T+\delta}$ associated with the reference asset $\delta B^{T+\delta}$, δ units of the bond maturing at time $T+\delta$. A popular model of the forward LIBOR is a geometric Brownian motion

$$\begin{aligned} dL(t,T) &= d\left([B^T - B^{T+\delta}]_{\delta B^{T+\delta}}(t)\right) \\ &= \sigma\left([B^T - B^{T+\delta}]_{\delta B^{T+\delta}}(t)\right) dW^{T+\delta}(t). \end{aligned}$$

In this case, we have

$$L(T,T) = L(t,T) \cdot \exp\left(\sigma W^{T+\delta}(T-t) - \tfrac{1}{2}\sigma^2(T-t)\right). \tag{4.4}$$

For pricing European contracts on the LIBOR, it is also important to consider dynamics of the inverse price $\frac{1}{L(t,T)}$ which has to be a martingale under probability measure associated with $[B^T - B^{T+\delta}]$ as a reference asset. We will call this measure $\mathbb{P}^{T,T+\delta}$; it corresponds to the measure $\mathbb{P}^X$ that was used in the previous sections. We have that

$$\begin{aligned} d\left(\tfrac{1}{L(t,T)}\right) &= d\left[\delta B^{T+\delta}\right]_{(B^T - B^{T+\delta})}(t) \\ &= \sigma\left[\delta B^{T+\delta}\right]_{(B^T - B^{T+\delta})}(t) dW^{T,T+\delta}(t). \end{aligned} \tag{4.5}$$

This gives us

$$L(T,T) = L(t,T) \cdot \exp\left(\sigma W^{T,T+\delta}(T-t) + \tfrac{1}{2}\sigma^2(T-t)\right). \tag{4.6}$$

4.1.1 Backset LIBOR

Backset LIBOR is a contract that pays off $L(T,T)$ units of a dollar \$ at time $T+\delta$

$$V_{T+\delta} = L(T,T) \cdot \$_{T+\delta}.$$

The payoff can be also expressed in terms of a $B^{T+\delta}$ bond as

$$V_{T+\delta} = L(T,T) \cdot B^{T+\delta}_{T+\delta} = L(T,T) \cdot \$_{T+\delta}. \tag{4.7}$$

When $t \leq T$, the price of this contract is given by

$$\begin{aligned} V_t &= \mathbb{E}^{T+\delta}_t \left[L(T,T) \cdot B^{T+\delta}_{B^{T+\delta}}(T+\delta)\right] \cdot B^{T+\delta}_t = L(t,T) \cdot B^{T+\delta}_t \\ &= \left[B^T - B^{T+\delta}\right]_{\delta B^{T+\delta}}(t) \cdot B^{T+\delta}_t = \tfrac{1}{\delta}\left[B^T_t - B^{T+\delta}_t\right]. \end{aligned} \tag{4.8}$$

We have used the fact that $L(t,T)$ is a $\mathbb{P}^{T+\delta}$ martingale. The hedge of this contract is obvious; one should long $\frac{1}{\delta}$ units of the bond B^T, and short $\frac{1}{\delta}$ units of the bond $B^{T+\delta}$. When the LIBOR $L(T,T)$ is already known at times $T \leq t \leq T+\delta$, the price of the contract is given by

$$V_t = \mathbb{E}_t^{T+\delta}\left[L(T,T)\cdot B^{T+\delta}_{B^{T+\delta}}(T+\delta)\right]\cdot B_t^{T+\delta} = L(T,T)\cdot B_t^{T+\delta}. \tag{4.9}$$

The hedge is to hold $L(T,T)$ units (now a known number) of the bond $B^{T+\delta}$.

4.1.2 Caplet

A **caplet** is a contract that pays off

$$V_{T+\delta} = (L(T,T)-K)^+ \cdot \$_{T+\delta} = \tfrac{1}{\delta}(L(T,T)-K)^+\cdot \delta B_{T+\delta}^{T+\delta} \tag{4.10}$$

at time $T+\delta$. It is a European call option on LIBOR settled in dollars. The price of caplet is given by the Black–Scholes formula

$$V^{EC}(X, K\cdot Y, T) = \tfrac{1}{\delta}\cdot\left[\mathbb{P}_t^X(X_Y(T)\geq K)\cdot X - K\cdot \mathbb{P}_t^Y(X_Y(T)\geq K)\cdot Y\right], \tag{4.11}$$

where $X = [B^T - B^{T+\delta}]$, and $Y = \delta B^{T+\delta}$. In this case we can write

$$\begin{aligned} V_t &= \tfrac{1}{\delta}\cdot\Big[\mathbb{P}_t^{T,T+\delta}(L(T,T)\geq K)\cdot[B^T - B^{T+\delta}] \\ &\qquad\qquad -K\cdot\mathbb{P}_t^{T+\delta}(L(T,T)\geq K)\cdot\delta B^{T+\delta}\Big] \\ &= \Big(L(t,T)\cdot\mathbb{P}_t^{T,T+\delta}(L(T,T)\geq K) - K\cdot\mathbb{P}_t^{T+\delta}(L(T,T)\geq K)\Big)\cdot B^{T+\delta} \end{aligned}$$

for $t \leq T$. For the geometric Brownian motion model, we have

$$\mathbb{P}_t^{T,T+\delta}(L(T,T)\geq K) = N\left(\tfrac{1}{\sigma\sqrt{T-t}}\cdot\log\left(\tfrac{1}{K}\cdot L(t,T)\right) + \tfrac{1}{2}\sigma\sqrt{T-t}\right),$$

and

$$\mathbb{P}_t^{T}(L(T,T)\geq K) = N\left(\tfrac{1}{\sigma\sqrt{T-t}}\cdot\log\left(\tfrac{1}{K}\cdot L(t,T)\right) - \tfrac{1}{2}\sigma\sqrt{T-t}\right).$$

The Black–Scholes formula for the price of the caplet is also known as the **Black caplet formula.**

The hedge of the European call option in a geometric Brownian motion model is given by

- $\Delta^X(t) = \mathbb{P}_t^X(X_Y(T)\geq K)$ units of an asset X, and
- $\Delta^Y(t) = -K\mathbb{P}_t^Y(X_Y(T)\geq K)$ units of an asset Y.

For the case of the caplet, one should have

- $\frac{1}{\delta}\mathbb{P}_t^{T,T+\delta}(L(T,T) \geq K)$ units of $[B^T - B^{T+\delta}]$, and

- $-K\mathbb{P}_t^{T+\delta}(L(T,T) \geq K)$ units of $B^{T+\delta}$.

The payoff of the caplet is scaled by a factor of $\frac{1}{\delta}$, so the hedging positions are scaled correspondingly. This is the same as having

- $\frac{1}{\delta}\mathbb{P}_t^{T,T+\delta}(L(T,T) \geq K)$ units of the bond B^T, and

- $-\frac{1}{\delta}\mathbb{P}_t^{T,T+\delta}(L(T,T) \geq K) - K\mathbb{P}_t^{T+\delta}(L(T,T) \geq K)$ units of the bond $B^{T+\delta}$.

4.2 Swaps and Swaptions

A **swap** is a series of payments of the form

$$\delta[L(T_{k-1}, T_{k-1}) - K] \cdot \$_{T_k} \tag{4.12}$$

for $k = 1, \cdots, n$, and $T_k = T_0 + \delta k$. The payments are paid at times T_k, where the time lag between the payments is given by $T_k - T_{k-1} = \delta$. In practice, δ is typically $\frac{1}{2}$ or $\frac{1}{4}$ years. The holder of the contract receives $L(T_{k-1}, T_{k-1})$ units of a dollar $\$_{T_k}$, which represents the LIBOR rates observed at the previous times T_{k-1}. Since the LIBOR rates are random variables, the corresponding payments are known as **floating payments.** On the other hand, the holder of the contract agrees to pay K units of a dollar $\$_{T_k}$ at time T_k, which corresponds to **fixed payments.** Note that the backset LIBOR is a special case of a swap with one payment only, $n = 1$.

Let V be the swap with the payments (4.12). We can rewrite the contract as taken at time t in a more compact form and in terms of no-arbitrage assets that have no time value as

$$V_t = \delta \cdot \sum_{k=1}^{n} [L(t, T_{k-1}) - K] \cdot B_t^{T_k}. \tag{4.13}$$

Furthermore, if we substitute for $L(t,T_{k-1}) = \frac{1}{\delta}\left[B^{T_{k-1}} - B^{T_k}\right]_{B^{T_k}}(t)$, we get

$$\begin{aligned}
V_t &= \delta \cdot \sum_{k=1}^{n} [L(t,T_{k-1}) - K] \cdot B_t^{T_k} \\
&= \delta \cdot \sum_{k=1}^{n} \left[\tfrac{1}{\delta}\left[B^{T_{k-1}} - B^{T_k}\right]_{B^{T_k}}(t) - K\right] \cdot B_t^{T_k} \\
&= \sum_{k=1}^{n} \left[\left(B_t^{T_{k-1}} - B_t^{T_k}\right) - \delta \cdot K \cdot B_t^{T_k}\right] \\
&= \left[B_t^{T_0} - B_t^{T_n}\right] - \delta \cdot K \cdot \sum_{k=1}^{n} B_t^{T_k}.
\end{aligned}$$

Therefore the swap is equal to

$$\boxed{V_t = \left[B_t^{T_0} - B_t^{T_n}\right] - \delta \cdot K \cdot \sum_{k=1}^{n} B_t^{T_k}.} \tag{4.14}$$

The **forward swap rate** $y(t)$ is the value of K in (4.14) which makes the value of the swap V equal to zero. Solving for $V_t = 0$, we get

$$\left[B_t^{T_0} - B_t^{T_n}\right] - \delta \cdot y(t) \cdot \sum_{k=1}^{n} B_t^{T_k} = 0.$$

Therefore

$$y(t) = \left[B^{T_0} - B^{T_n}\right]_{\sum_{k=1}^{n} \delta \cdot B^{T_k}}(t), \tag{4.15}$$

so the forward swap rate is also a special instance of a price, where the asset X is equal to $B^{T_0} - B^{T_n}$, and the reference asset Y is equal to $\sum_{k=1}^{n} \delta \cdot B^{T_k}$.

The price of the swap V in terms of the reference asset $\sum_{k=1}^{n} \delta \cdot B^{T_k}$ can be written as

$$V_{\sum_{k=1}^{n} \delta \cdot B^{T_k}}(t) = \left[B^{T_0} - B^{T_n}\right]_{\sum_{k=1}^{n} \delta \cdot B^{T_k}}(t) - K = [y(t) - K]. \tag{4.16}$$

A closely related contract is the right to enter the swap contract at time T_0, known as a **swaption.** The payoff of the swaption is given by

$$V_{T_0} = (y(T_0) - K)^+ \cdot \sum_{k=1}^{n} \delta \cdot B_{T_0}^{T_k} = \left(\left[B_{T_0}^{T_0} - B_{T_0}^{T_n}\right] - K \cdot \sum_{k=1}^{n} \delta \cdot B_{T_0}^{T_k}\right)^+. \tag{4.17}$$

Thus a swaption is a special case of an option with a payoff $(X_{T_0} - KY_{T_0})^+$, where $X = B^{T_0} - B^{T_n}$, and $Y = \sum_{k=1}^{n} \delta \cdot B^{T_k}$. The value of the swaption at time $t \leq T_0$ is given by the Black–Scholes formula (1.77) as

$$V_t = \mathbb{P}_t^{T_0,T_n}(y(T_0) \geq K) \cdot \left[B_t^{T_0} - B_t^{T_n}\right] - K \cdot \mathbb{P}_t^{\Sigma}(y(T_0) \geq K) \cdot \sum_{k=1}^{n} \delta \cdot B_t^{T_k}. \quad (4.18)$$

Here, the probability measure $\mathbb{P}^{T_0,T_n}$ corresponds to the reference asset $X = B^{T_0} - B^{T_n}$, and the probability measure $\mathbb{P}^{\Sigma}$ corresponds to the reference asset $Y = \sum_{k=1}^{n} \delta \cdot B^{T_k}$.

According to the First Fundamental Theorem of Asset Pricing, the forward swap rate $y(t)$ is a $\mathbb{P}^{\Sigma}$ martingale. A popular approach for modeling the evolution of the forward swap rate is to assume geometric Brownian motion dynamics

$$dy(t) = \sigma y(t) dW^{\Sigma}(t). \quad (4.19)$$

The Black–Scholes formula then simplifies to

$$V_t = N(d_+) \cdot \left[B_t^{T_0} - B_t^{T_n}\right] - K \cdot N(d_-) \cdot \sum_{k=1}^{n} \delta \cdot B_t^{T_k},$$

where

$$d_{\pm} = \frac{1}{\sigma\sqrt{T-t}} \cdot \log\left(\frac{y(t)}{K}\right) \pm \sigma\sqrt{T-t}.$$

4.3 Term Structure Models

Term structure models describe a no-arbitrage evolution of the prices of bonds with different maturities. Recall that a **bond** is a contract to deliver one dollar at time T. We denote this contract as B^T. Given that T can be chosen arbitrarily, we have in principle an infinite number of such contracts. A closely related product is a **money market** M, an investment account that gains interest.

The reader should note that bonds with different maturities represent different contracts only because the underlying asset to deliver, a dollar or a different currency, is an arbitrage asset. Imagine that the asset to be delivered is a no-arbitrage asset, such as a stock S. A contract to deliver a no-arbitrage asset is the no-arbitrage asset itself. So even if one had infinitely many possible delivery dates T, the corresponding contract to deliver coincides with the original underlying asset, and the term structure becomes trivial in such a case.

Let us focus on term structure models where the underlying asset, say a dollar \$, is an arbitrage asset. Besides the dollar \$ and the bonds B^T we also have another asset, a money market account M. Money market accounts are another no-arbitrage proxy asset for a dollar, where the dollars are invested in the bond with the shortest available maturity. When this bond expires, dollars are reinvested to the next bond with the shortest available maturity. Since the dollar is an arbitrage asset and the money market is a no-arbitrage asset, the price of the money market in terms of dollars appreciates (has a nonnegative dt term):

$$\boxed{dM_{\$}(t) = r(t)M_{\$}(t)dt.} \tag{4.20}$$

The First Fundamental Theorem of Asset Pricing implies that $B_M^T(t)$ should be $\mathbb{P}^M$ martingales for all maturities T, and that $B_{B^{T_1}}^{T_2}(t)$ or $M_{B^{T_1}}(t)$ should be $\mathbb{P}^{T_1}$ martingales. The pricing measure $\mathbb{P}^M$ that corresponds to the money market is also known as a **risk-neutral measure,** and the pricing measure $\mathbb{P}^T$ that corresponds to the bond with maturity T is known as a **T-forward measure.**

The key question is how one can capture the martingale evolution of the prices when there are infinitely many available assets. The bonds differ only in maturity T, and it is natural to expect that bonds with higher maturity are more volatile with respect to the money market than bonds with smaller maturity. Thus it is reasonable to assume that

$$dB_M^T(t) = \sigma^*(t,T)B_M^T(t)dW^M(t), \tag{4.21}$$

where $\sigma^*(t,T_1) \le \sigma^*(t,T_2)$ for $T_1 \le T_2$. Note that Equation (4.21) describes the price evolution of an unlimited number of assets B^T that differ in maturity time T. The monotonicity of $\sigma^*(t,T)$ can also be captured by expressing it as an integral

$$\sigma^*(t,T) = \int_t^T \sigma(t,u)du$$

for $\sigma(t,u) \ge 0$. Therefore we can write

$$\boxed{dB_M^T(t) = B_M^T(t)\left[\int_t^T \sigma(t,u)du\right] dW^M(t).} \tag{4.22}$$

The inverse price can be also considered; it is given by

$$\boxed{dM_{B^T}(t) = M_{B^T}(t)\left[\int_t^T \sigma(t,u)du\right] dW^T(t).} \tag{4.23}$$

The relationship between $W^M(t)$ and $W^T(t)$ is given by

$$dW^T(t) = -dW^M(t) + \left[\int_t^T \sigma(t,u)du\right] dt, \tag{4.24}$$

or equivalently by

$$dW^M(t) = -dW^T(t) + \left[\int_t^T \sigma(t,u)du\right] dt. \tag{4.25}$$

REMARK 4.1 Deterministic interest rates
It was noted in Section 1.4 that when the interest rate is deterministic, the bonds and the money market are just deterministic multiples of each other. In this case all T-forward measures $\mathbb{P}^T$ and the risk-neutral measure $\mathbb{P}^M$ are the same. This corresponds to the situation when $\sigma(t,T) \equiv 0$. □

Let us determine the continuously compounded interest rate $R(t,T,T+\epsilon)$ that corresponds to borrowing money at time T, returning it at time $T+\epsilon$ as seen at time t. Assume that 1 \$ is borrowed at time T. The amount to be returned at time $T+\epsilon$ is equal to $\exp(\epsilon R(t,T,T+\epsilon))$ units of a dollar. Having 1 \$ at time T is the same as having one bond B^T at time t. The bond B^T will deliver one dollar at time T. Similarly, having $\exp(\epsilon R(t,T,T+\epsilon))$ dollars at time $T+\epsilon$ is the same as having $\exp(\epsilon R(t,T,T+\epsilon))$ bonds $B^{T+\epsilon}$ at time t. The two assets should have the same price at time t, so we need

$$B_t^T = \exp(\epsilon R(t,T,T+\epsilon)) \cdot B_t^{T+\epsilon}. \tag{4.26}$$

We can rewrite the above equation in terms of dollar prices as

$$B_\$^T(t) = \exp(\epsilon R(t,T,T+\epsilon)) \cdot B_\$^{T+\epsilon}(t), \tag{4.27}$$

leading to the formula for $R(t,T,T+\epsilon)$

$$R(t,T,T+\epsilon) = -\frac{1}{\epsilon} \cdot \left[\log(B_\$^{T+\epsilon}(t)) - \log(B_\$^T(t))\right]. \tag{4.28}$$

The **forward rate** $f(t,T)$ is defined as a limiting case when $\epsilon \to 0$ which corresponds to an instantaneous interest rate at time T as seen from time t. Formally,

$$\boxed{f(t,T) = \lim_{\epsilon \to 0} R(t,T,T+\epsilon) = -\frac{\partial}{\partial T}\log(B_\$^T(t)).} \tag{4.29}$$

The goal is to determine the evolution of $f(t,T)$ as time t passes for a fixed T, or in other words the dynamics of $df(t,T)$. The strategy is to first determine the evolution of $\int_t^T f(t,u)du$. Note that

$$\int_t^T f(t,u)du = -\left[\log(B_\$^T(t)) - \log(B_\$^t(t))\right] = -\log(B_\$^T(t)). \tag{4.30}$$

Now the evolution follows from applying Ito's formula to $-\log(B_\$^T(t))$. The price $B_\$^T(t)$ is equal to $B_M^T(t) \cdot M_\$(t)$, and thus

$$-\log(B_\$^T(t)) = -\log(B_M^T(t) \cdot M_\$(t)) = -\log(B_M^T(t)) - \log(M_\$(t)).$$

Hence,

$$
\begin{aligned}
d\left[\int_t^T f(t,u)du\right] &= -d\left[\log(B_{\$}^T(t))\right] = -d\left[\log(B_M^T(t))\right] - d\left[\log(M_{\$}(t))\right] \\
&= -\frac{1}{B_M^T(t)}dB_M^T(t) + \tfrac{1}{2}\frac{1}{\left[B_M^T(t)\right]^2}(dB_M^T(t))^2 - \frac{1}{M_{\$}(t)}dM_{\$}(t) \\
&= -\left[\int_t^T \sigma(t,u)du\right]dW^M(t) + \tfrac{1}{2}\left[\int_t^T \sigma(t,u)du\right]^2 dt - r(t)dt. \qquad (4.31)
\end{aligned}
$$

Differentiating with respect to T leads to the following result.

THEOREM 4.1 Heath–Jarrow–Morton

The forward rate $f(t,T)$ has the dynamics

$$
\boxed{df(t,T) = \sigma(t,T)\cdot\left[\int_t^T \sigma(t,u)du\right]dt - \sigma(t,T)dW^M(t).} \qquad (4.32)
$$

Note that the above relationship (4.32) represents in fact infinitely many equations, one equation for each T. As time t passes, the forward rate $f(t,T)$ fluctuates according to (4.32). However, the forward rates $f(t,T)$ for different times T are driven by the same Brownian motion $W^M(t)$, a single noise factor. It is possible to generalize the Heath–Jarrow–Morton model to include more noise factors.

REMARK 4.2 Relationship to spot rate models

The evolution of the spot rate $r(t)$ is already determined by the HJM model. Note that

$$
\begin{aligned}
r(t) = f(t,t) &= \int_0^t df(u,t)du + f(0,t) \\
&= f(0,t) + \int_0^t \left[\sigma(u,t)\cdot\sigma^*(u,t)\right]du + \int_0^t \sigma(u,t)dW^M(u). \qquad (4.33)
\end{aligned}
$$

A more compact formula for $r(t)$ is possible for particular choices of the volatility function $\sigma(t,T)$. Assume for instance $\sigma(t,T) = \sigma e^{-\gamma(T-t)}$ for $\sigma > 0, \gamma > 0$, which means that the shorter forward rates fluctuate more than the longer forward rates, a model that is consistent with empirical data. Then

$$
\sigma^*(t,T) = \int_t^T \sigma(t,u)du = \int_t^T \sigma e^{-\gamma(T-u)}du = -\frac{\sigma}{\gamma}\cdot(e^{-\gamma(T-t)} - 1),
$$

and consequently

$$
df(t,T) = -\frac{\sigma^2}{\gamma}\cdot e^{-\gamma(T-t)}(e^{-\gamma(T-t)} - 1)dt + \sigma e^{-\gamma(T-t)}dW^M(t),
$$

and

$$r(t) = f(0,t) - \int_0^t \left[\sigma e^{-\gamma(T-t)} \cdot \frac{\sigma}{\gamma} \cdot (e^{-\gamma(T-t)} - 1) \right] du + \int_0^t \sigma e^{-\gamma(T-t)} dW^M(u)$$

$$= f(0,t) + \frac{\sigma^2}{2\gamma^2}(1 - e^{-\gamma t})^2 + \int_0^t \sigma e^{-\gamma(T-t)} dW^M(u).$$

Note that the random variable $\int_0^t \sigma e^{-\gamma(T-t)} dW^M(u)$ has a normal distribution as a stochastic integral of a deterministic function. The stochastic integral itself is a limit of the sum of normal increments, and the sum of normal random variables is normal. Therefore negative values of the interest rate $r(t)$ are not excluded by this model. The interest rate $r(t)$ satisfies the following stochastic differential equation

$$dr(t) = (a(t) - \gamma r(t))dt + \sigma dW^M(t), \tag{4.34}$$

where $a(t) = \gamma m(t) + m'(t)$, and $m(t) = f(0,t) + \frac{\sigma^2}{2\gamma^2}(1 - e^{-\gamma t})^2$. The evolution of the spot rate in (4.34) is known as **Vasicek interest rate model,** or **Hull–White interest rate model** which generalized Vasicek's model. The model is mean reverting to the value $\frac{a(t)}{\gamma}$ that makes the dt term zero in (4.34). Since this model allows for a negative interest rate $r(t)$, several modifications were suggested. Another popular spot rate model is

$$dr(t) = (a(t) - b(t)r(t))dt + \sigma\sqrt{r(t)}dW^M(t) \tag{4.35}$$

which cannot become negative. This is known as the **Cox–Ingersoll–Ross interest rate model.** □

References and Further Reading

Historically the mathematical treatment of the fixed income markets started with modeling the spot rate. The Vasicek model appears first in Vasicek (1977), and it was later generalized for time-varying coefficients by Hull and White (1990). Another popular spot rate model is due to Cox–Ingersoll–Ross which appeared in Cox et al. (1985). The Heath–Jarrow–Morton model in Heath et al. (1992) established the no-arbitrage evolution of the forward rates. Since the spot rate is a special case of the forward rate, it also determines all possible no-arbitrage evolutions of the spot rate.

The link between the HJM model and the no-arbitrage evolution of the LIBOR rates is due to Brace et al. (1997). The formula for caplets appeared earlier in Black (1976). Jamshidian (1997) extended these results to swap rates. The reader may refer to the monographs of Brigo and Mercurio (2006), Sadr (2009), Filipovic (2009), James and Webber (2000), Pellser (2000), or Musiela and Rutkowski (2008) for more information on the subject.

Exercises

4.1 Consider the evolution of the forward rate given by

$$df(t,T) = -\sigma dW^M(t).$$

Since the evolution violates the no-arbitrage condition in the HJM model, arbitrage is possible. Find the arbitrage portfolio.

Hint: The goal is to find a portfolio P consisting of one or several bonds and the money market such that

$$dP_M(t) = \mu(t)dt$$

for $\mu(t) > 0$. Then we can start with $P_0 = 0$ and end up with $P_T > 0$. Finding the arbitrage portfolio can be split in two steps: first determine $dB_M^T(t)$. It turns out that in this model, $dB_M^T(t)$ has a positive dt term, but it also has a noise term $dW^M(t)$, and thus the arbitrage is not locked considering just two assets B^T and M. Second, the arbitrage can be locked with another bond of a different maturity which would cancel the noise term $dW^M(t)$ in the evolution of the portfolio in the form

$$P_t = B^{T_2} + \Delta^{T_1}(t) \cdot B^{T_1} + \Delta^M(t) \cdot M,$$

where $T_2 > T_1$. Determine Δ^{T_1} such that

$$dP_M(t) = d\left(B_M^{T_2} + \Delta^{T_1}(t)B_M^{T_1}\right) = \mu(t)dt$$

for $\mu(t) > 0$.

Chapter 5

Barrier Options

A barrier option is a contract whose payoff depends on the event that the underlying price crosses a certain boundary. Typical variants of the barrier option depend only on the assets X and Y, and thus they can be regarded as contracts on two assets. Barrier options are cheaper than their corresponding plain vanilla counterparts. They may appeal to investors who want to have a higher exposure on the payoff of the plain vanilla option. Such investors may buy more units of the barrier option than the plain vanilla options for the same price, but with the risk that the barrier option may expire worthless in contrast to the plain vanilla option. Barrier options typically appear in foreign exchange markets.

The first section of this chapter describes different types of barrier options. Barrier options come in two flavors: **knock-out** and **knock-in**. Knock-out options expire worthless if the barrier is hit during the life of the option. On the other hand, knock-in options convert to a plain vanilla option at the time of the first hit of the barrier. The barrier is usually a constant in terms of dollar prices. However, in terms of no-arbitrage assets, such as the bond B^T, the barrier takes an exponential form because of the time value of money. This is an important observation for pricing such contracts since the price of the option has to be computed with respect to a no-arbitrage asset.

The second section of this chapter shows that a barrier contract has a corresponding plain vanilla "sibling" contract that has the same price up to the first time of hitting the barrier. This result is based on the exchangeability of the assets X and Y, meaning that the returns of the prices with respect to the reference asset Y and the reference asset X have the same distribution in the corresponding martingale measures. We can express this assumption mathematically as

$$\mathcal{L}_t^Y\left(\frac{X_Y(T)}{X_Y(t)}\right) = \mathcal{L}_t^X\left(\frac{Y_X(T)}{Y_X(t)}\right).$$

As a consequence, the holder of the barrier contract is ambivalent between having a contract that pays off when the price first hits a lower barrier and then goes up, and a contract that pays off when the price first hits a lower barrier and ends up even lower. The second contract is a plain vanilla type option since the barrier must be crossed in order to collect the payoff. As a side result of this approach, we can easily derive the distribution of the hitting time of the boundary. Determination of the distribution of the hitting time is an interesting problem on its own since it may have consequences on the value of a portfolio even if it does not include barrier options. For instance a very low boundary may be hit during a market crash.

When the interest rate r is positive, the barrier takes an exponential form, and the corresponding "sibling" contract depends on a power option R^α. A specific choice of α converts the exponential barrier problem for the assets X and Y to a constant barrier problem for the assets R^α and Y. For computing the price of the barrier option, we decompose its payoff in two Arrow–Debreu securities, and determine their "sibling" plain vanilla contracts in terms of the power options R^α. The third section illustrates how to compute the price of the down-and-in call option by combining the two Arrow–Debreu securities in order to express the barrier option as a plain vanilla European call option on two power options. This part of the book is mathematically demanding, in particular the transformation of the exponential boundary to the constant boundary for the asset R^α. The reader who is interested in the conceptual ideas rather than in the computational details may just follow the text which covers the case when $r = 0$, which is relatively simple.

5.1 Types of Barrier Options

Let X and Y be two arbitrary assets (both arbitrage or no-arbitrage). A barrier knock-out option pays off $f^Y(X_Y(T))$ units of an asset Y, subject to $\bar{f}^Y(t, X_Y(t)) \geq 0$ at all times $t \in [0, T]$ for some function $\bar{f}^Y(t, x)$. The last condition says that the price $X_Y(t)$ must stay in a certain region at all times in order to collect the option payoff. Similarly, the barrier knock-out option can be defined as a contract that pays off $f^X(Y_X(T))$ units of an asset X, subject to $\bar{f}^X(t, Y_X(t)) \geq 0$ at all times $t \in [0, T]$ for some function $\bar{f}^X(t, x)$. When the two definitions describe the same contract, the functions $\bar{f}^Y(t, x)$ and $\bar{f}^X(t, x)$ are related by

$$\bar{f}^Y(t, x) = \bar{f}^X\left(t, \tfrac{1}{x}\right) \cdot x, \quad \text{or} \quad \bar{f}^X(t, x) = \bar{f}^Y\left(t, \tfrac{1}{x}\right) \cdot x. \tag{5.1}$$

This is analogous to the relationship between the payoff functions f^Y and f^X.

Example 5.1
A down-and-out call option pays off

$$\boxed{(X_T - K \cdot Y_T)^+ \quad \text{if} \quad \min_{0 \leq t \leq T} X_Y(t) \geq L.} \tag{5.2}$$

The payoff function is the same as for a European call option, where $f^Y(x) = (x - K)^+$, and $f^X(x) = (1 - K \cdot x)^+$. The barrier condition is given by $\bar{f}^Y(t,x) = x - L$, or equivalently by $\bar{f}^X(t,x) = 1 - L \cdot x$.

Similarly, the up-and-out call option pays off

$$\boxed{(X_T - K \cdot Y_T)^+ \quad \text{if} \quad \max_{0 \le t \le T} X_Y(t) \le U.} \tag{5.3}$$

The barrier condition is given by $\bar{f}^Y(t,x) = U - x$, or equivalently by $\bar{f}^X(t,x) = U \cdot x - 1$. □

A barrier knock-in option pays off $f^Y(X_Y(T))$ units of an asset Y, subject to $\bar{f}^Y(t, X_Y(t)) < 0$ for at least one time $t \in [0,T]$ for some function $\bar{f}^Y(t,x)$. The last condition says that the price $X_Y(t)$ must enter a certain region for at least one time in order to collect the option payoff. Similarly, the barrier knock-in option can be defined as a contract that pays off $f^X(Y_X(T))$ units of an asset X, subject to $\bar{f}^X(t, Y_X(t)) < 0$ for at least one time $t \in [0,T]$ for some function $\bar{f}^X(t,x)$. When the two definitions describe the same contract, the functions $\bar{f}^Y$ and $\bar{f}^X$ are related by

$$\bar{f}^Y(t,x) = \bar{f}^X\left(t, \tfrac{1}{x}\right) \cdot x, \quad \text{or} \quad \bar{f}^X(t,x) = \bar{f}^Y\left(t, \tfrac{1}{x}\right) \cdot x.$$

Example 5.2
A down-and-in call option pays off

$$\boxed{(X_T - K \cdot Y_T)^+ \quad \text{if} \quad \min_{0 \le t \le T} X_Y(t) < L.} \tag{5.4}$$

The payoff function is the same as for a standard European call option, where $f^Y(x) = (x - K)^+$, and $f^X(x) = (1 - K \cdot x)^+$. The barrier condition is given by $\bar{f}^Y(t,x) = x - L$, or equivalently by $\bar{f}^X(t,x) = 1 - L \cdot x$.

Similarly, the up-and-in call option pays off

$$\boxed{(X_T - K \cdot Y_T)^+ \quad \text{if} \quad \max_{0 \le t \le T} X_Y(t) > U.} \tag{5.5}$$

The barrier condition is given by $\bar{f}^Y(t,x) = U - x$, or equivalently by $\bar{f}^X(t,x) = U \cdot x - 1$. □

If we denote by V^{KO} a barrier knock-out call option and by V^{KI} a barrier knock-in call option, we have a simple relationship with the plain vanilla European call option V^{EC}:

$$\boxed{V^{EC} = V^{KO} + V^{KI}.} \tag{5.6}$$

This relationship is easy to see. When the barrier is crossed, the knock-in option is activated but the corresponding knock-out option becomes worthless. Similarly, if the barrier is not crossed, the knock-out option is alive but the corresponding knock-in option is worthless.

REMARK 5.1 Exponential boundaries
Consider the situation when Y is an arbitrage asset \$, and X is a no-arbitrage asset S. Then the down-and-in call option pays off

$$(S_T - K \cdot \$_T)^+ \quad \text{if} \quad \min_{0\le t\le T} S_{\$}(t) < L.$$

Assuming a constant interest rate with $B_{\$}^T(t) = e^{-r(T-t)}$, we have $S_{\$}(t) = S_{B^T}(t) \cdot B_{\$}^T(t) = e^{-r(T-t)} \cdot S_{B^T}(t)$, and the above contract can be rewritten in terms of no-arbitrage assets S and B^T as

$$(S_T - K \cdot B_T^T)^+ \quad \text{if} \quad \min_{0\le t\le T} e^{-r(T-t)} S_{B^T}(t) < L.$$

In this case, the barrier condition is given by

$$\bar{f}^T(t,x) = x - L \cdot e^{r(T-t)}$$

when we express it in terms of the bond B^T. Therefore we will consider this boundary type in the next section. □

5.2 Barrier Option Pricing via Power Options

The basic idea of pricing barrier options is that the barrier option has a related plain vanilla option that has the same price. The pricing problem of the barrier option then reduces to the pricing problem of the corresponding plain vanilla option. Let us illustrate this technique in the simplest case when the interest rate r is zero, and the barrier function is in the form $\bar{f}^Y(t,x) = x - L$.

5.2.1 Constant Barrier

Consider a down-and-in call option that pays off

$$(X_T - K \cdot Y_T)^+ \quad \text{if} \quad \min_{0\le t\le T} X_Y(t) \le L.$$

The payoff of a European call option can be expressed as a combination of two Arrow–Debreu securities, one that pays off a unit of the asset X on the event

$$A = \{\min_{0\le t\le T} X_Y(t) \le L; X_Y(T) \ge K\}, \tag{5.7}$$

and one that pays off $-K$ units of the asset Y on the same event. Thus it is sufficient to find the price of the corresponding Arrow–Debreu securities. Let us denote by V the Arrow–Debreu security that pays off

$$V_T = \mathbb{I}_A(\omega) \cdot Y_T.$$

Note that when the price of $X_Y(t)$ is on the boundary L (meaning $X_Y(t) = L$), the price of the Arrow–Debreu security is simply

$$V_t = \mathbb{P}_t^Y(X_Y(T) \geq K) \cdot Y_t$$

since the barrier is already hit. Furthermore, we can write

$$\begin{aligned}
V_t &= \mathbb{P}_t^Y(X_Y(T) \geq K) \cdot Y_t \\
&= \mathbb{P}_t^Y\left(X_Y(t) \cdot \exp(\sigma W^Y(T-t) - \tfrac{1}{2}\sigma^2(T-t)) \geq K\right) \cdot Y_t \\
&= \mathbb{P}_t^X\left(X_Y(t) \cdot \exp(\sigma W^X(T-t) - \tfrac{1}{2}\sigma^2(T-t)) \geq K\right) \cdot \tfrac{1}{L} \cdot X_t \\
&= \mathbb{P}_t^X\left(X_Y(t) \cdot \frac{Y_X(T)}{Y_X(t)} \geq K\right) \cdot \tfrac{1}{L} \cdot X_t \\
&= \mathbb{P}_t^X\left(\tfrac{1}{K} \geq \frac{X_Y(T)}{[X_Y(t)]^2}\right) \cdot \tfrac{1}{L} \cdot X_t \\
&= \mathbb{P}_t^X\left(\tfrac{L^2}{K} \geq X_Y(T)\right) \cdot \tfrac{1}{L} \cdot X_t.
\end{aligned} \tag{5.8}$$

The equality between the second and the third line follows from the fact that $X_t = L \cdot Y_t$ on the boundary, and that the distribution of W^Y under $\mathbb{P}^Y$ is the same as the distribution of W^X under $\mathbb{P}^X$. In mathematical notation we can write

$$\mathcal{L}^Y(W^Y(T)) = \mathcal{L}^X(W^X(T)),$$

meaning that the probability laws of the corresponding Brownian motions agree. In a geometric Brownian motion model, this statement is equivalent to

$$\mathcal{L}_t^Y\left(\frac{X_Y(T)}{X_Y(t)}\right) = \mathcal{L}_t^X\left(\frac{Y_X(T)}{Y_X(t)}\right), \tag{5.9}$$

implying that the assets X and Y are exchangeable. We can also rewrite this relationship as

$$\mathcal{L}_t^Y(X_Y(T)) = \mathcal{L}_t^X\left(\frac{[X_Y(t)]^2}{X_Y(T)}\right).$$

In conclusion, we have proved that on the boundary $X_Y(t) = L$, the Arrow–Debreu security is equal to

$$V_t = \mathbb{P}_t^Y(X_Y(T) \geq K) \cdot Y_t = \mathbb{P}_t^X\left(\tfrac{L^2}{K} \geq X_Y(T)\right) \cdot \tfrac{1}{L} \cdot X_t.$$

But $\mathbb{P}_t^X\left(\tfrac{L^2}{K} \geq X_Y(T)\right) \cdot \tfrac{1}{L} \cdot X_t$ corresponds to $\tfrac{1}{L}$ units of the Arrow–Debreu security U with a payoff

$$U_T = \mathbb{I}_B(\omega) \cdot X_T,$$

where B is the event

$$B = \{\tfrac{L^2}{K} \geq X_Y(T)\}.$$

Let τ_L be the first hitting time of the boundary

$$\tau_L = \inf\{t \geq 0 : X_Y(t) \leq L\}. \tag{5.10}$$

We have already shown that $V_{\tau_L} = \frac{1}{L} U_{\tau_L}$. Moreover, if $\tau_L > T$, both U and V are worthless. Therefore we have also the relationship

$$V_{\tau_L \wedge T} = \tfrac{1}{L} \cdot U_{\tau_L \wedge T}, \tag{5.11}$$

where $x \wedge y = \min(x, y)$. Thus for all times $t \leq \tau_L$, we also have

$$V_Y(t) = \mathbb{E}_t^Y[V_Y(\tau_L \wedge T)] = \tfrac{1}{L}\mathbb{E}_t^Y[U_Y(\tau_L \wedge T)] = \tfrac{1}{L}U_Y(t).$$

Therefore we have $V_t = \frac{1}{L}U_t$ for all times $t \leq \tau_L$. The price of the Arrow–Debreu security with a barrier feature V can be computed from the price of the Arrow–Debreu security U that is a plain vanilla contract.

Figure 5.1 illustrates this situation. Before the price $X_Y(t)$ hits the barrier τ_L for the first time, the price of the Arrow–Debreu security V agrees with the price of $\frac{1}{L}U$. The price of the plain vanilla Arrow–Debreu security that pays one unit of an asset Y at time T if $X_Y(T) \geq K$ dominates the price of V which has the barrier feature up to the first hitting time τ_L. Once the barrier is hit, the security V agrees with its plain vanilla counterpart, and the price of $\frac{1}{L}U$ is no longer relevant.

Similar to an Arrow–Debreu security V that pays off a unit of an asset Y on the event A defined in (5.7), we can consider an Arrow–Debreu security $\widetilde{U}$ that pays off a unit of an asset X on the same event A

$$\widetilde{U}_T = \mathbb{I}_A(\omega) \cdot X_T.$$

Following the same arguments as in (5.8), we can show that when $X_Y(t) = L$, we have

$$\widetilde{U}_t = \mathbb{P}_t^X(X_Y(T) \geq K) \cdot X_t = \mathbb{P}_t^Y(\tfrac{L^2}{K} \geq X_Y(T)) \cdot LY_t.$$

Therefore the security $\widetilde{U}$ has the same price as L units of an Arrow–Debreu security $\widetilde{V}$ that pays off an asset Y when $\{\frac{L^2}{K} \geq X_Y(T)\}$ up to the first time τ_L when the barrier is hit.

In conclusion, the barrier option

$$(X_T - K \cdot Y_T)^+ \quad \text{if} \quad \min_{0 \leq t \leq T} X_Y(t) \leq L$$

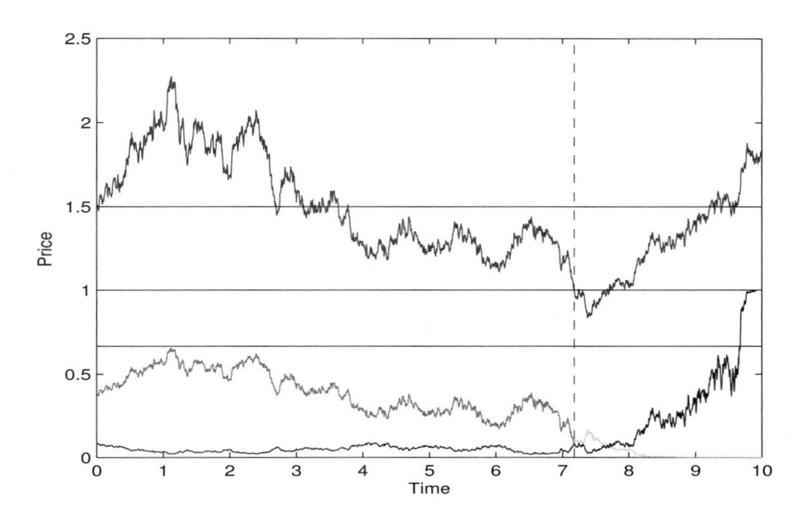

FIGURE 5.1: The evolution of the price $X_Y(t)$ (top graph). Consider an Arrow–Debreu security V that pays a unit of an asset Y at time T if the price of $X_Y(T)$ exceeds K, subject to the condition that $X_Y(t) \leq L$ for some $t \in [0, T]$. The parameters are $X_Y(0) = 1.5$, $K = 1.5$ (top solid line), $L = 1$ (middle solid line), $\frac{L^2}{K} \approx 0.666$ (bottom solid line), $\sigma = 0.2$, $T = 10$. The price of V agrees up to time τ_L with the price of $\frac{1}{L}$ units of a plain vanilla contract that pays a unit of X if $X_Y(T) \leq \frac{L^2}{K}$ (the bottom graph starting at 0.082 and ending at 0). The first hit happens at time $\tau_L = 7.175$ (vertical dash line). Compare the price of V with the price of the Arrow–Debreu security that pays a unit of an asset Y when $X_Y(T) \geq K$ (plain vanilla type, the graph starting at 0.376 and ending at 1); it dominates the price of V up to time τ_L. When the price $X_Y(t)$ hits the barrier $L = 1$, the prices of V and $\frac{1}{L}U$ agree: $V = \frac{1}{L}U$. Once the barrier is hit, the price of V agrees with its plain vanilla counterpart. The contract U becomes irrelevant once the barrier is hit.

has the same price as a plain vanilla option with a payoff

$$(LY_T - \tfrac{K}{L}X_T)^+$$

up to the first time τ_L when the barrier is hit.

REMARK 5.2 Distribution of the first hitting time of a geometric Brownian motion.
It is interesting to note that we can determine the distribution of the hitting time

$$\tau_L = \inf\{t \geq 0 : X_Y(t) \leq L\}$$

from the prices of two Arrow–Debreu securities. Consider a financial contract V that pays off a unit of Y at time τ_L

$$V_{\tau_L} = Y_{\tau_L}.$$

We can also rewrite this payoff as

$$V_T = \mathbb{I}(\tau_L \leq T) \cdot Y_T.$$

Define two Arrow–Debreu securities V^1 and V^2 by

$$V_T^1 = \mathbb{I}(\min_{t\in[0,T]} X_Y(t) \leq L, X_Y(T) > L) \cdot Y_T,$$

and

$$V_T^2 = \mathbb{I}(\min_{t\in[0,T]} X_Y(t) \leq L, X_Y(T) \leq L) \cdot Y_T.$$

Note that

$$V_t = V_t^1 + V_t^2.$$

As for the price of V^2, the condition $X_Y(T) \leq L$ already implies $\min_{t\in[0,T]} X_Y(t) \leq L$, and thus it is a plain vanilla security with price

$$V_t^2 = \mathbb{P}^Y(X_Y(T) \leq L) \cdot Y.$$

The security V^1 is a knock-in Arrow–Debreu security with the barrier L and strike $K = L$, and thus its price is equal to $\frac{1}{L}$ units of a plain vanilla Arrow–Debreu security with the payoff $\mathbb{I}(X_Y(T) \leq L) \cdot \frac{1}{L} \cdot X$ as explained in the previous text. Thus the price of V^1 is given by

$$V_t^1 = \mathbb{P}^X(X_Y(T) \leq L) \cdot \tfrac{1}{L} \cdot X$$

up to time τ_L. Thus we have

$$\begin{aligned} V &= \mathbb{P}^Y(\tau_L \leq T) \cdot Y \\ &= \mathbb{P}^X(X_Y(T) \leq L) \cdot \tfrac{1}{L} \cdot X + \mathbb{P}^Y(X_Y(T) \leq L) \cdot Y. \end{aligned}$$

Therefore

$$\begin{aligned}\mathbb{P}^Y(\tau_L \le T) &= \mathbb{P}^X(X_Y(T) \le L) \cdot \tfrac{X_Y(0)}{L} + \mathbb{P}^Y(X_Y(T) \le L) \\ &= N\left(\tfrac{1}{\sigma\sqrt{T}} \log\left(\tfrac{L}{X_Y(0)}\right) - \tfrac{1}{2}\sigma\sqrt{T}\right) \cdot \tfrac{X_Y(0)}{L} \\ &\qquad + N\left(\tfrac{1}{\sigma\sqrt{T}} \log\left(\tfrac{L}{X_Y(0)}\right) + \tfrac{1}{2}\sigma\sqrt{T}\right).\end{aligned}$$

Thus we have determined the cumulative distribution function of the stopping time τ_L. Its density is given by

$$\frac{\partial}{\partial T}\mathbb{P}^Y(\tau_L \le T) = \frac{\log\left(\frac{X_Y(0)}{L}\right)}{\sigma\sqrt{2\pi T^3}} \exp\left(-\frac{\left(\frac{1}{\sigma\sqrt{T}}\log\left(\frac{L}{X_Y(0)}\right) + \frac{1}{2}\sigma\sqrt{T}\right)^2}{2T}\right).$$

□

5.2.2 Exponential Barrier

When the interest rate r is greater than zero, the barrier takes an exponential form. We will show here that when $X_Y(t)$ hits an exponential boundary, there is a corresponding power option R^α whose price is hitting a constant boundary. Thus we are able to transform the problem of an exponential boundary to a constant boundary, although this requires a proxy asset R^α.

Recall that a power option R^α pays off $[X_Y(T)]^\alpha$ units of an asset Y at time T. Note that $R^0 = Y$, and $R^1 = X$, so the assets X and Y are just special cases of power options. The price of the power option is given by

$$R_Y^\alpha(t) = \exp(\tfrac{1}{2}\alpha(\alpha-1)\sigma^2(T-t)) \cdot [X_Y(t)]^\alpha. \tag{5.12}$$

The evolution of the power option price is given by

$$dR_Y^\alpha(t) = \alpha\sigma R_Y^\alpha(t) dW^Y(t) \tag{5.13}$$

with a closed form solution

$$R_Y^\alpha(T) = R_Y^\alpha(t) \cdot \exp\left(\alpha\sigma W^Y(T-t) - \tfrac{1}{2}\alpha^2\sigma^2(T-t)\right). \tag{5.14}$$

The volatility of the power option R^α is α times the volatility σ of the price X_Y. Since R^α itself is a no-arbitrage asset with a positive price, it can be used as a numeraire, leading to a pricing measure $\mathbb{P}^{(\alpha)}$ that is associated with R^α. From the exchangeability of the assets Y and R^α, the dynamics of the inverse price $Y_{R^\alpha}(t)$ are given by

$$dY_{R^\alpha}(t) = \alpha\sigma Y_{R^\alpha}(t) dW^{(\alpha)}(t). \tag{5.15}$$

Now consider an Arrow–Debreu security V that pays off a unit of an asset Y at time T on the event

$$A = \{\min_{0 \leq t \leq T}(e^{-r(T-t)} X_Y(t)) \leq L; X_Y(T) \geq K\}.$$

The first time of hitting the boundary is given by

$$\tau_L = \inf\{t \geq 0 : X_Y(t) \leq Le^{r(T-t)}\}.$$

Let us show that for all $t \leq \tau_L$,

$$V_t = \left(\tfrac{1}{L}\right)^{\alpha} \cdot V_t^{(\alpha)},$$

where $V^{(\alpha)}$ is an Arrow–Debreu security that pays off a unit of the power option R^{α} at time T on the event

$$B = \{X_Y(T) \leq \tfrac{L^2}{K}\}$$

for $\alpha = 1 - \frac{2r}{\sigma^2}$.

THEOREM 5.1
The Arrow–Debreu security with a barrier feature V that pays off

$$V_T = \mathbb{I}_A(\omega) \cdot Y_T$$

for

$$A = \{\min_{0 \leq t \leq T}(e^{-r(T-t)} X_Y(t)) \leq L; X_Y(T) \geq K\}$$

has the same price up to the first hitting time of the barrier

$$\tau_L = \inf\{t \geq 0 : X_Y(t) \leq Le^{r(T-t)}\}$$

as $\left(\frac{1}{L}\right)^{\alpha}$ units of a plain vanilla Arrow–Debreu security $V^{(\alpha)}$ that pays off

$$V_T^{(\alpha)} = \mathbb{I}_B(\omega) \cdot R_T^{\alpha}$$

for

$$B = \{X_Y(T) \leq \tfrac{L^2}{K}\},$$

with $\alpha = 1 - \frac{2r}{\sigma^2}$.

PROOF Let us first show that V and $\left(\frac{1}{L}\right)^{\alpha} \cdot V^{(\alpha)}$ have the same price when the price $X_Y(t)$ is on the barrier at time t, meaning $X_Y(t) = Le^{r(T-t)}$. Note that in this case, the price of the power option $R_Y^{\alpha}(t)$ is given by

$$\begin{aligned} R_Y^{\alpha}(t) &= \exp(\tfrac{1}{2}\alpha(\alpha-1)\sigma^2(T-t)) \cdot [X_Y(t)]^{\alpha} \\ &= \exp(\tfrac{1}{2}\alpha(\alpha-1)\sigma^2(T-t)) \cdot e^{\alpha r(T-t)} \cdot L^{\alpha} \\ &= \exp([r + \tfrac{1}{2}(\alpha-1)\sigma^2] \cdot \alpha(T-t)) \cdot L^{\alpha}. \end{aligned}$$

In particular, when $\alpha = 1 - \frac{2r}{\sigma^2}$, the above price simplifies to $L^{1-\frac{2r}{\sigma^2}}$, so we have

$$R_Y^{(1-\frac{2r}{\sigma^2})}(t) = L^{1-\frac{2r}{\sigma^2}}$$

when $X_Y(t)$ is on the barrier $Le^{r(T-t)}$. As for the value of V, for $\alpha = 1 - \frac{2r}{\sigma^2}$ we get (assuming $\alpha > 0$)

$$\begin{aligned}
V_t &= \mathbb{P}_t^Y(X_Y(T) \geq K) \cdot Y_t \\
&= \mathbb{P}_t^Y([X_Y(T)]^\alpha \geq K^\alpha) \cdot Y_t = \mathbb{P}_t^Y(R_Y^\alpha(T) \geq K^\alpha) \cdot Y_t \\
&= \mathbb{P}_t^Y\left(R_Y^\alpha(t) \cdot \exp\left(\alpha\sigma W^Y(T-t) - \tfrac{1}{2}\alpha^2\sigma^2(T-t)\right) \geq K^\alpha\right) \cdot Y_t \\
&= \mathbb{P}_t^{(\alpha)}\left(R_Y^\alpha(t) \cdot \exp\left(\alpha\sigma W^{(\alpha)}(T-t) - \tfrac{1}{2}\alpha^2\sigma^2(T-t)\right) \geq K^\alpha\right) \cdot \left(\tfrac{1}{L}\right)^\alpha \cdot R_t^\alpha \\
&= \mathbb{P}_t^{(\alpha)}\left(R_Y^\alpha(t) \cdot \frac{Y_{R^\alpha}(T)}{Y_{R^\alpha}(t)} \geq K^\alpha\right) \cdot \left(\tfrac{1}{L}\right)^\alpha \cdot R_t^\alpha \\
&= \mathbb{P}_t^{(\alpha)}\left(\left(\tfrac{[R_Y^\alpha(t)]^2}{K}\right)^\alpha \geq R_Y^\alpha(T)\right) \cdot \left(\tfrac{1}{L}\right)^\alpha \cdot R_t^\alpha \\
&= \mathbb{P}_t^{(1-\frac{2r}{\sigma^2})}\left(\left(\tfrac{L^2}{K}\right)^{(1-\frac{2r}{\sigma^2})} \geq R_Y^{(1-\frac{2r}{\sigma^2})}(T)\right) \cdot \left(\tfrac{1}{L}\right)^{(1-\frac{2r}{\sigma^2})} \cdot R_t^{(1-\frac{2r}{\sigma^2})} \\
&= \mathbb{P}_t^{(1-\frac{2r}{\sigma^2})}\left(\tfrac{L^2}{K} \geq X_Y(T)\right) \cdot \left(\tfrac{1}{L}\right)^{(1-\frac{2r}{\sigma^2})} \cdot R_t^{(1-\frac{2r}{\sigma^2})}.
\end{aligned}$$

When $\alpha \leq 0$, we obtain the same result. We have used the fact that $R_t^{(1-\frac{2r}{\sigma^2})} = L^{1-\frac{2r}{\sigma^2}} \cdot Y_t$ when $X_Y(t) = Le^{r(T-t)}$, and that the distribution of the Brownian motion W^Y under the probability measure $\mathbb{P}^Y$ is the same as the distribution of the Brownian motion $W^{(\alpha)}$ under the probability measure $\mathbb{P}^{(\alpha)}$. The latter statement is another way of saying that the assets Y and R^α are exchangeable, meaning that

$$\mathcal{L}_t^Y\left(\frac{R_Y^\alpha(T)}{R_Y^\alpha(t)}\right) = \mathcal{L}_t^{(\alpha)}\left(\frac{Y_{R^\alpha}(T)}{Y_{R^\alpha}(t)}\right).$$

We can also rewrite this relationship as

$$\mathcal{L}_t^Y(R_Y^\alpha(T)) = \mathcal{L}_t^{(\alpha)}\left(\frac{[R_Y^\alpha(t)]^2}{R_Y^\alpha(T)}\right).$$

We have shown that

$$V_t = \mathbb{P}_t^Y(X_Y(T) \geq K) \cdot Y_t = \mathbb{P}_t^{(1-\frac{2r}{\sigma^2})}\left(\tfrac{L^2}{K} \geq X_Y(T)\right) \cdot \left(\tfrac{1}{L}\right)^{(1-\frac{2r}{\sigma^2})} \cdot R_t^{(1-\frac{2r}{\sigma^2})}.$$

Thus when $X_Y(t)$ is on the barrier $Le^{r(T-t)}$, the price of the Arrow–Debreu security is also equal to $\mathbb{P}_t^{(1-\frac{2r}{\sigma^2})}\left(\frac{L^2}{K} \geq X_Y(T)\right) \cdot \left(\frac{1}{L}\right)^{(1-\frac{2r}{\sigma^2})} \cdot R_t^{(1-\frac{2r}{\sigma^2})}$. But this representation corresponds to $\left(\frac{1}{L}\right)^{(1-\frac{2r}{\sigma^2})}$ units of an Arrow–Debreu security

$V^{(\alpha)}$ that pays off

$$V_T^{(\alpha)} = \mathbb{I}_B(\omega) \cdot R_T^{(1-\frac{2r}{\sigma^2})},$$

where $B = \{X_Y(T) \leq \frac{L^2}{K}\}$. Since

$$V_{\tau_L \wedge T} = \left(\tfrac{1}{L}\right)^{(1-\frac{2r}{\sigma^2})} \cdot V_{\tau_L \wedge T}^{(\alpha)},$$

we also have

$$V_t = \left(\tfrac{1}{L}\right)^{(1-\frac{2r}{\sigma^2})} \cdot V_t^{(\alpha)}$$

for all $t \leq \tau_L$. $\square$

Figure 5.2 illustrates the price evolution of the Arrow–Debreu security V in more detail together with the corresponding prices of the asset X, power option R^α, the plain vanilla security $\left(\frac{1}{L}\right)^\alpha V^{(\alpha)}$ and a plain vanilla Arrow–Debreu security that pays off a unit of the asset Y at time T if $X_Y(T) \geq K$. Up to the first hitting time τ_L of the boundary, the prices of V and $\left(\frac{1}{L}\right)^\alpha V^{(\alpha)}$ agree, and they are dominated by the price of the plain vanilla Arrow–Debreu security that pays off a unit of the asset Y at time T if $X_Y(T) \geq K$. Once the barrier is hit, the security V agrees with its plain vanilla counterpart, and the security $V^{(\alpha)}$ is no longer relevant.

Similarly, the Arrow–Debreu security U that pays off a unit of an asset X on the event

$$A = \{\min_{0 \leq t \leq T} (e^{-r(T-t)} X_Y(t)) \leq L; X_Y(T) \geq K\}$$

has the same price up to time τ_L as an Arrow–Debreu security $U^{(\alpha)}$ that pays off $L^{1+\frac{2r}{\sigma^2}}$ units of a power option $R^{(-\frac{2r}{\sigma^2})}$ on the event

$$B = \{X_Y(T) \leq \tfrac{L^2}{K}\}.$$

See Exercise 5.2 for more details.

5.3 Price of a Down-and-In Call Option

Let us illustrate how to compute a price of a down-and-in call option using power options. The prices of other types of barrier options can be determined in an analogous way. The down-and-in barrier option with a payoff

$$(X_T - K \cdot Y_T)^+ \quad \text{if} \quad \min_{0 \leq t \leq T} (e^{-r(T-t)} X_Y(t)) \leq L$$

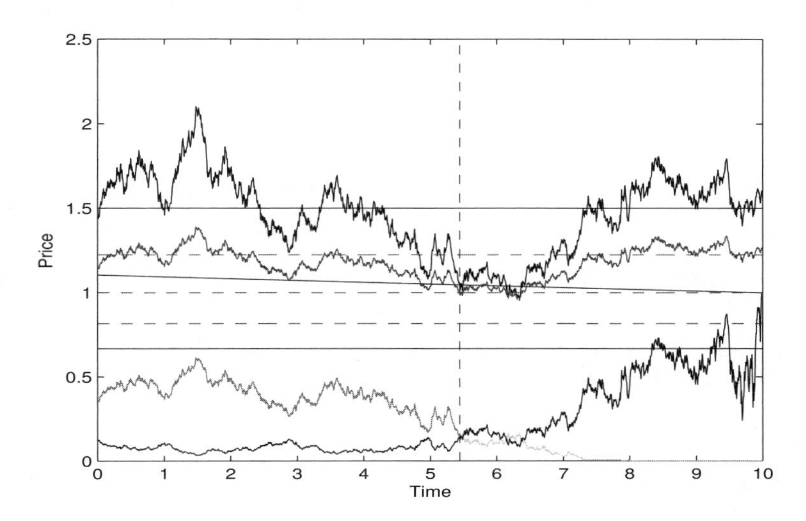

FIGURE 5.2: The evolution of the prices $X_Y(t)$ (top graph) and $R^{\alpha}_Y(t)$ (immediately below it) with $X_Y(0) = 1.5$, $K = 1.5$, $L = 1$, $r = 0.01$, $\sigma = 0.2$, $T = 10$. This gives the values of $\alpha = 1 - \frac{2r}{\sigma^2} = 0.5$, $K^{\alpha} \approx 1.225$, $\left(\frac{L^2}{K}\right)^{\alpha} \approx 0.816$. Consider an Arrow–Debreu security V that pays off one unit of an asset Y at time T when $X_Y(T) \geq K = 1.5$ (upper solid line) if the price $X_Y(t)$ gets below the barrier $Le^{r(T-t)}$ (middle solid curve). The security V is equivalent to the one that pays off a unit of Y when $R^{\alpha}_Y(T) \geq K^{\alpha} \approx 1.225$ (top dash line), subject to $R^{\alpha}_Y(t) \leq L^{\alpha} = 1$ (middle dash line) in terms of the price of the power option R^{α}. Note that each time the price $X_Y(t)$ hits the barrier $Le^{r(T-t)}$, the price of $R^{\alpha}_Y(t)$ is at $L^{\alpha} = 1$. The first hitting time of the barrier happens at time $\tau_L = 5.445$ (vertical dash line). Up to time τ_L, the security V has the same price as $(\frac{1}{L})^{\alpha}$ units of a security $V^{(\alpha)}$ that pays off one unit of R^{α} when $R^{\alpha}_Y(T) \leq \left(\frac{L^2}{K}\right)^{\alpha} \approx 0.816$ (bottom dash line). This event is equivalent to $X_Y(T) \leq \frac{L^2}{K} \approx 0.666$ (bottom solid line). Since $\frac{L^2}{K} < Le^{r(T-t)}$, the price of $X_Y(t)$ must cross the barrier so that $V^{(\alpha)}$ does not expire worthless. Therefore $V^{(\alpha)}$ is a plain vanilla contract. The price of $(\frac{1}{L})^{\alpha} \cdot V^{(\alpha)}$ is the bottom graph up to time τ_L; it agrees with the price of V. The graph immediately above corresponds to the price of a plain vanilla Arrow–Debreu security that pays off a unit of an asset Y if $X_Y(T) \geq K$. It is more expensive since it does not have a barrier feature in contrast to the security V. Once the price $X_Y(t)$ hits the barrier $Le^{r(T-t)}$ at time τ_L, the security V is knocked in, and from that moment on it agrees with its plain vanilla counterpart. Also at time τ_L, the prices of V (now a plain vanilla contract) and $\left(\frac{1}{L}\right)^{\alpha} V^{(\alpha)}$ agree. The price of V expires at 1 since $X_Y(T)$ ends up above K in this scenario. Compare the price of V from time τ_L with the price of $\left(\frac{1}{L}\right)^{\alpha} V^{(\alpha)}$ that expires worthless.

is equivalent up to time τ_L to a plain vanilla European call option V with a payoff

$$V_T = \left(L^{\frac{2r}{\sigma^2}+1} R_T^{(-\frac{2r}{\sigma^2})} - K L^{\frac{2r}{\sigma^2}-1} R_T^{(1-\frac{2r}{\sigma^2})}\right)^+$$
$$= L^{\frac{2r}{\sigma^2}} \cdot \left(L R_T^{(-\frac{2r}{\sigma^2})} - \tfrac{K}{L} R_T^{(1-\frac{2r}{\sigma^2})}\right)^+. \quad (5.16)$$

This is clear from the representation of an option payoff as two Arrow–Debreu securities. The security that pays off a unit of an asset X has a corresponding plain vanilla counterpart that is settled in units of $R^{(-\frac{2r}{\sigma^2})}$; the security that pays off a unit of an asset Y has a corresponding plain vanilla counterpart that is settled in units of $R^{(1-\frac{2r}{\sigma^2})}$.

The payoff that corresponds to power options can also be expressed in terms of the assets X and Y as

$$L^{\frac{2r}{\sigma^2}} \cdot \left(L R_T^{(-\frac{2r}{\sigma^2})} - \tfrac{K}{L} R_T^{(1-\frac{2r}{\sigma^2})}\right)^+$$
$$= L^{\frac{2r}{\sigma^2}} \left(L[X_Y(T)]^{-\frac{2r}{\sigma^2}} \cdot Y_T - \tfrac{K}{L}[X_Y(T)]^{1-\frac{2r}{\sigma^2}} \cdot Y_T\right)^+$$
$$= \left[\frac{L}{X_Y(T)}\right]^{\frac{2r}{\sigma^2}} \cdot \left(L \cdot Y_T - \tfrac{K}{L} \cdot X_T\right)^+.$$

Let us determine the price of the option V. From the Black–Scholes formula,

$$V_t = \left[L \mathbb{P}_t^{(-\frac{2r}{\sigma^2})}\left(R^{-2r/\sigma^2}_{R^{1-2r/\sigma^2}}(T) \geq \tfrac{K}{L}\right)\right] \cdot L^{\frac{2r}{\sigma^2}} \times R_t^{-\frac{2r}{\sigma^2}}$$
$$+ \left[-\tfrac{K}{L} \mathbb{P}_t^{(1-\frac{2r}{\sigma^2})}\left(R^{-2r/\sigma^2}_{R^{1-2r/\sigma^2}}(T) \geq \tfrac{K}{L}\right)\right] \cdot L^{\frac{2r}{\sigma^2}} \times R_t^{1-\frac{2r}{\sigma^2}}.$$

Since

$$R^{\alpha}_{R^{1+\alpha}}(T) = R^{\alpha}_{Y}(T) \cdot Y_{R^{\alpha+1}}(T) = [X_Y(T)]^{\alpha} \cdot [Y_X(T)]^{\alpha+1} = Y_X(T),$$

we can rewrite the above formula as

$$V_t = \left[L \mathbb{P}_t^{(-\frac{2r}{\sigma^2})}(X_Y(T) \leq \tfrac{L^2}{K})\right] \cdot L^{\frac{2r}{\sigma^2}} \cdot R_t^{-\frac{2r}{\sigma^2}}$$
$$+ \left[-\tfrac{K}{L} \mathbb{P}_t^{(1-\frac{2r}{\sigma^2})}(X_Y(T) \leq \tfrac{L^2}{K})\right] \cdot L^{\frac{2r}{\sigma^2}} \cdot R_t^{1-\frac{2r}{\sigma^2}}. \quad (5.17)$$

Let us determine $\mathbb{P}_t^{(\alpha)}(X_Y(T) \le \frac{L^2}{K})$ for $\alpha > 0$:

$$\begin{aligned}
\mathbb{P}_t^{(\alpha)} & (X_Y(T) \le \tfrac{L^2}{K}) = \mathbb{P}_t^{(\alpha)}\left([X_Y(T)]^\alpha \le \left(\tfrac{L^2}{K}\right)^\alpha\right) \\
&= \mathbb{P}_t^{(\alpha)}\left(R_Y^\alpha(T) \le \left(\tfrac{L^2}{K}\right)^\alpha\right) = \mathbb{P}_t^{(\alpha)}\left(\left(\tfrac{K}{L^2}\right)^\alpha \le Y_{R^\alpha}(T)\right) \\
&= \mathbb{P}_t^{(\alpha)}\left(\left(\tfrac{K}{L^2}\right)^\alpha \le Y_{R^\alpha}(t)\cdot\exp(\alpha\sigma W^{(\alpha)}(T-t) - \tfrac{1}{2}\alpha^2\sigma^2(T-t))\right) \\
&= \mathbb{P}_t^{(\alpha)}\left(\left(\tfrac{K}{L^2}\right)^\alpha \le [Y_X(t)]^\alpha\cdot\exp(\alpha\sigma W^{(\alpha)}(T-t) - \alpha(\alpha-\tfrac{1}{2})\sigma^2(T-t))\right) \\
&= \mathbb{P}_t^{(\alpha)}\left(-\tfrac{W^{(\alpha)}}{\sqrt{T-t}} \le \tfrac{1}{\sigma\sqrt{T-t}}\log\left(\tfrac{L^2}{X_Y(t)\cdot K}\right) - (\alpha-\tfrac{1}{2})\sigma\sqrt{T-t}\right) \\
&= N\left(\tfrac{1}{\sigma\sqrt{T-t}}\log\left(\tfrac{L^2}{X_Y(t)\cdot K}\right) - (\alpha-\tfrac{1}{2})\sigma\sqrt{T-t}\right).
\end{aligned}$$

We get the same result for $\alpha \le 0$. We can also express the power option R^α in terms of the asset Y as

$$R_t^\alpha = \exp(\tfrac{1}{2}\alpha(\alpha-1)\sigma^2(T-t))[X_Y(t)]^\alpha \cdot Y_t.$$

Therefore

$$\begin{aligned}
V_t &= \left[L\mathbb{P}_t^{(-\frac{2r}{\sigma^2})}(X_Y(T) \le \tfrac{L^2}{K})\right]\cdot L^{\frac{2r}{\sigma^2}}\cdot R_t^{-\frac{2r}{\sigma^2}} \\
&\qquad + \left[-\tfrac{K}{L}\mathbb{P}_t^{(1-\frac{2r}{\sigma^2})}(X_Y(T) \le \tfrac{L^2}{K})\right]\cdot L^{\frac{2r}{\sigma^2}}\cdot R_t^{1-\frac{2r}{\sigma^2}} \\
&= \left[N\left(\tfrac{1}{\sigma\sqrt{T-t}}\log\left(\tfrac{L^2}{X_Y(t)\cdot K}\right) + (\tfrac{2r}{\sigma^2}+\tfrac{1}{2})\sigma\sqrt{T-t}\right)\right] \qquad (5.18) \\
&\qquad \times\left[\tfrac{Le^{r(T-t)}}{X_Y(t)}\right]^{\frac{2r}{\sigma^2}+1}\cdot X_t \\
&\quad -K\left[N\left(\tfrac{1}{\sigma\sqrt{T-t}}\log\left(\tfrac{L^2}{X_Y(t)\cdot K}\right) + (\tfrac{2r}{\sigma^2}-\tfrac{1}{2})\sigma\sqrt{T-t}\right)\right] \\
&\qquad \times\left[\tfrac{Le^{r(T-t)}}{X_Y(t)}\right]^{\frac{2r}{\sigma^2}-1}\cdot Y_t.
\end{aligned}$$

This is a relationship of the down-and-in option V before its knock in with the no-arbitrage assets X and Y. If we define

$$u^Y(t,x) = \mathbb{E}^Y[V_Y(T)|X_Y(t) = x],$$

we can also write the price of the down-and-in barrier option as

$$\begin{aligned}
u^Y(t,x) &= \left[N\left(\tfrac{1}{\sigma\sqrt{T-t}}\log\left(\tfrac{L^2}{x\cdot K}\right) + (\tfrac{2r}{\sigma^2}+\tfrac{1}{2})\sigma\sqrt{T-t}\right)\right] \qquad (5.19) \\
&\qquad \times\left[\tfrac{Le^{r(T-t)}}{x}\right]^{\frac{2r}{\sigma^2}+1}\cdot x \\
&\quad -K\left[N\left(\tfrac{1}{\sigma\sqrt{T-t}}\log\left(\tfrac{L^2}{x\cdot K}\right) + (\tfrac{2r}{\sigma^2}-\tfrac{1}{2})\sigma\sqrt{T-t}\right)\right] \\
&\qquad \times\left[\tfrac{Le^{r(T-t)}}{x}\right]^{\frac{2r}{\sigma^2}-1}.
\end{aligned}$$

When the option is written on a stock S and on a dollar \$, the corresponding no-arbitrage assets are $X = S$, and $Y = B^T$. In order to get the price of V in terms of a dollar \$, we can use the change of numeraire formula. Note that

$$S_\$(t) = S_{B^T}(t) \cdot B^T_\$(t) = e^{-r(T-t)} \cdot S_{B^T}(t) = e^{-r(T-t)} X_Y(t).$$

Thus the dollar price $V_\$(t)$ of the contract takes the following form

$$\begin{aligned} V_\$(t) = {} & N\left(\tfrac{1}{\sigma\sqrt{T-t}} \log\left(\tfrac{L^2}{S_\$(t)\cdot K}\right) + (\tfrac{r}{\sigma^2} + \tfrac{1}{2})\sigma\sqrt{T-t}\right) \cdot \left[\tfrac{L}{S_\$(t)}\right]^{\frac{2r}{\sigma^2}+1} \cdot S_\$(t) \\ & - Ke^{-r(T-t)} N\left(\tfrac{1}{\sigma\sqrt{T-t}} \log\left(\tfrac{L^2}{S_\$(t)\cdot K}\right) + (\tfrac{r}{\sigma^2} - \tfrac{1}{2})\sigma\sqrt{T-t}\right) \cdot \left[\tfrac{L}{S_\$(t)}\right]^{\frac{2r}{\sigma^2}-1}. \end{aligned}$$

If we denote by $v^\$(t, S_\$(t)) = V_\$(t)$ the dollar price function, we can also write

$$\begin{aligned} v^\$(t,x) = {} & N\left(\tfrac{1}{\sigma\sqrt{T-t}} \log\left(\tfrac{L^2}{x\cdot K}\right) + (\tfrac{r}{\sigma^2} + \tfrac{1}{2})\sigma\sqrt{T-t}\right) \cdot \left(\tfrac{L}{x}\right)^{\frac{2r}{\sigma^2}+1} \cdot x \\ & - Ke^{-r(T-t)} N\left(\tfrac{1}{\sigma\sqrt{T-t}} \log\left(\tfrac{L^2}{x\cdot K}\right) + (\tfrac{r}{\sigma^2} - \tfrac{1}{2})\sigma\sqrt{T-t}\right) \cdot \left(\tfrac{L}{x}\right)^{\frac{2r}{\sigma^2}-1}. \end{aligned} \tag{5.20}$$

The functions $v^\$(t,x)$ and $u^Y(t,x)$ are linked by

$$v^\$(t,x) = e^{-r(T-t)} \cdot u^Y(t, e^{r(T-t)}x), \tag{5.21}$$

which is the same as equation (3.59). When comparing Equation (5.18) with Equation (5.20), the term inside the normal cumulative distribution function changes from $\frac{2r}{\sigma^2}$ to $\frac{r}{\sigma^2}$ due to the change of numeraire from the bond prices to the dollar prices.

Let us find the hedging portfolio for the down-and-in barrier call option. Since the Black–Scholes formula in (5.17) is written in a form of a self-financing portfolio, it is also a hedging portfolio. Thus one should have

$$\begin{aligned} & \mathbb{P}_t^{(-\frac{2r}{\sigma^2})}(X_Y(T) \leq \tfrac{L^2}{K}) \cdot L^{\frac{2r}{\sigma^2}+1} \\ & \qquad = N\left(\tfrac{1}{\sigma\sqrt{T-t}} \log\left(\tfrac{L^2}{X_Y(t)\cdot K}\right) + (\tfrac{2r}{\sigma^2} + \tfrac{1}{2})\sigma\sqrt{T-t}\right) \cdot L^{\frac{2r}{\sigma^2}+1} \end{aligned}$$

units of a power option $R^{-\frac{2r}{\sigma^2}}$, and

$$\begin{aligned} & - K\mathbb{P}_t^{(1-\frac{2r}{\sigma^2})}(X_Y(T) \leq \tfrac{L^2}{K}) \cdot L^{\frac{2r}{\sigma^2}-1} \\ & \qquad = -KN\left(\tfrac{1}{\sigma\sqrt{T-t}} \log\left(\tfrac{L^2}{X_Y(t)\cdot K}\right) + (\tfrac{2r}{\sigma^2} - \tfrac{1}{2})\sigma\sqrt{T-t}\right) \cdot L^{\frac{2r}{\sigma^2}-1} \end{aligned}$$

units of a power option $R^{1-\frac{2r}{\sigma^2}}$ at time t. The power options $R^{-\frac{2r}{\sigma^2}}$ and $R^{1-\frac{2r}{\sigma^2}}$ can be replicated by trading in the underlying assets X and Y according to the following self-financing representations

$$R_t^{-\frac{2r}{\sigma^2}} = \left[-\left(\tfrac{2r}{\sigma^2}\right)\left[\tfrac{e^{r(T-t)}}{X_Y(t)}\right]^{\frac{2r}{\sigma^2}+1}\right]\cdot X_t + \left[\left(1+\tfrac{2r}{\sigma^2}\right)e^{r(T-t)}\left[\tfrac{e^{r(T-t)}}{X_Y(t)}\right]^{\frac{2r}{\sigma^2}}\right]\cdot Y_t, \tag{5.22}$$

and

$$\begin{aligned} R_t^{(1-\frac{2r}{\sigma^2})} = {} & \left[\left(1-\tfrac{2r}{\sigma^2}\right)\left[\tfrac{e^{r(T-t)}}{X_Y(t)}\right]^{\frac{2r}{\sigma^2}-1}\cdot\tfrac{1}{X_Y(t)}\right]\cdot X_t \\ & + \left[\left(\tfrac{2r}{\sigma^2}\right)\left[\tfrac{e^{r(T-t)}}{X_Y(t)}\right]^{\frac{2r}{\sigma^2}-1}\right]\cdot Y_t. \end{aligned} \tag{5.23}$$

See Exercise 5.1 for more details. We conclude that the hedging positions are given up to time τ_L by

$$\begin{aligned} \Delta^X(t) = {} & N\left(\tfrac{1}{\sigma\sqrt{T-t}}\log\left(\tfrac{L^2}{X_Y(t)\cdot K}\right)+\left(\tfrac{2r}{\sigma^2}+\tfrac{1}{2}\right)\sigma\sqrt{T-t}\right) \\ & \times\left[-\left(\tfrac{2r}{\sigma^2}\right)\left[\tfrac{Le^{r(T-t)}}{X_Y(t)}\right]^{\frac{2r}{\sigma^2}+1}\right] \\ & -KN\left(\tfrac{1}{\sigma\sqrt{T-t}}\log\left(\tfrac{L^2}{X_Y(t)\cdot K}\right)+\left(\tfrac{2r}{\sigma^2}-\tfrac{1}{2}\right)\sigma\sqrt{T-t}\right) \\ & \times\left[\left(1-\tfrac{2r}{\sigma^2}\right)\left[\tfrac{Le^{r(T-t)}}{X_Y(t)}\right]^{\frac{2r}{\sigma^2}-1}\cdot\tfrac{1}{X_Y(t)}\right], \end{aligned} \tag{5.24}$$

and

$$\begin{aligned} \Delta^Y(t) = {} & N\left(\tfrac{1}{\sigma\sqrt{T-t}}\log\left(\tfrac{L^2}{X_Y(t)\cdot K}\right)+\left(\tfrac{2r}{\sigma^2}+\tfrac{1}{2}\right)\sigma\sqrt{T-t}\right) \\ & \times\left[\left(1+\tfrac{2r}{\sigma^2}\right)Le^{r(T-t)}\left[\tfrac{Le^{r(T-t)}}{X_Y(t)}\right]^{\frac{2r}{\sigma^2}}\right] \\ & -KN\left(\tfrac{1}{\sigma\sqrt{T-t}}\log\left(\tfrac{L^2}{X_Y(t)\cdot K}\right)+\left(\tfrac{2r}{\sigma^2}-\tfrac{1}{2}\right)\sigma\sqrt{T-t}\right) \\ & \times\left[\left(\tfrac{2r}{\sigma^2}\right)\left[\tfrac{Le^{r(T-t)}}{X_Y(t)}\right]^{\frac{2r}{\sigma^2}-1}\right]. \end{aligned} \tag{5.25}$$

5.4 Connections with the Partial Differential Equations

The price function u^Y in Equation (5.19) satisfies the Black–Scholes partial differential equation (3.52)

$$u_t^Y(t,x)+\tfrac{1}{2}\sigma^2x^2u_{xx}^Y(t,x)=0$$

for times before the knock-in and before the expiration time T. This is clear from the fact that the price process $u^Y(t, X_Y(t))$ must be a $\mathbb{P}^Y$ martingale, and thus the corresponding dt term must be equal to zero. The form of the partial differential equation agrees with its plain vanilla counterpart, and it is the same for all types of the barrier options that depend on the price process $X_Y(t)$.

The only difference is in the boundary and the terminal conditions. The knock-in barrier option has the terminal condition equal to

$$u^Y(T, x) = 0 \tag{5.26}$$

for times before the knock-in and before the expiration time T. This corresponds to the situation that the barrier is not hit by the expiration, and thus the option expires worthless. On the other hand, when the barrier is hit, the price of the barrier option agrees with the price of its European option counterpart. The barrier is given by the relationship $x = Le^{r(T-t)}$, and thus the boundary condition is

$$\begin{aligned} u^Y(t, Le^{r(T-t)}) &= \mathbb{E}^Y[(X_Y(T) - K)^+ | X_Y(t) = Le^{r(T-t)}] \\ &= N(d_+)Le^{r(T-t)} - N(d_-), \end{aligned} \tag{5.27}$$

where

$$\begin{aligned} d_\pm &= \tfrac{1}{\sigma\sqrt{T-t}} \log\left(\tfrac{X_Y(t)}{K}\right) \pm \tfrac{1}{2}\sigma\sqrt{T-t} \\ &= \tfrac{1}{\sigma\sqrt{T-t}} \log\left(\tfrac{L}{K}\right) + \left(\tfrac{r}{\sigma^2} \pm \tfrac{1}{2}\right)\sigma\sqrt{T-t}. \end{aligned} \tag{5.28}$$

One can easily check that the price of the barrier option in (5.19) satisfies these boundary conditions. A straightforward computation also confirms that the price satisfies the Black–Scholes partial differential equation.

Similarly, the price function $v^\$$ from Equation (5.20) satisfies the same Black–Scholes partial differential equation as in (3.61):

$$-rv^\$(t, x) + v^\$_t(t, x) + rxv^\$_x(t, x) + \tfrac{1}{2}\sigma^2 x^2 v^\$_{xx}(t, x) = 0.$$

The terminal condition for $v^\$$ agrees with the terminal condition for u^Y:

$$v^\$(T, x) = 0, \tag{5.29}$$

but the exponential boundary $Le^{r(T-t)}$ for the function u^Y transforms into a constant boundary L for the function $v^\$$:

$$v^\$(t, L) = N(d_+)L - N(d_-), \tag{5.30}$$

where

$$d_\pm = \tfrac{1}{\sigma\sqrt{T-t}} \log\left(\tfrac{L}{K}\right) + \left(\tfrac{r}{\sigma^2} \pm \tfrac{1}{2}\right)\sigma\sqrt{T-t}. \tag{5.31}$$

The price functions $u^X(t,x) = u^Y(t,\frac{1}{x}) \cdot x$ and $v^S(t,x) = v^\$(t,\frac{1}{x}) \cdot x$ satisfy the same partial differential equations as their European option counterparts (Equations (3.55), and (3.69)). The terminal and the boundary conditions can be determined from the perspective mapping.

References and Further Reading

The price of the barrier option in the geometric Brownian motion model was computed in Rubinstein and Reiner (1991). Kunitomo and Ikeda (1992) determined prices of barrier options with exponential boundaries. Our approach is related to the work of Carr and Chou (1997) who found several connections between barrier and plain vanilla options based on the symmetry argument. Broadie et al. (1997) studied the difference between the prices of discretely and continuously monitored barrier options. Other papers related to pricing barrier options include Brown et al. (2001), Zvan et al. (2000), Figlewski and Gao (1999), or Carr et al. (1998).

Exercises

5.1 Consider a contract that pays off

$$R_T^\alpha = [X_Y(T)]^\alpha \cdot Y_T.$$

(a) Compute the price of this contract in a geometric Brownian motion model from the stochastic representation

$$u^Y(t,x) = \mathbb{E}_t^Y\left[[X_Y(T)]^\alpha | X_Y(t) = x\right].$$

(b) Compute the price of this contract in the same model by solving the partial differential equation

$$u_t^Y(t,x) + \tfrac{1}{2}\sigma^2 x^2 u_{xx}^Y(t,x) = 0$$

with the terminal condition $u^Y(T,x) = x^\alpha$.
Hint: Find the solution of the form $u^Y(t,x) = h(t) \cdot x^\alpha$.

(c) Determine the hedge of the contract: $\Delta^X(t)$ and $\Delta^Y(t)$.

(d) Use Ito's formula to show

$$dR_Y^\alpha(t) = \alpha\sigma R_Y^\alpha(t)dW^Y(t).$$

5.2 Show that the price of an Arrow–Debreu security U that pays off a unit of an asset X on the event

$$A = \{\min_{0\le t\le T}(e^{-r(T-t)}X_Y(t)) \le L; X_Y(T) \ge K\}$$

has the same price up to time τ_L as an Arrow–Debreu security $U^{(\alpha)}$ that pays off $L^{1+\frac{2r}{\sigma^2}}$ units of a power option $R^{(-\frac{2r}{\sigma^2})}$ on the event

$$B = \{X_Y(T) \le \tfrac{L^2}{K}\}.$$

5.3 Determine

$$\mathbb{P}^{(\alpha)}(X_Y(T) \ge K),$$

where $\mathbb{P}^{(\alpha)}$ is a martingale measure associated with a power option R^α. Check that your result agrees with the special cases $\mathbb{P}^{(0)}(X_Y(T) \ge K) = \mathbb{P}^Y(X_Y(T) \ge K) = N(d_-)$, and $\mathbb{P}^{(1)}(X_Y(T) \ge K) = \mathbb{P}^X(X_Y(T) \ge K) = N(d_+)$.

5.4 Find the price and the hedge of the contract V with the payoff

$$V_T = (X_Y(T) - K)^+ \cdot X_T.$$

Hint: This is like a standard European call option, only settled in a "wrong" asset X instead of Y. Note that $X_Y(T)\cdot X_T = [X_Y(T)]^2 \cdot Y_T$, which represents a power option R^2. Thus

$$V_T = (R_T^2 - K \cdot X_T)^+.$$

You can apply Black–Scholes formula using the assets R^2 and X.

5.5 Consider a perpetual barrier option that pays off a unit of Y when X_Y hits a level $L > X_Y(0)$ in a geometric Brownian motion model. Let

$$\tau = \inf\{t \ge 0 : X_Y(t) = L\}$$

be the first time that the price hits the barrier. Obviously, the price of the perpetual barrier option is given by

$$\mathbb{E}^Y[\mathbb{I}(\tau < \infty)] = \mathbb{P}^Y(\tau < \infty).$$

Determine $\mathbb{P}^Y(\tau < \infty)$.
Hint: Define an auxiliary stopping time τ_ϵ as

$$\tau_\epsilon = \inf\{t \ge 0 : X_Y(t) = L \text{ or } X_Y(t) = \epsilon\}$$

for $0 < \epsilon < X_Y(0)$.

(a) Use the Optional Sampling Theorem (Theorem A.1: $\mathbb{E}^Y[X_Y(\tau_\epsilon)] = X_Y(0)$) to determine $\mathbb{P}^Y(X_Y(\tau_\epsilon) = L)$.

(b) Compute

$$\mathbb{P}^Y(\tau < \infty) = \lim_{\epsilon \to 0} \mathbb{P}^Y(X_Y(\tau_\epsilon) = L).$$

You may easily generalize the formula to determine $\mathbb{P}_t^Y(\tau < \infty)$ when $X_Y(t) < L$.

(c) Determine the hedging portfolio of this contract. Show that your hedging portfolio delivers a unit of Y when the barrier is hit.

Chapter 6

Lookback Options

This chapter studies **lookback options.** The payoff of a lookback option depends on either the maximum price of $X_Y(t)$ or the minimum price of $X_Y(t)$. We describe the contract on the maximum price. In order to collect the payoff that depends on the maximum, all the intermediate price levels between zero and the maximum have to be reached during the lifetime of the option, and thus one can think of the lookback option as a combination of knock-in option contracts. Once a particular level K is reached, the price may end up either above or below this level. The first case corresponds to a plain vanilla European option; the second case corresponds to a knock-in barrier option. Using the results from the previous sections, the second contract can also be expressed as a plain vanilla European option. Thus we obtain a representation of the lookback option price in terms of two plain vanilla European call options. We also give partial differential equations that correspond to the pricing of the lookback option together with the characterization of the hedging portfolio. The last section introduces the maximum drawdown, a widely used portfolio performance measure.

6.1 Connections of Lookbacks with Barrier Options

A **lookback option** is a contract whose payoff depends on the maximal or the minimal price of $X_Y(t)$. Let us illustrate how the **maximal asset** M^* defined as

$$M_t^* = \left[\max_{0\leq s\leq t} X_Y(s)\right] \cdot Y_t$$

is related to barrier contracts. The maximal asset is also known as a **high watermark** in hedge fund management. Note that M^* is an arbitrage asset, but a contract V that agrees to deliver this asset at time T is a no-arbitrage asset. The most typical contracts monitor the maximum of the price in terms of a dollar, so let us choose Y to be a dollar \$, and X to be a stock S.

Note that the contract V that pays off

$$V_T = M_T^* = \left[\max_{0\leq t\leq T} S_\$(t)\right] \cdot \$_T = \left[\max_{0\leq t\leq T} S_\$(t)\right] \cdot B_T^T$$

can be also expressed as

$$V_T = \left[\int_0^\infty \mathbb{I}(\tau_L \leq T)dL\right] \cdot B_T^T. \tag{6.1}$$

The hitting time τ_L is defined as the first time the price of the stock $S_\$$ reaches level L

$$\tau_L = \inf\{t \geq 0 : S_\$(t) \geq L\}.$$

The price level L is reached from below rather than from above in contrast to the situations studied in the previous sections, but all the analysis remains the same. The insight of the relationship (6.1) is the following. When the maximal price is equal to m, i.e., $M_\$^*(T) = m$, the price process $S_\$(t)$ must cross all levels $L \leq m$ by time T, while the higher levels are not reached by time T.

A contract that pays off $\mathbb{I}(\tau_L \leq T)$ units of B^T can be split in the following way

$$\mathbb{I}(\tau_L \leq T) \cdot B_T^T = \mathbb{I}(\tau_L \leq T, S_\$(T) \geq L) \cdot B_T^T + \mathbb{I}(\tau_L \leq T, S_\$(T) < L) \cdot B_T^T. \tag{6.2}$$

Since the event $S_\$(T) \geq L$ already implies that $\tau_L \leq T$, we have

$$\mathbb{I}(\tau_L \leq T, S_\$(T) \geq L) \cdot B_T^T = \mathbb{I}(S_\$(T) \geq L) \cdot B_T^T.$$

The second contract with a payoff $\mathbb{I}(\tau_L \leq T, S_\$(T) < L) \cdot B_T^T$ is a knock-in Arrow–Debreu security with a barrier equal to L in terms of a dollar price, and a payoff on the event $S_\$(T) < L$, so the strike is equal to the barrier. We have seen that this security has its plain vanilla counterpart that has the same price up to time τ_L. Let us determine the corresponding plain vanilla contract. First, we need to express the barrier in terms of a no-arbitrage asset B^T in the following way

$$S_\$(t) \geq L \Longleftrightarrow S_{B^T}(t) \geq Le^{r(T-t)},$$

assuming $B_\$^T(t) = e^{-r(T-t)}$. Thus the problem corresponds to hitting an exponential barrier $Le^{r(T-t)}$. The knock-in contract on the barrier has the same price as $\left(\frac{1}{L}\right)^\alpha$ units of a plain vanilla Arrow–Debreu security that pays off a power option R^α at time T when $S_\$(T) \geq L$ for $\alpha = 1 - \frac{2r}{\sigma^2}$.

Let $M_\$^*(t) = m$ be the current running maximum of the stock price $S_\$(t)$. Based on the previous arguments, the payoff of V_T is equivalent to

$$V_T = m \cdot B_T^T + \left[\int_m^\infty \mathbb{I}(S_\$(T) \geq L)dL\right] \cdot B_T^T \\ + \left[\int_m^\infty \left(\tfrac{1}{L}\right)^\alpha \mathbb{I}(S_\$(T) \geq L)dL\right] \cdot R_T^\alpha. \quad (6.3)$$

Since

$$\left[\int_m^\infty \mathbb{I}(S_\$(T) \geq L)dL\right] \cdot B_T^T = \left[\int_m^{S_\$(T)\vee m} dL\right] \cdot B_T^T \\ = \left[(S_\$(T) - m)^+\right] \cdot B_T^T = \left[(S_{B^T} - m)^+\right] \cdot B_T^T = (S_T - mB_T^T)^+, \quad (6.4)$$

the contract with a payoff $\left[\int_m^\infty \mathbb{I}(S_\$(T) \geq L)dL\right] \cdot B_T^T$ is a plain vanilla call option with strike m

$$(S_T - m \cdot B_T^T)^+.$$

We use the notation $x \vee y = \max(x, y)$. The form of the contract that pays off $\left[\int_m^\infty \left(\frac{1}{L}\right)^\alpha \mathbb{I}(S_\$(T) \geq L)dL\right] \cdot S_T$ depends on the value of α.

6.1.1 Case $\alpha = 1$

When $\alpha = 1$, or equivalently when $r = 0$, we get

$$\left[\int_m^\infty \tfrac{1}{L} \cdot \mathbb{I}(S_\$(T) \geq L)dL\right] \cdot S_T = \left[\int_m^{S_\$(T)\vee m} \tfrac{1}{L}dL\right] \cdot S_T \\ = \left[\log\left(\tfrac{S_\$(T)}{m}\right)^+\right] \cdot S_T = \left[\log\left(\tfrac{S_{B^T}(T)}{m}\right)^+\right] \cdot S_T. \quad (6.5)$$

Therefore the the lookback option V has the same value as a combination of two plain vanilla options.

$$\boxed{V_T = m \cdot B_T^T + (S_T - m \cdot B_T^T)^+ + \left[\log\left(\tfrac{S_{B^T}(T)}{m}\right)^+\right] \cdot S_T.} \quad (6.6)$$

Let

$$u^T(t, x, y) = \mathbb{E}^T[V_{B^T}(T)|S_{B^T}(t) = x, M_{B^T}^*(t) = y]$$

be the price of the lookback option V with respect to the bond B^T. In the case when $r = 0$, it also agrees with the dollar price. Furthermore, $M_{B^T}^*(t) = M_\$^*(t) = m$, so we also have $y = m$. The price of the first contract with the payoff $(S_T - y \cdot B_T^T)^+$ is simply given by the Black–Scholes formula; the price of the second contract follows from a straightforward computation, and it is a subject of Exercise 6.1. Thus we get

$$\boxed{u^T(t, x, y) = y + xN(d_+) - yN(d_-) + x\sigma\sqrt{T-t}\left[d_+N(d_+) + \phi(d_+)\right],} \quad (6.7)$$

where

$$d_{\pm} = \tfrac{1}{\sigma\sqrt{T-t}} \cdot \log(\tfrac{x}{y}) \pm \tfrac{1}{2}\sigma\sqrt{T-t},$$

and where $\phi(x) = \frac{1}{\sqrt{2\pi}} \cdot e^{-\frac{x^2}{2}}$ is a density of a standard normal random variable. The hedging portfolio is given by

$$\Delta^S(t) = u_x^T(t,x,y) = 2N(d_+) + \sigma\sqrt{T-t}\left[d_+N(d_+) + \phi(d_+)\right], \quad (6.8)$$

$$\Delta^T(t) = u_y^T(t,x,y) = yN(-d_-) - xN(d_+). \quad (6.9)$$

6.1.2 Case $\alpha < 1$

When $\alpha = 1 - \frac{2r}{\sigma^2} < 1$, which is the case when $r > 0$, we have

$$\begin{aligned}
\left[\int_m^\infty \left(\tfrac{1}{L}\right)^\alpha \cdot \mathbb{I}(S_\$(T) \geq L)dL\right] \cdot R_T^\alpha &= \left[\int_m^{S_\$(T)\vee m} \left(\tfrac{1}{L}\right)^\alpha dL\right] \cdot R_T^\alpha \\
&= \left[\frac{1}{1-\alpha} \cdot \left[[S_\$(T)]^{1-\alpha} - m^{1-\alpha}\right]^+\right] \cdot R_T^\alpha \\
&= \left[\frac{1}{1-\alpha} \cdot \left[[S_{B^T}(T)]^{1-\alpha} - m^{1-\alpha}\right]^+\right] \cdot R_T^\alpha \\
&= \frac{1}{1-\alpha} \cdot \left[S_T - m^{1-\alpha}R_T^\alpha\right]^+ \\
&= \frac{\sigma^2}{2r} \cdot \left[S_T - m^{\frac{2r}{\sigma^2}} R_T^{(1-\frac{2r}{\sigma^2})}\right]^+. \quad (6.10)
\end{aligned}$$

The second-to-the last relationship follows from

$$[S_{B^T}(T)]^{1-\alpha} \cdot R_T^\alpha = [S_{B^T}(T)]^{1-\alpha} \cdot [S_{B^T}(T)]^\alpha \cdot B_T^T = S_{B^T}(T) \cdot B_T^T = S_T.$$

Thus the lookback option V has the same price as a combination of two plain vanilla options, one expressed in terms of the assets S and B^T that pays off

$$(S_T - m \cdot B_T^T)^+,$$

and one expressed in terms of the assets S and R^α that pays off

$$\tfrac{1}{1-\alpha}[S_T - m^{1-\alpha}R_T^\alpha]^+ = \tfrac{\sigma^2}{2r} \cdot \left[S_T - m^{\frac{2r}{\sigma^2}} R_T^{(1-\frac{2r}{\sigma^2})}\right]^+.$$

Thus we can write

$$\boxed{V_T = m \cdot B_T^T + (S_T - m \cdot B_T^T)^+ + \tfrac{\sigma^2}{2r} \cdot \left[S_T - m^{\frac{2r}{\sigma^2}} R_T^{(1-\frac{2r}{\sigma^2})}\right]^+.} \quad (6.11)$$

In order to express the price of the lookback option V in terms of the prices with respect to the bond B^T, we must substitute $M_{B^T}^*(t)$ for $M_\$^*(t) = m$ that appears in the above formula. Let $M_{B^T}^*(t) = y$. Then from

$$M_{B^T}^*(t) = M_\$^*(t) \cdot \$_{B^T}(t) = me^{r(T-t)} = y,$$

we have that $m = ye^{-r(T-t)}$, and thus we can also write

$$V_T = ye^{-r(T-t)} \cdot B_T^T + (S_T - ye^{-r(T-t)} \cdot B_T^T)^+ \\ + \frac{\sigma^2}{2r} \cdot \left[S_T - \left(ye^{-r(T-t)} \right)^{\frac{2r}{\sigma^2}} R_T^{(1-\frac{2r}{\sigma^2})} \right]^+ . \quad (6.12)$$

Let $u^T(t, x, y)$ be the price of V in terms of B^T defined as

$$u^T(t, x, y) = \mathbb{E}^T[V_{B^T}(T) | S_{B^T}(t) = x, M^*_{B^T}(t) = y].$$

We can easily obtain the closed form solution using the fact that the price of V agrees with the sum of the prices of two plain vanilla options. Recall that the price of a plain vanilla option with a payoff $(X_T - K \cdot Y_T)^+$ is given by the Black–Scholes formula

$$\mathbb{P}_t^X(X_Y(T) \geq K) \cdot X_t - K \mathbb{P}_t^Y(X_Y(T) \geq K) \cdot Y_t,$$

which also represents the hedging portfolio for this option. Thus the price of the first option with a payoff $(S_T - ye^{-r(T-t)} \cdot B_T^T)^+$ is simply

$$\mathbb{P}_t^S \left(S_{B^T}(T) \geq ye^{-r(T-t)} \right) \cdot S_{B^T}(t) - ye^{r(T-t)} \cdot \mathbb{P}_t^T \left(S_{B^T}(T) \geq ye^{-r(T-t)} \right) \\ = xN(d_+) - ye^{-r(T-t)} N(d_-) \quad (6.13)$$

according to the Black–Scholes formula, where

$$d_\pm = \frac{1}{\sigma\sqrt{T-t}} \log \left(\frac{x}{ye^{-r(T-t)}} \right) \pm \frac{1}{2} \sigma \sqrt{T-t}.$$

The price of the second option with a payoff $\left[S_T - \left(ye^{-r(T-t)} \right)^{\frac{2r}{\sigma^2}} R_T^{(1-\frac{2r}{\sigma^2})} \right]^+$ is also given by the Black–Scholes formula

$$\mathbb{P}_t^S \left[S_{R^\alpha}(T) \geq \left(ye^{-r(T-t)} \right)^{\frac{2r}{\sigma^2}} \right] \cdot S_{B^T}(t) \\ - \left(ye^{-r(T-t)} \right)^{\frac{2r}{\sigma^2}} \cdot \mathbb{P}_t^{(\alpha)} \left[S_{R^\alpha}(T) \geq \left(ye^{-r(T-t)} \right)^{\frac{2r}{\sigma^2}} \right] \cdot R_{B^T}^{(1-\frac{2r}{\sigma^2})}(t).$$

But since

$$S_{R^\alpha}(T) = S_{B^T}(T) \cdot B_{R^\alpha}^T(T) \\ = S_{B^T}(T) \cdot [S_{B^T}(T)]^{-\alpha} = [S_{B^T}(T)]^{1-\alpha} = [S_{B^T}(T)]^{\frac{2r}{\sigma^2}},$$

the event $S_{R^\alpha}(T) \geq \left(ye^{-r(T-t)} \right)^{\frac{2r}{\sigma^2}}$ is equivalent to the event $S_{B^T}(T) \geq ye^{-r(T-t)}$. Furthermore, the price of the power option R^α in terms of the bond B^T is given by

$$R^{\alpha}_{B^T}(t) = \exp(\tfrac{1}{2}\alpha(\alpha-1)\sigma^2(T-t)) \cdot [S_{B^T}(t)]^{\alpha}$$
$$= \left(xe^{-r(T-t)}\right)^{\alpha} = \left(xe^{-r(T-t)}\right)^{1-\frac{2r}{\sigma^2}}.$$

Thus the price of the second option can be expressed as

$$\begin{aligned}
&\mathbb{P}^S_t\left[S_{B^T}(T) \geq ye^{-r(T-t)}\right] \cdot S_{B^T}(t) \\
&\quad - \left(ye^{-r(T-t)}\right)^{\frac{2r}{\sigma^2}} \cdot \mathbb{P}^{(\alpha)}_t\left[S_{B^T}(T) \geq ye^{-r(T-t)}\right] \cdot R^{(1-\frac{2r}{\sigma^2})}_{B^T}(t) \\
&= xN(d_+) - \left(ye^{-r(T-t)}\right)^{\frac{2r}{\sigma^2}} \cdot \left(xe^{-r(T-t)}\right)^{1-\frac{2r}{\sigma^2}} \cdot N\left(d_{\left(1-\frac{2r}{\sigma^2}\right)}\right), \qquad (6.14)
\end{aligned}$$

where

$$\mathbb{P}^{(\alpha)}_t\left[S_{B^T}(T) \geq ye^{-r(T-t)}\right] = N(d_{\alpha})$$

for

$$d_{\alpha} = \tfrac{1}{\sigma\sqrt{T-t}}\log\left(\tfrac{x}{ye^{-r(T-t)}}\right) + (\alpha - \tfrac{1}{2})\sigma\sqrt{T-t}.$$

See Exercise 5.3 for more details on how to determine $\mathbb{P}^{(\alpha)}(X_Y(T) \geq K)$. Combining (6.13) and (6.14), we conclude that

$$\begin{aligned}
u^T(t,x,y) &= ye^{-r(T-t)} + xN(d_+) - ye^{-r(T-t)}N(d_-) \\
&\quad + \frac{1}{1-\alpha} \cdot \left[xN(d_+) - \left(ye^{-r(T-t)}\right)^{1-\alpha} \cdot \left(xe^{-r(T-t)}\right)^{\alpha} \cdot N(d_{\alpha})\right] \\
&= ye^{-r(T-t)} + xN(d_+) - ye^{-r(T-t)}N(d_-) + \tfrac{\sigma^2}{2r} \cdot xN(d_+) \\
&\quad - \tfrac{\sigma^2}{2r} \cdot \left(ye^{-r(T-t)}\right)^{\frac{2r}{\sigma^2}} \cdot \left(xe^{-r(T-t)}\right)^{1-\frac{2r}{\sigma^2}} \cdot N\left(d_{\left(1-\frac{2r}{\sigma^2}\right)}\right) \\
&= ye^{-r(T-t)}N(-d_-) + (1 + \tfrac{\sigma^2}{2r})xN(d_+) \\
&\quad - \tfrac{\sigma^2}{2r} \cdot \left(\tfrac{y}{x}\right)^{\frac{2r}{\sigma^2}} \cdot \left(xe^{-r(T-t)}\right) \cdot N\left(d_{\left(1-\frac{2r}{\sigma^2}\right)}\right). \qquad (6.15)
\end{aligned}$$

We have shown the following result.

THEOREM 6.1

The price of a contract V that delivers the maximal asset M^* at time T defined as

$$V_{B^T}(t) = u^T(t,x,y) = \mathbb{E}^T[M^*_{B^T}(T) | S_{B^T}(t) = x, M^*_{B^T}(t) = y]$$

is given by

$$\begin{aligned}
u^T(t,x,y) &= ye^{-r(T-t)}N(-d_-) + (1 + \tfrac{\sigma^2}{2r})xN(d_+) \\
&\quad - \tfrac{\sigma^2}{2r} \cdot \left(\tfrac{y}{x}\right)^{\frac{2r}{\sigma^2}} \cdot \left(xe^{-r(T-t)}\right) \cdot N\left(d_{\left(1-\frac{2r}{\sigma^2}\right)}\right). \qquad (6.16)
\end{aligned}$$

This corresponds to the terminal condition

$$f^T(x,y) = y.$$

REMARK 6.1 The payoff of the lookback option V written as

$$V_T = ye^{-r(T-t)} \cdot B_T^T + (S_T - ye^{-r(T-t)} \cdot B_T^T)^+ + \frac{\sigma^2}{2r} \cdot \left[S_T - \left(ye^{-r(T-t)}\right)^{\frac{2r}{\sigma^2}} R_T^{(1-\frac{2r}{\sigma^2})}\right]^+$$

can be expressed in a more compact form by observing that both options expire in the money on the same event $S_{B^T}(T) \geq ye^{-r(T-t)}$, and thus the two option payoffs can be combined into one

$$V_T = ye^{-r(T-t)} \cdot B_T^T + \left[(1+\frac{\sigma^2}{2r})S_T - ye^{-r(T-t)} \cdot B_T^T - \frac{\sigma^2}{2r}\left(ye^{-r(T-t)}\right)^{\frac{2r}{\sigma^2}} R_T^{(1-\frac{2r}{\sigma^2})}\right]^+. \tag{6.17}$$

Furthermore, since $R^{(0)} = B^T$, $R^{(1)} = S$, we can also write

$$V_T = ye^{-r(T-t)} \cdot R_T^{(0)} + \left[(1+\frac{\sigma^2}{2r})R_T^{(1)} - ye^{-r(T-t)} \cdot R_T^{(0)} - \frac{\sigma^2}{2r}\left(ye^{-r(T-t)}\right)^{\frac{2r}{\sigma^2}} R_T^{(1-\frac{2r}{\sigma^2})}\right]^+. \tag{6.18}$$

□

REMARK 6.2 We can also express the price of the contract V that delivers the maximal asset M^* at time T in terms of a dollar \$ from the change of numeraire formula:

$$V_\$(t) \cdot \$_t = V_{B^T}(t) \cdot B^T.$$

If we denote $v^\$(t, S_\$(t), M_\$^*(t)) = V_\$(t)$, we can obtain the relationship to the price function $u^T(t, S_{B^T}(t), M_{B^T}^*(t)) = V_{B^T}(t)$, which is given by

$$v^\$(t,x,y) = e^{-r(T-t)} \cdot u^T(t, e^{r(T-t)}x, e^{r(T-t)}y). \tag{6.19}$$

This formula generalizes Equation (3.59). Note that for the particular payoff we have been considering, $f^T(x,y) = y$, the function $u^T(t,x,y)$ is homogeneous in the variables x and y, meaning that

$$u^T(t, ax, ay) = a \cdot u^T(t,x,y). \tag{6.20}$$

If the prices of $S_{B^T}(t)$ and $M_{B^T}^*(t)$ are multiplied by a factor of a, the whole contract becomes a times more expensive. This is also seen directly from Equation (6.16). Therefore we have

$$v^\$(t,x,y) = e^{-r(T-t)} \cdot u^T(t, e^{r(T-t)}x, e^{r(T-t)}y) = u^T(t,x,y)$$

for the case of the contract that delivers the maximal asset at time T. This is true only for special payoff functions $f^T(x,y)$, not in general. □

6.1.3 Hedging

We have already determined the hedging portfolio for a contract that agrees to deliver the maximal asset M^* in the case when $r = 0$, which corresponds to the case of $\alpha = 1$. Let us show the hedging portfolio for the case of $\alpha < 1$. Recall that the contract that delivers the maximal asset corresponds to a combination of two European options

$$V_T = ye^{-r(T-t)} \cdot B_T^T + (S_T - ye^{-r(T-t)} \cdot B_T^T)^+ + \tfrac{\sigma^2}{2r} \cdot \left[S_T - \left(ye^{-r(T-t)}\right)^{\frac{2r}{\sigma^2}} R_T^{(1-\frac{2r}{\sigma^2})}\right]^+.$$

The hedging portfolio for the option with the payoff $(S_T - ye^{-r(T-t)} \cdot B_T^T)^+$ is given by $N(d_+)$ units of the stock S and $-ye^{-r(T-t)}N(d_-)$ units of the bond B^T. The hedging portfolio for the option with the payoff $\frac{\sigma^2}{2r}\left[S_T - \left(ye^{-r(T-t)}\right)^{\frac{2r}{\sigma^2}} R_T^{(1-\frac{2r}{\sigma^2})}\right]^+$ is given by $\frac{\sigma^2}{2r}N(d_+)$ units of the stock S and $-\frac{\sigma^2}{2r}\left(ye^{-r(T-t)}\right)^{\frac{2r}{\sigma^2}} N\left(d_{\left(1-\frac{2r}{\sigma^2}\right)}\right)$ units of the power option $R^{(1-\frac{2r}{\sigma^2})}$. The power option $R^{(1-\frac{2r}{\sigma^2})}$ itself has the hedging portfolio given in Equation (5.23), which corresponds to the $(1-\frac{2r}{\sigma^2})\left[\frac{e^{r(T-t)}}{S_{B^T}(t)}\right]^{\frac{2r}{\sigma^2}-1}\frac{1}{S_{B^T}(t)}$ units of the stock S and $\left(\frac{2r}{\sigma^2}\right)\left[\frac{e^{r(T-t)}}{S_{B^T}(t)}\right]^{\frac{2r}{\sigma^2}-1}$ units of the bond B^T. Combining these expressions, we get the following result.

THEOREM 6.2
The hedging portfolio for the contract V that delivers the maximal asset M^* at time T is given by

$$\Delta^S(t) = (1+\tfrac{\sigma^2}{2r}) \cdot N(d_+) + (1-\tfrac{\sigma^2}{2r}) \cdot \left(\tfrac{y}{x}\right)^{\frac{2r}{\sigma^2}} \cdot e^{-r(T-t)} \cdot N\left(d_{\left(1-\frac{2r}{\sigma^2}\right)}\right), \quad (6.21)$$

$$\Delta^T(t) = ye^{-r(T-t)}N(-d_-) - \left(\tfrac{y}{x}\right)^{\frac{2r}{\sigma^2}} \cdot \left(xe^{-r(T-t)}\right) \cdot N\left(d_{\left(1-\frac{2r}{\sigma^2}\right)}\right), \quad (6.22)$$

where $x = S_{B^T}(t) = e^{r(T-t)}S_\$(t)$, and $y = M^*_{B^T}(t) = e^{r(T-t)}M^*_\$(t)$.

REMARK 6.3 Hedging positions and the hitting times
The price of the contract V that delivers a maximal asset M^* at time T can

be written as

$$u^T(t,x,y) = \mathbb{E}^T[M^*_{B^T}(T)|S_{B^T}(t) = x, M^*_{B^T}(t) = y].$$

We have seen that the function $u^T(t,x,y)$ is homogeneous in the variables x and y, meaning that

$$u^T(t,ax,ay) = a \cdot u^T(t,x,y).$$

Homogeneous functions satisfy the following relationship

$$u^T(t,x,y) = xu^T_x(t,x,y) + yu^T_y(t,x,y),$$

which follows from differentiating (6.20) with respect to a. Substituting the prices of $S_{B^T}(t)$ and $M^*_{B^T}(t)$ for the variables x and y, we get

$$\begin{aligned} u^T(t,S_{B^T}(t),M^*_{B^T}(t)) = {} & S_{B^T}(t)u^T_x(t,S_{B^T}(t),M^*_{B^T}(t)) \\ & + M^*_{B^T}(t)u^T_y(t,S_{B^T}(t),M^*_{B^T}(t)). \end{aligned}$$

Note that we also have

$$u^T(t,S_{B^T}(t),M^*_{B^T}(t)) = \Delta^S(t)S_{B^T}(t) + \Delta^T(t)$$

from the hedging representation of the contract V. Therefore we must have

$$\Delta^T(t) = M^*_{B^T}(t) \cdot u^T_y(t,S_{B^T}(t),M^*_{B^T}(t)).$$

Let us determine $u^T_y(t,S_{B^T}(t),M^*_{B^T}(t))$. Recall that we can write

$$M^*_\$(T) = M^*_\$(t) + \int_{M^*_\$(t)}^{\infty} \mathbb{I}(\tau_L \le T)dL,$$

so the future dollar price of the maximal asset consists of two components: the present dollar price of the maximal asset plus all the price levels that will be crossed between time t and T. We can also rewrite the above relationship in the terms of the bond prices as

$$M^*_{B^T}(T) = e^{-r(T-t)}M^*_{B^T}(t) + \int_{e^{-r(T-t)}M^*_{B^T}(t)}^{\infty} \mathbb{I}(\tau_L \le T)dL.$$

Thus we have

$$\begin{aligned} u^T(t,S_{B^T}(t),M^*_{B^T}(t)) &= \mathbb{E}^T_t[M^*_{B^T}(T)] \\ &= \mathbb{E}^T_t\left[e^{-r(T-t)}M^*_{B^T}(t) + \int_{e^{-r(T-t)}M^*_{B^T}(t)}^{\infty} \mathbb{I}(\tau_L \le T)dL\right] \\ &= e^{-r(T-t)}M^*_{B^T}(t) + \int_{e^{-r(T-t)}M^*_{B^T}(t)}^{\infty} \mathbb{P}^T_t(\tau_L \le T)dL. \end{aligned}$$

Taking the derivative with respect to the variable $y = M^*_{BT}(t)$, we get

$$\begin{aligned} u_y^T(t, S_{BT}(t), M^*_{BT}(t)) &= e^{-r(T-t)} \cdot \left(1 - \mathbb{P}_t^T(\tau_{M^*_\$(t)} \le T)\right) \\ &= e^{-r(T-t)} \cdot \mathbb{P}_t^T(\tau_{M^*_\$(t)} > T) = e^{-r(T-t)} \cdot \mathbb{P}_t^T(M^*_\$(T) = M^*_\$(t)). \end{aligned}$$

Therefore

$$\boxed{\Delta^T(t) = e^{-r(T-t)} M^*_{BT}(t) \cdot \mathbb{P}_t^T(M^*_\$(T) = M^*_\$(t)).} \tag{6.23}$$

Thus the hedging position in the asset Y is between zero and the current level of $M^*_\$(t)$. The event $M^*_\$(T) = M^*_\$(t)$ means that the current level of $M^*_\$(t)$ will remain the same until the expiration of the contract. In particular, when $S_\$(t) = M^*_\$(t)$, which means that the price of $S_\$(t)$ is at the historical maximum, $\mathbb{P}_t^T(M^*_\$(T) = M^*_\$(t)) = 0$ as the price of $S_\$(t)$ will fluctuate to a higher level for sure. Let us determine the hedging position $\Delta^S(t)$. The hedging portfolio can be expressed as

$$V_t = \Delta^S(t) \cdot S_t + \Delta^T(t) \cdot B_t^T,$$

and thus we have

$$\begin{aligned} \Delta^S(t) &= V_S(t) - \Delta^T(t) \cdot B_S^T(t) \\ &= V_S(t) - e^{-r(T-t)} M^*_{BT}(t) \cdot \mathbb{P}_t^T(M^*_\$(T) = M^*_\$(t)) \cdot B_S^T(t). \end{aligned}$$

We conclude that

$$\boxed{\Delta^S(t) = V_S(t) - e^{-r(T-t)} M^*_S(t) \cdot \mathbb{P}_t^T(M^*_\$(T) = M^*_\$(t)).} \tag{6.24}$$

□

6.2 Partial Differential Equation Approach for Lookbacks

The previous section showed the relationship of the maximal asset M^* defined as

$$M_t^* = \left[\max_{0 \le s \le t} S_\$(s)\right] \cdot \$_t \tag{6.25}$$

to European options. However, the payoff that depends on the asset M^* can be more complicated. In general, **a lookback option** is a contract that depends on underlying assets S and \$ and upon the maximum of the price process $S_\$(t)$. The asset M^* itself is an arbitrage asset. It can still be used as a numeraire, but there is no probability measure that would have M^* as a reference asset. More formally, we can define:

DEFINITION 6.1 *A lookback option is a contract that pays off one of the following:*

- $f^T\left(S_{B^T}(T), M^*_{B^T}(T)\right)$ *units of an asset* B^T,
- $f^S\left(B^T_S(T), M^*_S(T)\right)$ *units of an asset* S,
- $f^*\left(S_{M^*}(T), B^T_{M^*}(T)\right)$ *units of an asset* M^*.

We can substitute a dollar \$ with the corresponding bond B^T in the payoff of the lookback option. Should the different settlements represent the same contract, we must have

$$f^T(x,y) = f^S(\tfrac{1}{x}, \tfrac{y}{x}) \cdot x, \quad f^S(x,y) = f^T(\tfrac{1}{x}, \tfrac{y}{x}) \cdot x,$$

$$f^T(x,y) = f^*(\tfrac{x}{y}, \tfrac{1}{y}) \cdot y, \quad f^*(x,y) = f^T(\tfrac{x}{y}, \tfrac{1}{y}) \cdot y,$$

$$f^S(x,y) = f^*(\tfrac{1}{y}, \tfrac{x}{y}) \cdot y, \quad f^*(x,y) = f^S(\tfrac{y}{x}, \tfrac{1}{x}) \cdot x.$$

This is a generalization of the **perspective mapping** for three assets. Let us show for instance the relationship of the two payoff functions f^S and f^*. In order for them to represent the same contract, we must have

$$f^S\left(B^T_S(T), M^*_S(T)\right) \cdot S = f^*\left(S_{M^*}(T), B^T_{M^*}(T)\right) \cdot M^*,$$

or in other words

$$f^S\left(B^T_S(T), M^*_S(T)\right) = f^*\left(S_{M^*}(T), B^T_{M^*}(T)\right) \cdot M^*_S(T).$$

Therefore the x variable in f^S stands for $B^T_S(T)$, and the y variable stands for $M^*_S(T)$. It is easy to see that $S_{M^*}(T)$ is now $\frac{1}{y}$, and $B^T_{M^*}(T) = B^T_S(T) \cdot S_{M^*}(T)$ is represented by $\frac{x}{y}$.

Example 6.1

Consider a lookback option contract with a payoff

$$M^*_T - S_T. \tag{6.26}$$

This is known as a **drawdown.** One can think of the drawdown as a difference of two assets, indicating how far the asset X is from the maximal asset M^*. Expressing the contract in terms of the payoff functions, we get

$$f^T(x,y) = y - x, \qquad f^S(x,y) = y - 1, \qquad f^*(x,y) = 1 - x.$$

Note that $M^*_T - S_T \geq 0$. □

Let V be the lookback option. We can write

$$V = V_{B^T}(t) \cdot B^T = V_S(t) \cdot S = V_{M*}(t) \cdot M^*.$$

In the Markovian case, we can also write

$$V = u^T\left(t, S_{B^T}(t), M^*_{B^T}(t)\right) \cdot B^T = u^S\left(t, B^T_S(t), M^*_S(t)\right) \cdot S$$
$$= u^*\left(t, S_{M^*}(t), B^T_{M^*}(t)\right) \cdot M^*,$$

giving us the following relationships between u^T, u^S, and u^* in terms of a perspective mapping:

$$u^T(t,x,y) = u^S(t, \tfrac{1}{x}, \tfrac{y}{x}) \cdot x, \quad u^S(t,x,y) = u^T(t, \tfrac{1}{x}, \tfrac{y}{x}) \cdot x, \tag{6.27}$$

$$u^T(t,x,y) = u^*(t, \tfrac{x}{y}, \tfrac{1}{y}) \cdot y, \quad u^*(t,x,y) = u^T(t, \tfrac{x}{y}, \tfrac{1}{y}) \cdot y, \tag{6.28}$$

$$u^S(t,x,y) = u^*(t, \tfrac{1}{y}, \tfrac{x}{y}) \cdot y, \quad u^*(t,x,y) = u^S(t, \tfrac{y}{x}, \tfrac{1}{x}) \cdot x. \tag{6.29}$$

In this situation, we have only two no-arbitrage assets: S and B^T. The asset M^* is an arbitrage asset. The hedging of the lookback option must be done in no-arbitrage assets S and B^T only, and no position can be taken in the asset M^*. The First Fundamental Theorem of Asset Pricing gives just two possible martingale measures that can be used for pricing: $\mathbb{P}^T$ and $\mathbb{P}^S$. The corresponding price functions u^T and u^S take the following forms:

$$u^T(t,x,y) = \mathbb{E}^T\left[V_{B^T}(T) | S_{B^T}(t) = x, M^*_{B^T}(t) = y\right]$$
$$= \mathbb{E}^T\left[f^T\left(S_{B^T}(T), M^*_{B^T}(T)\right) | S_{B^T}(T) = x, M^*_{B^T}(t) = y\right], \tag{6.30}$$

$$u^S(t,x,y) = \mathbb{E}^S\left[V_S(T) | B^T_S(t) = x, M^*_S(t) = y\right]$$
$$= \mathbb{E}^S\left[f^S\left(B^T_S(T), M^*_S(T)\right) | B^T_S(t) = x, M^*_S(t) = y\right]. \tag{6.31}$$

The price function u^* does not have a stochastic representation with respect to $\mathbb{P}^*$ since no such measure exists. But we can still compute the price of the contract with respect to M^* chosen as a numeraire using the perspective mapping:

$$u^*(t,x,y) = u^T(t, \tfrac{x}{y}, \tfrac{1}{y}) \cdot y = u^S(t, \tfrac{y}{x}, \tfrac{1}{x}) \cdot x. \tag{6.32}$$

Let us again assume a geometric Brownian motion model with

$$dS_{B^T}(t) = \sigma S_{B^T}(t) dW^T(t),$$

and

$$dB^T_S(t) = \sigma B^T_S(t) dW^S(t).$$

The evolution of the price of the maximum asset M^* is given by

$$\begin{aligned} dM^*_{B^T}(t) &= d\left(M^*_\$(t)\cdot \$_{B^T}(t)\right) \\ &= \$_{B^T}(t)\cdot dM^*_\$(t) + M^*_\$(t)\cdot d\$_{B^T}(t) \\ &= \$_{B^T}(t)\cdot d\left[\max_{0\le s\le t} S_\$(t)\right] - rM^*_\$(t)\$_{B^T}dt \\ &= \$_{B^T}(t)\cdot d\left[\max_{0\le s\le t} S_\$(t)\right] - rM^*_{B^T}(t)dt, \end{aligned}$$

or by

$$\begin{aligned} dM^*_S(t) &= d\left[M^*_{B^T}(t)\cdot B^T_S(t)\right] \\ &= M^*_{B^T}(t)\cdot dB^T_S(t) + B^T_S(t)\cdot dM^*_{B^T}(t) \\ &= M^*_{B^T}(t)\cdot \sigma B^T_S(t)dW^S(t) \\ &\quad + B^T_S(t)\cdot\left[\$_{B^T}(t)\cdot d\left[\max_{0\le s\le t} S_\$(t)\right] - rM^*_{B^T}(t)dt\right] \\ &= \sigma M^*_S(t)dW^S(t) + \$_S(t)\cdot d\left[\max_{0\le s\le t} S_\$(t)\right] - rM^*_S(t)dt. \end{aligned}$$

THEOREM 6.3

The price function

$$u^T(t,x,y) = \mathbb{E}^T\left[f^T\left(S_{B^T}(T), M^*_{B^T}(T)\right)|S_{B^T}(t)=x, M^*_{B^T}(t)=y\right],$$

satisfies the partial differential equation

$$u^T_t(t,x,y) - ryu^T_y(t,x,y) + \tfrac{1}{2}\sigma^2x^2u^T_{xx}(t,x,y) = 0, \tag{6.33}$$

with the boundary condition

$$u^T_y(t,x,x) = 0, \tag{6.34}$$

and the terminal condition

$$u^T(T,x,y) = f^T(x,y). \tag{6.35}$$

The price function

$$u^S(t,x,y) = \mathbb{E}^S\left[f^S\left(B^T_S(T), M^*_S(T)\right)|B^T_S(t)=x, M^*_S(t)=y\right],$$

satisfies the partial differential equation

$$\begin{aligned} &u^S_t(t,x,y) - ryu^S_y(t,x,y) \\ &\quad + \tfrac{1}{2}\sigma^2\left(x^2u^S_{xx}(t,x,y) + 2xyu^S_{xy}(t,x,y) + y^2u^S_{yy}(t,x,y)\right) = 0, \end{aligned} \tag{6.36}$$

for $y \geq 1$ with the boundary condition

$$u_y^S(t, x, 1) = 0, \tag{6.37}$$

and the terminal condition

$$u^S(T, x, y) = f^S(x, y). \tag{6.38}$$

The price function

$$u^*(t, x, y) = u^T(t, \tfrac{x}{y}, \tfrac{1}{y}) \cdot y = u^S(t, \tfrac{y}{x}, \tfrac{1}{x}) \cdot x.$$

satisfies the partial differential equation

$$\begin{aligned} -ru^*(t, x, y) + u_t^*(t, x, y) + rxu_x^*(t, x, y) \\ + ryu_y^*(t, x, y) + \tfrac{1}{2}\sigma^2 x^2 u_{xx}^*(t, x, y) = 0, \end{aligned} \tag{6.39}$$

for $x \leq 1$, with the boundary condition

$$u^*(t, 1, y) - u_x^*(t, 1, y) - \tfrac{1}{x} u_y^*(t, 1, y) = 0, \tag{6.40}$$

and the terminal condition

$$u^*(T, x, y) = f^*(x, y). \tag{6.41}$$

PROOF The process $u^T(t, S_{B^T}(t), M_{B^T}^*(t))$ is a $\mathbb{P}^T$ martingale. From Ito's formula, we have

$$\begin{aligned} du^T(t, S_{B^T}(t), M_{B^T}^*(t)) &= u_t^T(t, S_{B^T}(t), M_{B^T}^*(t))dt \\ &+ u_x^T(t, S_{B^T}(t), M_{B^T}^*(t))dS_{B^T}(t) + u_y^T(t, S_{B^T}(t), M_{B^T}^*(t))dM_{B^T}^*(t) \\ &+ \tfrac{1}{2}u_{xx}^T(t, S_{B^T}(t), M_{B^T}^*(t))(dS_{B^T}(t))^2 = \Big[u_t^T(t, S_{B^T}(t), M_{B^T}^*(t)) \\ &- rM_{B^T}^*(t)u_y^T(t, S_{B^T}(t), M_{B^T}^*(t)) + \tfrac{1}{2}\sigma^2 (S_{B^T}(t))^2 u_{xx}^T(t, S_{B^T}(t), M_{B^T}^*(t))\Big] dt \\ &+ u_x^T(t, S_{B^T}(t), M_{B^T}^*(t))dS_{B^T}(t) \\ &+ \$_{B^T}(t)u_y^T(t, S_{B^T}(t), M_{B^T}^*(t)) \cdot d\left[\max_{0 \leq s \leq t} S_\$(t)\right]. \end{aligned}$$

The martingale part in the above evolution corresponds to

$$u_x^T(t, S_{B^T}(t), M_{B^T}^*(t))dS_{B^T}(t),$$

which also gives us the hedging position in the stock S

$$\Delta^S(t) = u_x^T(t, S_{B^T}(t), M_{B^T}^*(t)).$$

The dt term must also vanish, giving us the partial differential equation

$$u_t^T(t,x,y) - ryu_y^T(t,x,y) + \tfrac{1}{2}\sigma^2 x^2 u_{xx}^T(t,x,y) = 0.$$

The part that corresponds to the term

$$\$_{B^T}(t) u_y^T(t, S_{B^T}(t), M_{B^T}^*(t)) \cdot d\left[\max_{0\le s\le t} S_\$(t)\right]$$

must also disappear, as it does not correspond to a martingale, and it also does not correspond to a dt term. The $d\left[\max_{0\le s\le t} S_\$(t)\right]$ term is zero with probability one as the probability that the stock price is at its maximum is zero, $\mathbb{P}^T(S_\$(t) = M_\$^*(t)) = 0$. However, $d\left[\max_{0\le s\le t} S_\$(t)\right]$ is not zero when the stock price is at its maximum. In order to have

$$\$_{B^T}(t) u_y^T(t, S_{B^T}(t), M_{B^T}^*(t)) \cdot d\left[\max_{0\le s\le t} S_\$(t)\right] = 0$$

at all times, $u_y^T(t, S_{B^T}(t), M_{B^T}^*(t))$ must be zero when $S_\$(t) = M_\$^*(t)$, or equivalently, when $S_{B^T}(t) = M_{B^T}^*(t)$. This implies the boundary condition

$$u_y^T(t,x,x) = 0.$$

The partial differential equation for u^S follows from the relationship

$$u^T(t,x,y) = u^S(t, \tfrac{1}{x}, \tfrac{y}{x}) \cdot x,$$

and similarly, the partial differential equation for u^* follows from the relationship

$$u^T(t,x,y) = u^*(t, \tfrac{x}{y}, \tfrac{1}{y}) \cdot y.$$

$\square$

THEOREM 6.4
The hedging portfolio P_t of the lookback option satisfies each of the following equivalent relationships:

$$\begin{aligned} P_t = &\left[u_x^T\left(t, S_{B^T}(t), M_{B^T}^*(t)\right)\right] \cdot S \\ &+\left[u^T\left(t, S_{B^T}(t), M_{B^T}^*(t)\right) - S_{B^T}(t) \cdot u_x^T\left(t, S_{B^T}(t), M_{B^T}^*(t)\right)\right] \cdot B^T, \end{aligned} \tag{6.42}$$

$$\begin{aligned} P_t = &\Big[u^S\left(t, B_S^T(t), M_S^*(t)\right) - B_S^T(t) \cdot u_x^S\left(t, B_S^T(t), M_S^*(t)\right) \\ &\qquad -M_S^*(t) \cdot u_y^S\left(t, B_S^T(t), M_S^*(t)\right)\Big] \cdot S \\ &+\left[u_x^S\left(t, B_S^T(t), M_S^*(t)\right) + M_{B^T}^*(t) \cdot u_y^S\left(t, B_S^T(t), M_S^*(t)\right)\right] \cdot B^T, \end{aligned} \tag{6.43}$$

$$
\begin{aligned}
P_t = & \left[u_x^*\left(t, S_{M^*}(t), B_{M^*}^T(t)\right)\right] \cdot S \\
& + \Big[M_{B^T}^*(t) \cdot u^*\left(t, S_{M^*}(t), B_{M^*}^T(t)\right) \\
& \qquad - S_{B^T}(t) \cdot u_x^*\left(t, S_{M^*}(t), B_{M^*}^T(t)\right)\Big] \cdot B^T .
\end{aligned} \tag{6.44}
$$

PROOF The hedging position $\Delta^S(t)$ in terms of the price function u^T was already determined in the proof of Theorem 6.3. The hedging position $\Delta^T(t)$ satisfies

$$P_t = u^T\left(t, S_{B^T}(t), M_{B^T}^*(t)\right) \cdot B^T = \Delta^S(t) \cdot S + \Delta^T(t) \cdot B^T,$$

and thus

$$\Delta^T(t) = u^T\left(t, S_{B^T}(t), M_{B^T}^*(t)\right) - \Delta^S(t) \cdot S_{B^T}(t).$$

The hedging position in terms of the function u^S follows from the relationship

$$u^T(t, x, y) = u^S(t, \tfrac{1}{x}, \tfrac{y}{x}) \cdot x,$$

and the hedging position in terms of the function u^* follows from the relationship

$$u^T(t, x, y) = u^*(t, \tfrac{x}{y}, \tfrac{1}{y}) \cdot y.$$

□

Note that hedging is done only in the assets S and B^T, and no position is taken in the arbitrage asset M^*.

REMARK 6.4 Reduction of the pricing equations
When the payoff of the lookback option depends only on the assets M^* and S, it is possible to obtain simpler pricing equations. This is, for instance, the case of a drawdown with a payoff $M_T^* - S_T$. We have seen that it corresponds to the payoff functions

$$f^T(x, y) = y - x, \qquad f^S(x, y) = y - 1, \qquad f^*(x, y) = 1 - x.$$

The payoff is a function of two variables when B^T is taken as a reference asset, but only one variable when S or M^* is taken as a reference asset. This makes sense, when we use B^T as a reference asset; we must consider both prices S_{B^T} and $M_{B^T}^*$. But the asset B^T does not enter the contract directly, and thus both S and M^* are more natural numeraires for this problem. When the payoff does not depend on B^T, the pricing problem does not depend on the price S_{B^T}, and thus the partial differential equations for u^S and u^* reduce by one dimension that represents this price. The price function u^S does not depend on the variable x, and thus we get

$$u_t^S(t,y) - ryu_y^S(t,y) + \tfrac{1}{2}\sigma^2 y^2 u_{yy}^S(t,y) = 0, \tag{6.45}$$

for $y \geq 1$ with the boundary condition

$$u_y^S(t,1) = 0, \tag{6.46}$$

and the terminal condition

$$u^S(T,y) = f^S(y). \tag{6.47}$$

The representation of the hedging portfolio simplifies to

$$\begin{aligned} P_t = &\left[u^S\left(t, M_S^*(t)\right) - M_S^*(t) \cdot u_y^S\left(t, M_S^*(t)\right)\right] \cdot S \\ &+ \left[M_{B^T}^*(t) \cdot u_y^S\left(t, M_S^*(t)\right)\right] \cdot B^T. \end{aligned} \tag{6.48}$$

The price function u^* does not depend on the variable y, and thus we get

$$-ru^*(t,x) + u_t^*(t,x) + rxu_x^*(t,x) + \tfrac{1}{2}\sigma^2 x^2 u_{xx}^*(t,y) = 0, \tag{6.49}$$

for $x \leq 1$, with the boundary condition

$$u^*(t,1) - u_x^*(t,1) = 0, \tag{6.50}$$

and the terminal condition

$$u^*(T,x) = f^*(x). \tag{6.51}$$

The hedging portfolio is of the form

$$\begin{aligned} P_t = &\left[u_x^*\left(t, S_{M^*}(t)\right)\right] \cdot S \\ &+ \left[M_{B^T}^*(t) \cdot u^*\left(t, S_{M^*}(t)\right) - S_{B^T}(t) \cdot u_x^*\left(t, S_{M^*}(t)\right)\right] \cdot B^T. \end{aligned} \tag{6.52}$$

□

6.3 Maximum Drawdown

One of the popular portfolio performance measures is the **maximum drawdown.** It is related to the drawdown asset defined as the difference between the maximal asset M^* and the stock S. A large value of the price of the drawdown $M_\$^*(t) - S_\$(t)$ indicates that the price of the asset is far from its running maximum, which is negatively regarded if the stock is a part of an investment portfolio. Drawdown also plays a role in the compensation of hedge fund portfolio managers which is based in part on the performance of the price of the portfolio. A typical hedge fund charges 20% of the returns above the so-called high watermark, which is another name for the price of the maximal asset. Let us illustrate this compensation scheme on the following example. If the initial price of the fund is 100 million, and at the end of the first year the price of the fund is 110 million, the return is 10 million, and the compensation of the hedge fund manager is 2 million. If the price of the fund drops back to 100 million at the end of the year 2, the return will be negative, and the performance compensation will be zero for that year. If the price of the fund will increase to 115 million at the end of the third year, the return for the year will be 15 million, but the compensation is computed only from the return above the price of the maximal asset, which is only 5 million, the difference between 115 million and the previous maximum at 110 million. Thus the performance fee of a portfolio manager is conditional on the situation that the drawdown is zero, and the value of the portfolio reaches a new maximum.

For both investors and hedge fund managers it is important to have the value of the drawdown as small as possible. Since this is not possible to achieve at all times, one can keep the track of the historical value of the largest drawdown. More formally, consider two basic assets: Y which plays the role of the reference asset in the economy, and X which represents the hedge fund portfolio. In particular, we can consider X to be a single primary asset S representing a stock or a stock index. The maximal asset M^* is defined as

$$M_t^* = \left[\max_{0 \leq s \leq t} X_Y(s)\right] \cdot Y_t.$$

The drawdown is defined as an asset

$$M_T^* - X_T,$$

not as a price. For the price, we have three natural numeraires to consider: Y, X, and M^*. The maximal drawdown can be defined in three ways:

$$D_T^Y = \left[\max_{0 \leq t \leq T} (M_Y^*(t) - X_Y(t))\right] \cdot Y_T, \tag{6.53}$$

which is known as an absolute maximum drawdown, or

$$D_T^X = \left[\max_{0 \leq t \leq T} (M_X^*(t) - 1)\right] \cdot X_T, \tag{6.54}$$

or

$$D_T^{M^*} = \left[\max_{0 \leq t \leq T} (1 - X_{M^*}(t))\right] \cdot M_T^*, \tag{6.55}$$

which is known as a relative maximum drawdown. For hedge fund managers, the maximum drawdown indicates the largest distance they have been from being able to collect the performance fee. In fact, when the price of the maximum drawdown is large, it may lead to a liquidation of the fund itself.

Since D^Y, D^X and D^{M^*} can be regarded as assets, one can find prices of the contracts that depend on them from the First Fundamental Theorem of Asset Pricing. For instance, we can consider a contract V to deliver the maximum drawdown D^Y at time T. The price of this contract is given by

$$V_Y(t) = \mathbb{E}^Y[D^Y_Y(T)]. \tag{6.56}$$

In general, we can define a contract that depends on the maximum drawdown as a contract that pays off $f^Y(X_Y(T), M^*_Y(T), D^Y_Y(T))$ units of the asset Y, or equivalently using the remaining three assets X, M^* D^Y as alternative reference assets. The price of such a contract now depends on three price processes, $X_Y(T)$, $M^*_Y(T)$, $D^Y_Y(T))$, and thus the most general form of the corresponding partial differential equation has three spatial variables. For special payoff functions, such as for the case of a contract to deliver the maximum drawdown $f^Y(x, y, z) = z$, it is possible to reduce the dimensionality of the problem to two spatial dimensions, and solve the pricing problem numerically.

References and Further Reading

The price of the lookback option was first given in Goldman et al. (1979), and later studied, for instance, by Hobson (1998), Buchen and Konstandatos (2005) or Eberlein and Papapantoleon (2005). The connection between the hedging positions of lookback options and hitting times was presented in Pospisil and Vecer (2010). Maximum drawdown serves as an important performance measure for hedge or investment funds. Magdon-Ismail and Atiya (2004) and Magdon-Ismail et al. (2004) found the distribution of the maximum drawdown of the Brownian motion. The maximum drawdown as an asset was introduced in Vecer (2006) and the corresponding partial differential equation was solved numerically in Pospisil and Vecer (2008).

Exercises

6.1 Find the price and the hedge of the contract V with the payoff

$$V_T = \left[\log\left(\frac{X_Y(T)}{K}\right)^+\right] \cdot X_T.$$

6.2 Consider a contract V to deliver the maximal asset M^* at time T in the situation when $r = 0$. We have seen that its u^Y price is given by Equation (6.7).

(a) Determine the u^S price of the contract V.
Hint: Use perspective mapping from (6.27). Note that the price function u^S depends only on one spatial variable y.

(b) Show that the u^S function satisfies the partial differential equation (6.45) in the reduced form ($r = 0$). Verify the boundary conditions, and determine the hedge using the function u^S. Check that the hedging portfolio agrees with the previously obtained representation in (6.8) and (6.9).

(c) Do the same analysis for the price u^* of the contract V.

6.3 Determine the probability

$$\mathbb{P}_t^T(\tau_{M_\$^*(t)} > T) = \mathbb{P}_t^T(M_\$^*(T) = M_\$^*(t))$$

that appears in Equation (6.23) and verify that it leads to the same representation of the hedge $\Delta^T(t)$ as in (6.22).

Chapter 7

American Options

American options have the same payoff as their European option counterparts but an American option can be settled at any time τ before the maturity time T. American options pay off either $f^Y(X_Y(\tau))$ units of an asset Y, or $f^X(Y_X(\tau))$ units of an asset X at the exercise time $\tau \in [0,T]$. The exercise time is chosen by the holder of the option. When these two payoffs correspond to the same contract, they are related by the perspective mapping

$$f^Y(x) = f^X\left(\tfrac{1}{x}\right) \cdot x, \quad \text{or} \quad f^X(x) = f^Y\left(\tfrac{1}{x}\right) \cdot x. \tag{7.1}$$

We will distinguish two cases: when the underlying assets are no-arbitrage assets, and when one of the underlying assets is an arbitrage asset.

The holder of the American option has at each moment two choices: either exercise the option immediately and obtain $f^Y(X_Y(\tau))$ units of an asset Y (or $f^X(Y_X(\tau))$ units of an asset X), or keep the option and exercise it at some later time. The holder has to compare the value of the immediate exercise (known as the intrinsic value of the option) with the continuation value. The option should be exercised the first time its value coincides with its intrinsic value. This is a situation when the continuation value of the option is not larger than the intrinsic value.

The first section shows that when an American option with a convex payoff function is written on two no-arbitrage assets, the optimal exercise strategy is to wait until the maturity time T, and never exercise the option early. If the option holder wants to liquidate the contract earlier, he should sell it rather than exercise it earlier. However, typical American option contracts have at least one arbitrage asset as an underlying asset, so we focus our attention on this case.

When the contract is a call or a put, we can get relatively tight bounds for its price in terms of the price of its European counterpart. The payoff of the American call or put option cannot be expressed as two Arrow–Debreu securities and priced separately since the optimal exercise time also needs to be considered. The problem is that the resulting Arrow–Debreu securities must be exercised at exactly the same moment, and thus we would not gain any computational advantage by splitting the payoff. When we compare the prices of the American put option, it is within the price of its European put option counterpart plus a term that corresponds to the time value of the strike price. The holder of the American put option has to consider two competing effects: holding it for a longer time and improving the option value because of the convexity of the payoff, but the option holder is on the receiving side of the arbitrage asset (currency) that is deteriorating in time. When the loss of the value of money dominates the convexity factor of the option, it is better to exercise the contract. On the other hand, the two factors (convexity and time value of money) play both in favor to the holder of the American call option, and thus it is never optimal to exercise this option early. The price is the same as its European call option counterpart.

Therefore the most interesting case is the American put option written on the stock and the currency. In general, there is no known analytical solution, and the pricing problem has to be solved numerically. The reason why there is not too much hope for analytical methods is that American put option is technically a contract on three assets: the stock, the currency, and the money market which appears implicitly in hedging the option. Moreover, the American put option has a barrier feature; the option should be exercised as soon as the dollar price of the stock reaches a certain region. One can think of the American put option as a barrier option on three assets (stock, currency, money market), where the barrier is determined by the optimal actions of the option holder.

A closed form solution exists in the case of the perpetual American put option. The reason is that the barrier which determines the exercise region in terms of the dollar price should not depend on time, and thus it must be a constant. This greatly simplifies the pricing problem. We apply power assets to compute the perpetual option price and its hedging portfolio. When the option has a finite maturity time T, we derive the partial differential equation that corresponds to the pricing problem.

7.1 American Options on No-Arbitrage Assets

Let us first consider the situation when both assets X and Y are no-arbitrage assets. It turns out that when the payoff functions f^Y and f^X are convex, the option should never be exercised early ($\tau^* = T$), as the following result suggests. Convexity of the payoff function f^Y implies that the payoff function f^X is convex, and vice versa. The perspective mapping $f^Y(x) = f^X(\frac{1}{x}) \cdot x$ preserves convexity.

THEOREM 7.1
The optimal exercise time τ^* of the American option written on two no-arbitrage assets X and Y that pays off either $f^Y(X_Y(\tau))$ units of an asset Y, or $f^X(Y_X(\tau))$ units of an asset X at the exercise time $\tau \in [0,T]$, where f^Y and f^X are convex functions, is given by

$$\tau^* = T.$$

PROOF This is a result that follows from Jensen's inequality that states

$$\mathbb{E}[f(X)] \geq f(\mathbb{E}[X])$$

for a random variable X and a convex function f. For any time t, the value of the European option V with the same payoff function as its American counterpart can be written as

$$V_Y(t) = \mathbb{E}_t^Y\left[f\left(X_Y(T)\right)\right].$$

But according to Jensen's inequality, we also have

$$\mathbb{E}_t^Y[f^Y(X_Y(T))] \geq f^Y\left(\mathbb{E}_t^Y[X_Y(T)]\right) = f^Y(X_Y(t))$$

Thus we conclude that at any time $t \leq T$, the value of the European option V_t dominates the intrinsic value of the American option $f^Y(X_Y(t))$, and thus it is never optimal to exercise the American option early for convex payoff functions f^Y and f^X. ▯

REMARK 7.1 Convexity of the payoff functions f^Y and f^X is necessary for Theorem 7.1 to hold. Consider for instance the payoff function $f^Y(x) = \mathbb{I}(x \geq K)$, so that the holder of the option can receive an asset Y if the price $X_Y(t)$ is greater or equal to K. The indicator function is not convex, and clearly the option should be exercised the first time the price reaches K, even when both underlying assets are no-arbitrage assets. The corresponding payoff function f^X is given by $f^X(x) = f^Y(\frac{1}{x}) \cdot x = \frac{1}{K} \cdot \mathbb{I}(x = \frac{1}{K})$. ▯

Theorem 7.1 implies that American puts and calls should never be exercised early when both underlying assets are no-arbitrage assets. Call options have a payoff function $f(x) = (x-K)^+$, put options have a payoff function $f(x) = (K-x)^+$, and both payoff functions are convex. In particular, an American put option with a payoff $(K \cdot Y_\tau - X_\tau)^+$ should never be exercised before maturity. This may be slightly surprising because an American put option written on a stock S and the dollar \$ with a payoff $(K \cdot \$_\tau - S_\tau)^+$ can be optimally exercised before the maturity of the contract, but that does not contradict our result since currencies are arbitrage assets. However, if the payoff of the American put option written on the stock and the dollar is slightly modified, when the dollar is replaced with a no-arbitrage asset B^T so that the payoff becomes $(K \cdot B^T_\tau - S_\tau)^+$, such an option should not be exercised early. We will use this fact to obtain tight bounds on the price of American options written on one arbitrage asset.

7.2 American Call and Puts on Arbitrage Assets

The most traded types of American options involve typically one arbitrage asset represented by the dollar \$. For instance, an American put option written on a stock S and the dollar \$ pays off $(K \cdot \$_\tau - S_\tau)^+$ at the exercise time τ. Since the hedging of the contract must be done in no-arbitrage assets, we also need to include a no-arbitrage proxy asset for the dollar \$, such as a bond B^T that matures at the expiration time of the American option T.

The following theorem shows that an American call option written on a stock S and on a dollar \$ should be exercised at maturity, and thus the price of the American call option coincides with the corresponding European call option. The price of the American put option written on a stock S and on \$ is constrained by relatively tight bounds. We denote by V^{AC}, V^{AP}, V^{EC}, and V^{EP} the American call option, American put option, European call option, and European put option, respectively. The first two arguments represent the assets that settle the contract; the third argument is the maturity of the contract. Note that $V^{EP}(S, K \cdot B^T, T) = V^{EP}(S, K \cdot \$, T)$ since the arbitrage asset \$ can be replaced by a corresponding contract to deliver B^T.

THEOREM 7.2

The price of an American put option written on a stock S and on the dollar \$ is bounded by

$$V^{EP}_t(S, K \cdot B^T, T) \leq V^{AP}_t(S, K \cdot \$, T) \leq V^{EP}_t(S, K \cdot B^T, T) + (\$_t - B^T_t) \cdot K. \tag{7.2}$$

The price of an American call option written on a stock S and on the dollar \$ is identical to its European call option counterpart

$$V^{AC}_t(S, K \cdot \$, T) = V^{EC}_t(S, K \cdot B^T, T). \tag{7.3}$$

PROOF The left hand side of the inequality for the American put option is trivial. Let us show the right hand side of the inequality. Since

$$(K \cdot \$_t - S_t)^+ \leq (K \cdot B^T_t - S_t)^+ + (\$_t - B^T_t) \cdot K,$$

we also have that

$$\begin{aligned} V_t^{AP}(S, K \cdot \$, T) &\leq V_t^{AP}(S, K \cdot B^T, T) + (\$_t - B_t^T) \cdot K \\ &= V_t^{EP}(S, K \cdot B^T, T) + (\$_t - B_t^T) \cdot K. \end{aligned}$$

Similarly for the American call option we have

$$V_t^{EC}(S, K \cdot B^T, T) \leq V_t^{AC}(S, K \cdot \$, T),$$

so the price of the American call option dominates the price of the European call option. On the other hand we have

$$(S_t - K \cdot \$_t)^+ \leq (S_t - K \cdot B_t^T)^+ + (B_t^T - \$_t)^+ \cdot K = (S_t - K \cdot B_t^T)^+.$$

Therefore

$$V_t^{AC}(S, K \cdot \$, T) \leq V_t^{EC}(S, K \cdot B^T, T),$$

and thus the price of the American call option written on the stock S and on the dollar \$ coincides with the price of its European call option counterpart. □

COROLLARY 7.1
When $B_t^T \equiv 1 \cdot \$_t$ (or in other words when $r = 0$), we have that

$$V_t^{AP}(S, K \cdot \$, T) = V_t^{EP}(S, K \cdot B^T, T),$$

and thus it is not optimal to exercise the American put option early in the situation when the interest rate is zero.

7.3 Perpetual American Put

Consider a contract V with a payoff

$$V_\tau = (K \cdot \$_\tau - S_\tau)^+ = (K - S_\$(\tau))^+ \cdot \$_\tau, \tag{7.4}$$

where τ is a stopping time with no upper bound. This contract is known as a **perpetual American put** since it has no expiration. Obviously, the optimal exercise strategy should be independent of time; the holder should decide when to exercise based only on the price level of $S_\$$. Time is not an issue; there is always an infinite amount of time left.

Since the perpetual American option has an infinite time horizon, the exercise strategy is independent of time. The only strategy that is independent of

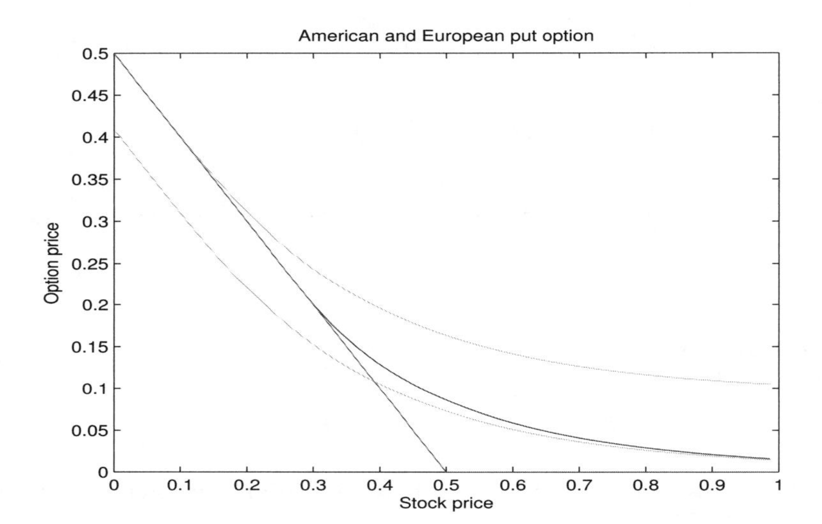

FIGURE 7.1: The price $V_{\$}^{EP}(0)$ of a European put option with a payoff $(\$_T - K \cdot S_T)^+$ (bottom curve), the price $V_{\$}^{AP}(0)$ of the corresponding American put option counterpart (middle curve), and the price of the European put option $V_{\$}^{EP}(0)$ plus $(1 - B_{\$}^{T}(0)) \cdot K$ (top curve), together with the option payoff $(1 - K \cdot S_{\$})^+$ as a function of the stock price $S_{\$}$. The parameters are $r = 0.02$, $K = \frac{1}{2}$, $\sigma = 0.2$, $T = 10$.

time is a stationary strategy. Let us first compute the price of the perpetual American put that is associated with the exercise strategy

$$\tau_L = \inf\{t \geq 0 : S_\$(t) \leq L\} \tag{7.5}$$

for a given value of $L < K$. Next we will find the value of L^* that maximizes the price of the perpetual American option among different choices of the exercise levels L.

In order to compute the price of the perpetual American put using the First Fundamental Theorem of Asset Pricing, we need to express the the contract in terms of no arbitrage assets. Since the contract has an infinite time horizon, there is no bond with an infinite maturity that can be used as a no-arbitrage proxy to a dollar \$, and thus we should use a money market M instead. The relationship between dollars and the money market is

$$M_t = e^{rt} \cdot \$_t.$$

In particular, the stopping time τ_L can be written as

$$\tau_L = \inf\{t \geq 0 : S_M(t) \leq L \cdot e^{-rt}\},$$

and the payoff of the option can be expressed as

$$(K - S_\$(\tau_L))^+ \cdot \$_\tau = (K - L)^+ \cdot e^{-r\tau_L} \cdot M_\tau.$$

The price can be computed from the Optional Sampling Theorem (Theorem A.1) as

$$V_M(0) = \mathbb{E}^M[(K - L) \cdot e^{-r\tau_L}] = (K - L) \cdot \mathbb{E}^M[e^{-r\tau_L}].$$

The Optional Sampling Theorem states that under certain conditions, the martingale keeps its expected value even when it is stopped at a stopping time τ. However, computation of $\mathbb{E}^M[e^{-r\tau_L}]$ requires one to determine the distribution of the hitting time τ_L, which is a nontrivial task. A more elegant solution uses properties of a **power asset**. The power asset is analogous to the power option that was studied in the previous chapters. The fact that the perpetual option has an infinite horizon prevents us from defining the power option as a contract with a particular payoff at some finite time T. Instead, one should define the initial value of a power asset as

$$R_M^\alpha(0) = \left[\tfrac{S_M(0)}{L}\right]^\alpha, \tag{7.6}$$

and assume that

$$dR_M^\alpha(t) = \alpha\sigma R_M^\alpha(t) dW^M(t), \tag{7.7}$$

which is the same evolution as (5.13). The price of the perpetual American put can be computed by finding α such that the assets R^α and $\frac{1}{L}$ units of S agree on the exercise boundary, in which case

$$R_M^\alpha(\tau_L) = \frac{S_M(\tau_L)}{L} = e^{-r\tau_L},$$

and the value of the perpetual American put would follow from

$$\mathbb{E}^M[e^{-r\tau_L}] = \mathbb{E}^M[R_M^\alpha(\tau_L)] = R_M^\alpha(0) = \left[\tfrac{S_M(0)}{L}\right]^\alpha.$$

Let us find the corresponding α. Since $R_M^\alpha(t)$ has a geometric Brownian motion evolution, it can be written as

$$\begin{aligned} R_M^\alpha(t) &= R_M^\alpha(0)\cdot \exp(\alpha\sigma W^M(t) - \tfrac{1}{2}\alpha^2\sigma^2 t) \\ &= \left[\tfrac{S_M(0)}{L}\right]^\alpha \cdot \exp(\alpha\sigma W^M(t) - \tfrac{1}{2}\alpha^2\sigma^2 t) \\ &= \left[\tfrac{S_M(t)}{L}\right]^\alpha \cdot \exp(-\tfrac{1}{2}\alpha(\alpha-1)\sigma^2 t). \end{aligned} \tag{7.8}$$

The last equality follows from

$$\left[\tfrac{S_M(t)}{L}\right]^\alpha = \left[\tfrac{S_M(0)}{L}\right]^\alpha \cdot \exp(\alpha\sigma W^M(t) - \tfrac{1}{2}\alpha\sigma^2 t).$$

Should $R_M^\alpha(\tau_L) = \frac{S_M(\tau_L)}{L}$, we must have

$$R_M^\alpha(\tau_L) = \left[\tfrac{S_M(\tau_L)}{L}\right]^\alpha \cdot \exp(-\tfrac{1}{2}\alpha(\alpha-1)\sigma^2\tau_L) = \tfrac{S_M(\tau_L)}{L}. \tag{7.9}$$

Since $\frac{S_M(\tau_L)}{L} = e^{-r\tau_L}$, the above equality is satisfied when

$$\exp((\alpha-1)(-r-\tfrac{1}{2}\alpha\sigma^2)\tau_L) = 1, \tag{7.10}$$

which is true either when $\alpha = 1$, or when $\alpha = -\frac{2r}{\sigma^2}$. Only the second solution represents a suitable power asset with the price given by

$$R_M^{-\frac{2r}{\sigma^2}}(t) = \left[\tfrac{S_M(t)}{L}\right]^{-\frac{2r}{\sigma^2}} \cdot \exp(-r(1+\tfrac{2r}{\sigma^2})t). \tag{7.11}$$

Let $S_M(t) \geq Le^{-rt}$. The value of the perpetual American put with the exercise boundary L at time t is given by

$$\begin{aligned} V_M(t) = (K-L)\cdot\mathbb{E}^M[e^{-r\tau_L}] &= (K-L)\cdot\mathbb{E}^M\left[R_M^{-\frac{2r}{\sigma^2}}(\tau_L)|S_M(t)\right] \\ &= (K-L)\cdot\left[\tfrac{S_M(t)}{L}\right]^{-\frac{2r}{\sigma^2}} \cdot \exp(-r(1+\tfrac{2r}{\sigma^2})t), \end{aligned} \tag{7.12}$$

which corresponds to the price function $u^M(t,x)$. Thus

$$u^M(t,x) = (K-L)\cdot\left(\tfrac{x}{L}\right)^{-\frac{2r}{\sigma^2}}\cdot\exp(-r(1+\tfrac{2r}{\sigma^2})t) = (K-L)\cdot e^{-rt}\cdot\left(\tfrac{e^{rt}x}{L}\right)^{-\frac{2r}{\sigma^2}}. \tag{7.13}$$

This is valid for $x \geq Le^{-rt}$. When $x \leq Le^{-rt}$, the perpetual American option should be exercised immediately, collecting K units of a dollar $\$_t$, and shorting a unit of a stock S_t. Thus $V_t = Ke^{-rt}\cdot M_t - S_t$, or equivalently,

$$V_M(t) = e^{-rt}K - S_M(t).$$

In terms of the price function u^M, we have

$$u^M(t,x) = e^{-rt}K - x. \tag{7.14}$$

The price of the perpetual American put in dollar terms is given by

$$V_\$(t) = V_M(t)\cdot M_\$(t) = (K-L)\cdot\left[\tfrac{S_\$(t)}{L}\right]^{-\frac{2r}{\sigma^2}}. \tag{7.15}$$

If we define the price function $v^\$$ as $v^\$(t,S_\$(t)) = V_\$(t)$, the prices of $v^\$$ and u^M are related by

$$v^\$(t,x) = e^{rt}\cdot u^M(t,e^{-rt}x). \tag{7.16}$$

Thus

$$v^\$(t,x) = (K-L)\cdot\left(\tfrac{x}{L}\right)^{-\frac{2r}{\sigma^2}} \tag{7.17}$$

when $x \geq L$. When $x \leq L$, we have

$$v^\$(t,x) = K - x. \tag{7.18}$$

The optimal choice of L^* maximizes the function

$$h(L) = u^M(0,x) = v^\$(t,x) = (K-L)\cdot\left(\tfrac{x}{L}\right)^{-\frac{2r}{\sigma^2}}.$$

It is straightforward to show that the function $h(L)$ is maximized for the value

$$L^* = \frac{2r}{2r+\sigma^2}\cdot K, \tag{7.19}$$

which can be seen from $h'(L^*) = 0$, and $h''(L^*) < 0$.

7.4 Partial Differential Equation Approach

This section studies the American stock option in a geometric Brownian motion model that pays off either $f^\$\,(S_\$(\tau))$ units of the dollar \$, or $f^S\,(\$_S(\tau))$

units of a stock S at the exercise time $\tau \in [0,T]$. Let us assume that the dynamics of the bond price with respect to the dollar are given by

$$dB_{\$}^{T}(t) = rB_{\$}^{T}(t)dt, \tag{7.20}$$

implying the dynamics of the stock price with respect to the dollar are

$$\begin{aligned} dS_{\$}(t) &= d[S_{B^T}(t) \cdot B_{\$}^{T}(t)] \\ &= S_{B^T}(t) \cdot dB_{\$}^{T}(t) + B_{\$}^{T}(t) \cdot dS_{B^T}(t) \\ &= rS_{\$}(t)dt + \sigma S_{\$}(t)dW^T(t). \end{aligned}$$

We can price the American stock option contract with respect to the dollar \$, or with respect to the stock S. When the dollar \$ is chosen as a reference asset, we can define the dollar price of the American option contract as:

$$v^{\$}(t,x) = \max_{\tau} \mathbb{E}_t^T \left[e^{-r(\tau-t)} V_{\$}(\tau) | S_{\$}(t) = x \right].$$

Then $e^{-rt}v^{\$}(t, S_{\$}(t))$ is a $\mathbb{P}^T$ martingale. Using Ito's formula, we get

$$\begin{aligned} d\left(e^{-rt}v^{\$}\left(t, S_{\$}(t)\right)\right) &= -re^{-rt}v^{\$}\left(t, S_{\$}(t)\right)dt + e^{-rt}dv^{\$}\left(t, S_{\$}(t)\right) \\ &= -re^{-rt}v^{\$}\left(t, S_{\$}(t)\right)dt + e^{-rt}\Big[v_t^{\$}\left(t, S_{\$}(t)\right)dt \\ &\quad + v_x^{\$}\left(t, S_{\$}(t)\right)dS_{\$}(t) + \tfrac{1}{2}v_{xx}^{\$}\left(t, S_{\$}(t)\right)d^2S_{\$}(t)\Big] \\ &= e^{-rt}\Big[-rv^{\$}\left(t, S_{\$}(t)\right) + v_t^{\$}\left(t, S_{\$}(t)\right) \\ &\quad + rS_{\$}(t)v_x^{\$}\left(t, S_{\$}(t)\right) + \tfrac{1}{2}\sigma^2 S_{\$}(t)^2 v\$_{xx}\left(t, S_{\$}(t)\right)\Big]dt \\ &\quad + \sigma S_{\$}(t)v_x^{\$}\left(t, S_{\$}(t)\right)dW^T(t). \end{aligned}$$

Since the dt term must be zero, we get the following partial differential equation:

$$-rv^{\$}(t,x) + v_t^{\$}(t,x) + rxv_x^{\$}(t,x) + \tfrac{1}{2}\sigma^2 x^2 v_{xx}^{\$}(t,x) = 0.$$

This partial differential equation is valid as long as the option should not be exercised, which is true when the option value exceeds its intrinsic value $v^{\$}(t,x) > f^{\$}(x)$. When the option should be exercised, we have that the option value coincides with its intrinsic value:

$$u^{\$}(t,x) = f^{\$}(x).$$

In case that the option is not exercised when it should be, the contract starts to lose the value and the dt term becomes negative. In this situation, the holder of the American option is creating an arbitrage opportunity at his own expense for the benefit of the seller of the option. The First Fundamental Theorem of Asset Pricing states that when there is no arbitrage, the prices

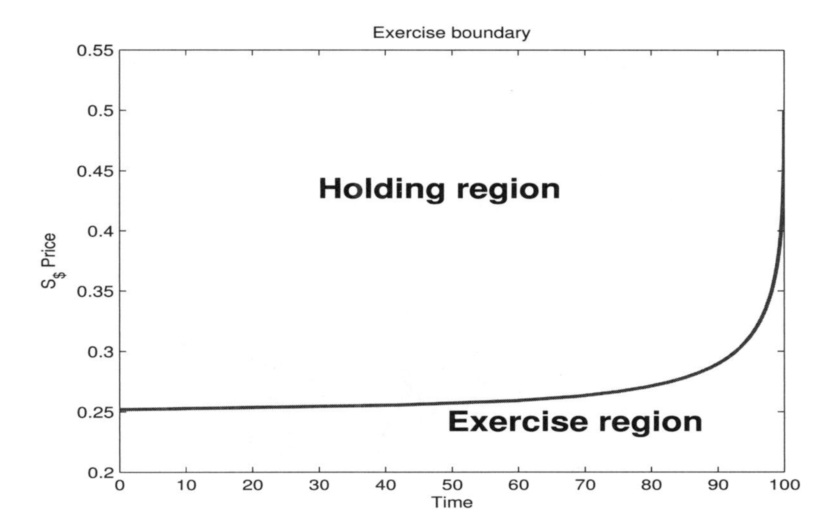

FIGURE 7.2: Optimal exercise boundary for the American put option with a payoff $(K \cdot \$_\tau - S_\tau)^+$ with parameters $r = 0.02$, $K = \frac{1}{2}$, $\sigma = 0.2$, $T = 100$ as a function of time.

are martingales under the probability measure that is associated with the reference asset. However, the holder of the American option may fail to take an optimal action that leads to the martingale price of the contract. In this case, the price of the contract becomes a **supermartingale**, which is a process that is nonincreasing in expectation:

$$\mathbb{E}_s[\mathcal{S}(t)] \leq \mathcal{S}(s). \tag{7.21}$$

Thus we have

$$-rv^\$(t,x) + v^\$_t(t,x) + rxv^\$_x(t,x) + \tfrac{1}{2}\sigma^2 x^2 v^\$_{xx}(t,x) < 0$$

in the region where the option should be exercised. We conclude that the American option is characterized by linear complementarity conditions:

$$\boxed{-rv^\$(t,x) + v^\$_t(t,x) + rxv^\$_x(t,x) + \tfrac{1}{2}\sigma^2 x^2 v^\$_{xx}(t,x) \leq 0,} \tag{7.22}$$

and

$$\boxed{v^\$(t,x) \geq f^\$(x).} \tag{7.23}$$

The above inequalities do not lead to a closed form solution, and the price of the contract has to be computed numerically.

Figure 7.2 shows the exercise boundary for the American put option with parameters $r = 0.02$, $K = \frac{1}{2}$, $\sigma = 0.2$. The price of the option above

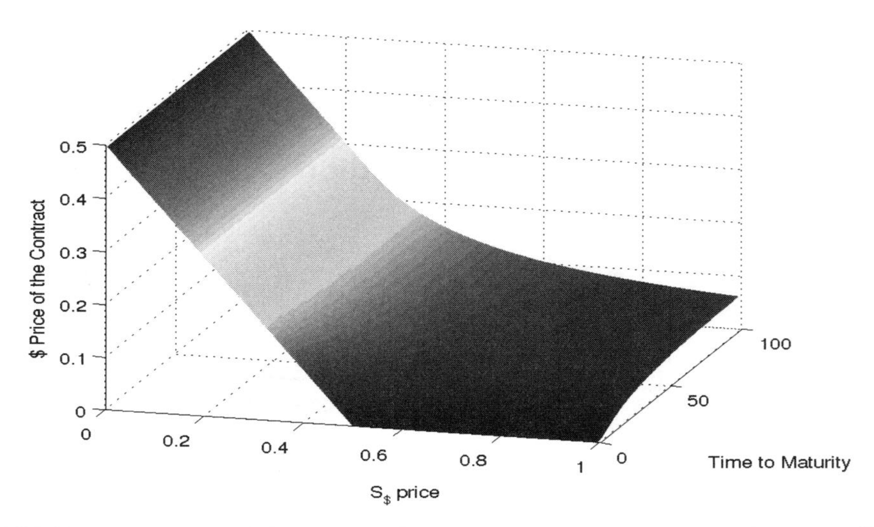

FIGURE 7.3: The dollar price of the American put option with a payoff $(K \cdot \$_\tau - S_\tau)^+$ with parameters $r = 0.02$, $K = \frac{1}{2}$, $\sigma = 0.2$ as a function of the stock price $S_\$$ and time to maturity $T - t$.

the boundary satisfies the Black-Scholes partial differential equation for the function $v^\$$, and the value of the option dominates its intrinsic value. The option should be exercised if the stock price is below the exercise boundary, where the option price agrees with its intrinsic value, and the term $-rv^\$(t,x) + v^\$_t(t,x) + rxv^\$_x(t,x) + \frac{1}{2}\sigma^2 x^2 v^\$_{xx}(t,x)$ is smaller than zero. Note that the exercise boundary flattens for large time to maturity, and the limiting level corresponds to $L^* = \frac{2r}{2r+\sigma^2} \cdot K = \frac{0.04}{0.04+0.04} \cdot \frac{1}{2} = 0.25$, the exercise boundary for the perpetual American put.

Figure 7.3 shows the dollar price of the American put option as a function of the price and time to maturity. For large times to maturity, the price of the contract approaches the price of the perpetual American put option, which is given by $v^\$(t,x) = K - x$ for $x \leq L^* = 0.25$, and $v^\$(t,x) = (K - L^*)\left(\frac{x}{L^*}\right)^{-1} = \frac{1}{16x}$ for $x \geq 0.25$.

Figures 7.4 and 7.5 show the hedging positions in the assets S and M. The hedging positions are similar to the hedging positions of its European option counterpart with the exception that the hedging position in the stock should be set to -1 in the exercise region. When the contract should be exercised, $K \cdot \$_t - S_t = K \cdot M_t - S_t$ should be collected. Thus the hedging becomes trivial (short the stock, and long K units of the money market). The exercise region is clearly visible in both graphs.

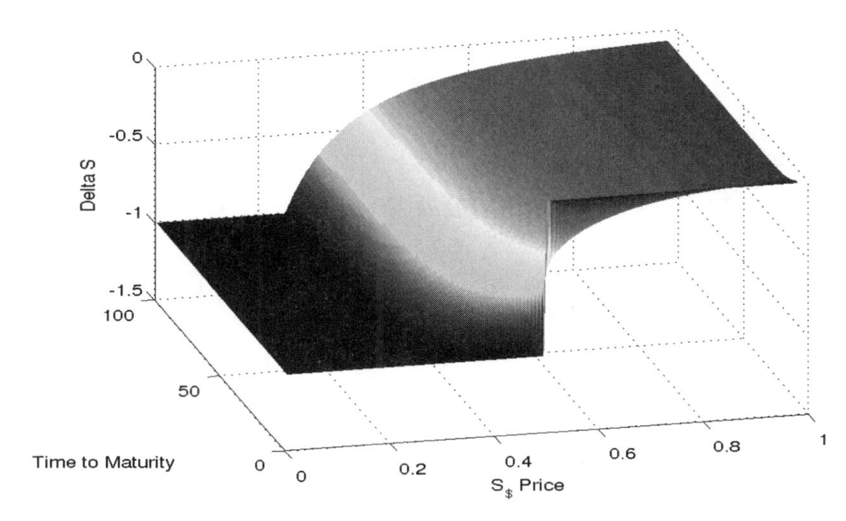

FIGURE 7.4: The hedging position in the stock S for the American put option contract with a payoff $(K \cdot \$_\tau - S_\tau)^+$ with parameters $r = 0.02$, $K = \frac{1}{2}$, $\sigma = 0.2$ as a function of the stock price $S_\$$ and time to maturity $T - t$.

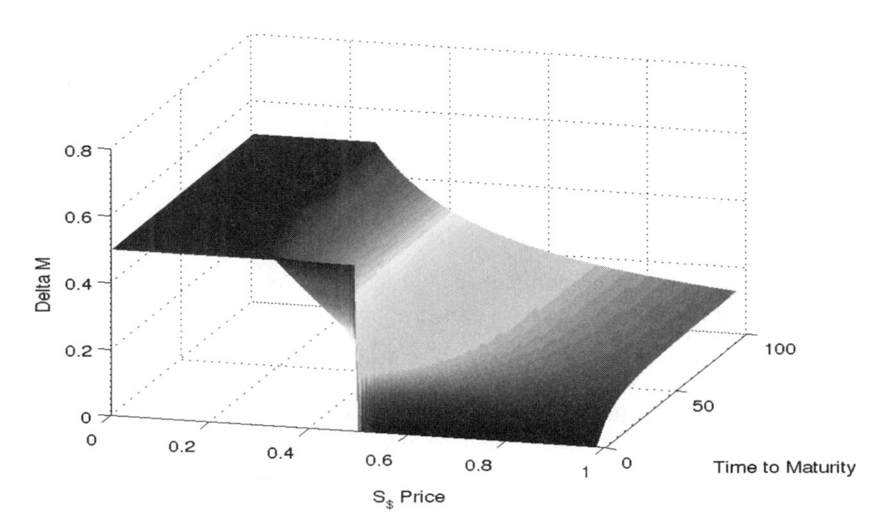

FIGURE 7.5: The hedging position in the money market M for the American put option contract with a payoff $(K \cdot \$_\tau - S_\tau)^+$ with parameters $r = 0.02$, $K = \frac{1}{2}$, $\sigma = 0.2$ as a function of the stock price $S_\$$ and time to maturity $T - t$.

REMARK 7.2 Stock as a reference asset
When the stock S is chosen as a reference asset, we can define the price of the American option with respect to the stock as

$$v^S(t,x) = \max_{\tau} \mathbb{E}_t^S \left[V_S(\tau) | \$_S(t) = x \right].$$

We have that $v^S(t, \$_S(t))$ is a $\mathbb{P}^S$ martingale. From Ito's formula we have

$$\begin{aligned} dv^S\left(t, \$_S(t)\right) &= v_t^S\left(t, \$_S(t)\right) dt + +v_x^S\left(t, \$_S(t)\right) d\$_S(t) \\ &\quad + \tfrac{1}{2} v_{xx}^S\left(t, \$_S(t)\right) d^2\$_S(t) \\ &= \Big[v_t^S\left(t, S_\$(t)\right) - r\$_S(t) v_x^S\left(t, \$_S(t)\right) \\ &\quad + \tfrac{1}{2}\sigma^2 \$_S(t)^2 v_{xx}^S\left(t, \$_S(t)\right) \Big] dt \\ &\quad + \sigma \$_S(t) v_x^S\left(t, \$_S(t)\right) dW^S(t). \end{aligned}$$

We have used the dynamics of the price of the dollar with respect to the stock

$$\begin{aligned} d\$_S(t) &= d[B_S^T(t) \cdot \$_{B^T}(t)] \\ &= B_S^T(t) \cdot d\$_{B^T}(t) + \$_{B^T}(t) \cdot dB_S^T(t) \\ &= \$_S(t) \left(-rdt + \sigma \cdot dW^S(t) \right). \end{aligned}$$

Since the price of the contract with respect to the stock is a martingale, the dt term must be zero and thus we obtain

$$v_t^S(t,x) - rxv_x^S(t,x) + \tfrac{1}{2}\sigma^2 x^2 v_{xx}^S(t,x) = 0.$$

This is true when the option value exceeds its intrinsic value $v^S(t,x) > f^S(x)$ and the option should not be exercised. This partial differential equation has a slight numerical advantage over the classical Black-Scholes since it contains only 3 terms. In the region where the option should be exercised, the option value coincides with its intrinsic value

$$v^S(t,x) = f^S(x),$$

but keeping the American contract would lead to a loss, and the corresponding dt term would become negative. Thus we have

$$v_t^S(t,x) - rxv_x^S(t,x) + \tfrac{1}{2}\sigma^2 x^2 v_{xx}^S(t,x) \leq 0$$

when the option should be exercised. We conclude that the option price with respect to the reference asset S satisfies the linear complementarity conditions

$$\boxed{v_t^S(t,x) - rxv_x^S(t,x) + \tfrac{1}{2}\sigma^2 x^2 v_{xx}^S(t,x) \geq 0,} \tag{7.24}$$

and

$$\boxed{v^S(t,x) \geq f^S(x).} \tag{7.25}$$

□

References and Further Reading

The price of the perpetual American put was first computed by McKean (1965), who also gave an analytic characterization of the American put price with a finite expiration. Merton (1973) observed that the price of the American call option agrees with its European option counterpart for a no-dividend-paying stock, which in our notation means that the asset X is a no-arbitrage asset. The bounds for the American put option price appeared in Carr et al. (1992). The probabilistic approach for pricing the American put option is given in Bensoussan (1984) and in Karatzas (1988). Longstaff and Schwartz (2001) and Tsitsiklis and Van Roy (2001) suggested the method of the least squares to estimate the continuation value and the optimal exercise boundary for the American put option. See also Broadie and Glasserman (1997) and Rogers (2002). For other references on pricing American options, see for instance, Barone-Adesi and Whaley (1987), Geske and Johnson (1984), Boyle et al. (1997), Jaillet et al. (1990), Haugh and Kogan (2004) Broadie and Detemple (1996), Zhu (2006), Carr (1998), Zvan et al. (1998), Clement et al. (2002), Boyarchenko and Levendorskii (2002), Levendorskii (2004), Pham (1997), and Zhang (1997).

Exercises

7.1 Show that the price $u^M(t,x)$ in (7.13) and (7.14) satisfies the Black–Scholes partial differential equation

$$u_t^M(t,x) + \tfrac{1}{2}\sigma^2 x^2 u_{xx}^M(t,x) = 0$$

when $x \geq Le^{-rt}$. When $x \leq Le^{-rt}$, the price $u^M(t,x)$ satisfies the partial differential inequality

$$u_t^M(t,x) + \tfrac{1}{2}\sigma^2 x^2 u_{xx}^M(t,x) \leq 0.$$

7.2 Determine the hedging portfolio for the perpetual American put option. Hint: $\Delta^S(t) = u_x^M(t,x)$.

7.3 Show that the price $v^\$(t,x)$ in (7.17) and (7.18) satisfies the partial differential equation

$$-rv^\$(t,x) + rxv_x^\$(t,x) + \tfrac{1}{2}\sigma^2 x^2 v_{xx}^\$(t,x) = 0$$

when $x \geq L$. When $x \leq L$, the price $v^\$(t,x)$ satisfies the partial differential inequality

$$-rv^\$(t,x) + rxv_x^\$(t,x) + \tfrac{1}{2}\sigma^2 x^2 v_{xx}^\$(t,x) \leq 0.$$

Chapter 8

Contracts on Three or More Assets: Quantos, Rainbows and "Friends"

Exotic options are contracts that depend on three or more underlying assets. We have already studied lookback options that depend on a stock S, a bond B^T (or equivalently a dollar \$), and on the maximal asset M^*. The maximal asset itself depends on the stock and the bond, so it is not a free asset that exists on its own. This chapter studies contracts on three "full" underlying assets; let us call them X, Y, and Z. There are several variants of these contracts. A **quanto** is a contract that pays off some function of the price of $X_Y(T)$ units of the third asset Z. The underlying assets for quanto contracts are usually currencies, but the definition is not limited to them. For example, a quanto forward pays off $X_Y(T) - K$ units of Z, and a quanto call option pays off $(X_Y(T) - K)^+$ units of Z. The quanto contracts have only one underlying price process, but the option is settled in a "wrong" asset. A contract that depends on three or more assets and two or more price processes is known as a **rainbow option.** The price of a rainbow option thus depends on the joint distribution of two or more price processes, and they are sensitive to the correlation structure of the prices. Some rainbow option contracts are known under a specific name, for instance a call option on the maximum of two assets, etc. We call them "the friends of quantos and rainbows."

An option on three underlying assets can be formally defined as:

DEFINITION 8.1 *An option on three underlying assets is a contract that pays off one of the following:*

- $f^Y(X_Y(T), Z_Y(T))$ *units of an asset* Y,
- $f^X(Y_X(T), Z_X(T))$ *units of an asset* X,
- $f^Z(X_Z(T), Y_Z(T))$ *units of an asset* Z.

Example 8.1

We have seen in Example 3.1 that the contract that pays off the best of two assets is completely symmetric (and dual to itself). The contract on the best of two assets is directly linked to European options via

$$(X_T - K \cdot Y_T)^+ = \max(X_T, K \cdot Y_T) - K \cdot Y_T,$$

and

$$(K \cdot Y_T - X_T)^+ = \max(X_T, K \cdot Y_T) - X_T,$$

and thus most of the traded contracts are related to it. Therefore an analogous contract that pays off the best of three assets

$$\max(X_T, K_1 Y_T, K_2 Z_T)$$

is a natural candidate whose variants would generalize the concept of most traded European options. Indeed, when we consider, for instance, the difference of the best of three assets and one of the assets itself, we get

$$\max(X_T, K_1 Y_T, K_2 Z_T) - K_2 Z_T = (\max(X_T, K_1 Y_T) - K_2 Z_T)^+, \qquad (8.1)$$

which represents the payoff of the call option on the maximum of two assets. □

As we saw earlier for the case of the general European contract, it is possible to settle the payoff in each of the underlying assets. Should the different settlements represent the same contract, the payoff functions must be linked by the **perspective mapping**

$$f^Y(x,y) = f^X(\tfrac{1}{x}, \tfrac{y}{x}) \cdot x, \quad f^X(x,y) = f^Y(\tfrac{1}{x}, \tfrac{y}{x}) \cdot x,$$
$$f^Y(x,y) = f^Z(\tfrac{x}{y}, \tfrac{1}{y}) \cdot y, \quad f^Z(x,y) = f^Y(\tfrac{x}{y}, \tfrac{1}{y}) \cdot y,$$
$$f^X(x,y) = f^Z(\tfrac{1}{y}, \tfrac{x}{y}) \cdot y, \quad f^Z(x,y) = f^X(\tfrac{y}{x}, \tfrac{1}{x}) \cdot x.$$

Example 8.2
Consider a quanto call option that pays off $(X_Y(T) - K)^+ \cdot Z$. In terms of the reference asset Y, the payoff takes the following form

$$(X_Y(T) - K)^+ \cdot Z_Y(T) \cdot Y,$$

representing $(X_Y(T) - K)^+ \cdot Z_Y(T)$ units of the asset Y. Thus the payoff function f^Y is given by

$$f^Y(x,y) = (x - K)^+ \cdot y.$$

Similarly, when the reference asset is chosen to be X, the payoff is of the form

$$(X_Y(T) - K)^+ \cdot Z_X(T) \cdot X$$

which is represented by the payoff function f^X given by

$$f^X(x,y) = (\tfrac{1}{x} - K)^+ \cdot y.$$

Finally, when the reference asset is Z the payoff is simply

$$(X_Y(T) - K)^+ \cdot Z$$

which is represented by the payoff function f^Z given by

$$f^Z(x, y) = (\tfrac{x}{y} - K)^+.$$

□

8.1 Pricing in the Geometric Brownian Motion Model

Let V denote a contract on three assets. We can use the change of numeraire formula to compute the price of V, this time using all three available assets:

$$V = V_Y(t) \cdot Y = V_X(t) \cdot X = V_Z(t) \cdot Z.$$

In a Markovian pricing model, we can represent the prices V_Y, V_X, and V_Z by functions u^Y, u^X, and u^Z:

$$\begin{aligned} V = u^Y(t, X_Y(t), Z_Y(t)) \cdot Y \\ &= u^X(t, Y_X(t), Z_X(t)) \cdot X = u^Z(t, X_Z(t), Y_Z(t)) \cdot Z, \end{aligned}$$

giving us the following relationships known as a perspective mapping between price functions u^Y, u^X, and u^Z:

$$u^Y(t, x, y) = u^X(t, \tfrac{1}{x}, \tfrac{y}{x}) \cdot x, \quad u^X(t, x, y) = u^Y(t, \tfrac{1}{x}, \tfrac{y}{x}) \cdot x, \tag{8.2}$$

$$u^Y(t, x, y) = u^Z(t, \tfrac{x}{y}, \tfrac{1}{y}) \cdot y, \quad u^Z(t, x, y) = u^Y(t, \tfrac{x}{y}, \tfrac{1}{y}) \cdot y, \tag{8.3}$$

$$u^X(t, x, y) = u^Z(t, \tfrac{1}{y}, \tfrac{x}{y}) \cdot y, \quad u^Z(t, x, y) = u^X(t, \tfrac{y}{x}, \tfrac{1}{x}) \cdot x. \tag{8.4}$$

When X, Y, and Z are no-arbitrage assets, we also have a stochastic representation of the prices:

$$\begin{aligned} u^Y(t, x, y) &= \mathbb{E}^Y\left[V_Y(T) | X_Y(t) = x, Z_Y(t) = y\right] \\ &= \mathbb{E}^Y\left[f^Y\left(X_Y(T), Z_Y(T)\right) | X_Y(t) = x, Z_Y(t) = y\right], \end{aligned} \tag{8.5}$$

$$\begin{aligned} u^X(t, x, y) &= \mathbb{E}^X\left[V_X(T) | Y_X(t) = x, Z_X(t) = y\right] \\ &= \mathbb{E}^X\left[f^X\left(Y_X(T), Z_X(T)\right) | Y_X(t) = x, Z_X(t) = y\right], \end{aligned} \tag{8.6}$$

$$
\begin{aligned}
u^Z(t,x,y) &= \mathbb{E}^Z\left[V_Z(T)|X_Z(t)=x, Y_Z(t)=y\right] \\
&= \mathbb{E}^Z\left[f^Z\left(X_Z(T),Y_Z(T)\right)|X_Z(t)=x, Y_Z(t)=y\right]. \quad (8.7)
\end{aligned}
$$

In order to get a more specific representation of the price of the contract on three assets, let us assume a diffusion model of the prices of the underlying assets. There are six possible prices to consider:

$$dX_Y(t) = \sigma_{xy}X_Y(t)dW^{1,Y}(t), \quad dZ_Y(t) = \sigma_{yz}Z_Y(t)dW^{2,Y}(t), \tag{8.8}$$

$$dY_X(t) = \sigma_{xy}Y_X(t)dW^{1,X}(t), \quad dZ_X(t) = \sigma_{xz}Z_X(t)dW^{2,X}(t), \tag{8.9}$$

$$dX_Z(t) = \sigma_{xz}X_Z(t)dW^{1,Z}(t), \quad dY_Z(t) = \sigma_{yz}Y_Z(t)dW^{2,Z}(t). \tag{8.10}$$

Since there are two price processes for each reference asset, they are driven by two Brownian motions W^1 and W^2 that can be correlated. Let us assume for instance

$$dW^{1,Y}(t)\cdot dW^{2,Y}(t) = \rho dt.$$

The value of ρ represents the **correlation.** The formula for $dW^{1,Y}(t)$ and $dW^{2,Y}(t)$ determines the correlation of $W^{1,X}(t)$ and $W^{2,X}(t)$, and the correlation of $W^{1,Z}(t)$ and $W^{2,Z}(t)$. For instance, from Ito's formula we get

$$
\begin{aligned}
dZ_X(t) &= d\left(Z_Y(t)\cdot Y_X(t)\right) \\
&= Z_Y(t)\cdot dY_X(t) + Y_X(t)\cdot dZ_Y(t) + dZ_Y(t)\cdot dY_X(t) \qquad (8.11)\\
&= Z_X(t)\left[\sigma_{xy}dW^{1,X}(t) + \sigma_{yz}dW^{2,Y}(t)\right] + Z_X(t)^2\sigma_{xy}\sigma_{yz}\left[-\rho\right]dt.
\end{aligned}
$$

Therefore

$$
\begin{aligned}
dW^{1,X}(t)\cdot dW^{2,X}(t) &= \frac{dY_X(t)}{\sigma_{xy}Y_X(t)}\cdot\frac{dZ_X(t)}{\sigma_{xz}Z_X(t)} \\
&= \frac{\sigma_{xy}^2 - \rho\sigma_{xy}\sigma_{yz}}{\sigma_{xy}\cdot\sigma_{xz}} = \frac{\sigma_{xy} - \rho\sigma_{yz}}{\sigma_{xz}}.
\end{aligned}
$$

Similarly we obtain

$$dW^{1,Z}(t)\cdot dW^{2,Z}(t) = \frac{\sigma_{yz} - \rho\sigma_{xy}}{\sigma_{xz}}.$$

From (8.11) we also get

$$\frac{d^2Z_X(t)}{Z_X(t)^2} = (\sigma_{xy}^2 - 2\rho\sigma_{xy}\sigma_{yz} + \sigma_{yz}^2)dt,$$

and therefore the volatility σ_{xz} of $Z_X(t)$ is constrained to satisfy

$$\sigma_{xz}^2 = \sigma_{xy}^2 - 2\rho\sigma_{xy}\sigma_{yz} + \sigma_{yz}^2.$$

THEOREM 8.1
The price function

$$u^Y(t,x,y) = \mathbb{E}^Y\Big[f\left(X_Y(T), Z_Y(T)\right)|X_Y(t) = x, Z_Y(t) = y\Big],$$

satisfies partial differential equation

$$u_t^Y(t,x,y) + \tfrac{1}{2}\sigma_{xy}^2 x^2 \cdot u_{xx}^Y(t,x,y) \\ + \rho\sigma_{xy}\sigma_{yz}xy \cdot u_{xy}^Y(t,x,y) + \tfrac{1}{2}\sigma_{yz}^2 y^2 \cdot u_{yy}^Y(t,x,y) = 0, \quad (8.12)$$

with the terminal condition

$$u^Y(T,x,y) = f^Y(x,y), \quad (8.13)$$

the price function

$$u^X(t,x,y) = \mathbb{E}^X\Big[f^X\left(Y_X(T), Z_X(T)\right)|Y_X(t) = x, Z_X(t) = y\Big],$$

satisfies partial differential equation

$$u_t^X(t,x,y) + \tfrac{1}{2}\sigma_{xy}^2 x^2 \cdot u_{xx}^X(t,x,y) + \sigma_{xy}(\sigma_{xy} - \rho\sigma_{yz})xy \cdot u_{xy}^X(t,x,y) \\ + \tfrac{1}{2}(\sigma_{xy}^2 - 2\rho\sigma_{xy}\sigma_{yz} + \sigma_{yz}^2)y^2 \cdot u_{yy}^X(t,x,y) = 0, \quad (8.14)$$

with the terminal condition

$$u^X(T,x,y) = f^X(x,y), \quad (8.15)$$

and the price function

$$u^Z(t,x,y) = \mathbb{E}^Z\Big[f^Z\left(X_Z(T), Y_Z(T)\right)|X_Z(t) = x, Y_Z(t) = y\Big]$$

satisfies partial differential equation

$$u_t^Z(t,x,y) + \tfrac{1}{2}(\sigma_{xy}^2 - 2\rho\sigma_{xy}\sigma_{yz} + \sigma_{yz}^2)x^2 \cdot u_{xx}^Z(t,x,y) \\ + \sigma_{yz}(\sigma_{yz} - \rho\sigma_{xy})xy \cdot u_{xy}^Z(t,x,y) + \tfrac{1}{2}\sigma_{yz}^2 y^2 \cdot u_{yy}^Z(t,x,y) = 0, \quad (8.16)$$

with the terminal condition

$$u^Z(T,x,y) = f^Z(x,y). \quad (8.17)$$

PROOF Let us derive the partial differential equation for u^X; the other partial differential equations can be proved in a similar fashion, or using the perspective mapping relationship between functions u^Y, u^X and u^Z. The process $u^X(t, Y_X(T), Z_X(T)) = \mathbb{E}^X[V_X(T)|Y_X(t), Z_X(t)]$ is a martingale. Using Ito's formula, we get

$$\begin{aligned}
du^X(t, Y_X(t), Z_X(t)) &= u_t^X dt + u_x^X dY_X(t) + u_y^X dZ_X(t) \\
&\quad + \tfrac{1}{2}u_{xx}^X d^2Y_X(t) + u_{xy}^X dY_X(t)dZ_X(t) + \tfrac{1}{2}u_{yy}^X d^2Z_X(t) \\
&= \Big[u_t^X + \tfrac{1}{2}\sigma_{xy}^2 x^2 \cdot u_{xx}^X + \sigma_{xy}(\sigma_{xy} - \rho\sigma_{yz})xy \cdot u_{xy}^X \\
&\quad + \tfrac{1}{2}(\sigma_{xy}^2 - 2\rho\sigma_{xy}\sigma_{yz} + \sigma_{yz}^2)y^2 \cdot u_{yy}^X\Big] dt \\
&\quad + u_x^X dY_X(t) + u_y^X dZ_X(t).
\end{aligned}$$

The dt term of a martingale has to be zero, giving us the partial differential equation for u^X. □

Example 8.3

Consider the quanto forward with a payoff

$$(X_Y(T) - K) \cdot Z$$

that corresponds to payoff functions $f^Y(x, y) = (x - K) \cdot y$, $f^X(x, y) = (\frac{1}{x} - K) \cdot y$, and $f^Z(x, y) = (\frac{x}{y} - K)$. It is not difficult to show that the price of the quanto forward is given by

$$u^Y(t, x, y) = \left(x \cdot e^{\rho\sigma_{xy}\sigma_{yz}(T-t)} - K\right) \cdot y$$

when Y is used as a reference asset. We can immediately obtain the price of the quanto forward with respect to the reference asset X

$$u^X(t, x, y) = u^Y(t, \tfrac{1}{x}, \tfrac{y}{x}) \cdot x = \left(\tfrac{1}{x} \cdot e^{\rho\sigma_{xy}\sigma_{yz}(T-t)} - K\right) \cdot y,$$

and the price of the quanto forward with respect to the reference asset Z

$$u^Z(t, x, y) = u^Y(t, \tfrac{x}{y}, \tfrac{1}{y}) \cdot y = \left(\tfrac{x}{y} \cdot e^{\rho\sigma_{xy}\sigma_{yz}(T-t)} - K\right).$$

One can check that the functions u^Y, u^X and u^Z satisfy the partial differential equations in Theorem 8.1 (Exercise 8.1). □

Example 8.4

Consider the quanto call option with a payoff

$$(X_Y(T) - K)^+ \cdot Z$$

that corresponds to payoff functions $f^Y(x,y) = (x-K)^+ \cdot y$, $f^X(x,y) = (\frac{1}{x} - K)^+ \cdot y$, and $f^Z(x,y) = (\frac{x}{y} - K)^+$. A slightly more complicated computation shows that

$$
\begin{aligned}
&u^Y(t,x,y) = \\
&= x \cdot e^{\rho\sigma_{xy}\sigma_{yz}(T-t)} \cdot y \cdot N\left(\tfrac{1}{\sigma_{xy}\sqrt{T-t}}\left[\log\left(\tfrac{x\cdot\exp(\rho\sigma_{xy}\sigma_{yz}(T-t))}{K}\right) + \tfrac{1}{2}\sigma_{xy}^2(T-t)\right]\right) \\
&\quad - K \cdot y \cdot N\left(\tfrac{1}{\sigma_{xy}\sqrt{T-t}}\left[\log\left(\tfrac{x\cdot\exp(\rho\sigma_{xy}\sigma_{yz}(T-t))}{K}\right) - \tfrac{1}{2}\sigma_{xy}^2(T-t)\right]\right) \qquad (8.18)
\end{aligned}
$$

when Y is used as a reference asset. We can immediately obtain the price of the quanto call option with respect to the reference asset X

$$
\begin{aligned}
&u^X(t,x,y) = u^Y(t, \tfrac{1}{x}, \tfrac{y}{x}) \cdot x = \\
&= \tfrac{1}{x} \cdot e^{\rho\sigma_{xy}\sigma_{yz}(T-t)} \cdot y \cdot N\left(\tfrac{1}{\sigma_{xy}\sqrt{T-t}}\left[\log\left(\tfrac{\exp(\rho\sigma_{xy}\sigma_{yz}(T-t))}{K\cdot x}\right) + \tfrac{1}{2}\sigma_{xy}^2(T-t)\right]\right) \\
&\quad - K \cdot y \cdot N\left(\tfrac{1}{\sigma_{xy}\sqrt{T-t}}\left[\log\left(\tfrac{\exp(\rho\sigma_{xy}\sigma_{yz}(T-t))}{K\cdot x}\right) - \tfrac{1}{2}\sigma_{xy}^2(T-t)\right]\right) \qquad (8.19)
\end{aligned}
$$

and the price of the quanto call option with respect to the reference asset Z

$$
\begin{aligned}
&u^Z(t,x,y) = u^Y(t, \tfrac{x}{y}, \tfrac{1}{y}) \cdot y = \\
&= \tfrac{x}{y} \cdot e^{\rho\sigma_{xy}\sigma_{yz}(T-t)} \cdot N\left(\tfrac{1}{\sigma_{xy}\sqrt{T-t}}\left[\log\left(\tfrac{x\cdot\exp(\rho\sigma_{xy}\sigma_{yz}(T-t))}{K\cdot y}\right) + \tfrac{1}{2}\sigma_{xy}^2(T-t)\right]\right) \\
&\quad - K \cdot N\left(\tfrac{1}{\sigma_{xy}\sqrt{T-t}}\left[\log\left(\tfrac{x\cdot\exp(\rho\sigma_{xy}\sigma_{yz}(T-t))}{K\cdot y}\right) - \tfrac{1}{2}\sigma_{xy}^2(T-t)\right]\right). \qquad (8.20)
\end{aligned}
$$

Functions u^Y, u^X and u^Z satisfy the partial differential equations in Theorem 8.1. □

8.2 Hedging

Contracts on three assets admit a perfect hedge in the geometric Brownian motion model driven by two Brownian motions. The hedging is done in all three underlying assets, and has the form summarized in the next theorem.

THEOREM 8.2

The hedging portfolio P of the contract on three assets is given by each of the *following equivalent relationships:*

$$
\begin{aligned}
P_t = & \left[u_x^Y\left(t, X_Y(t), Z_Y(t)\right)\right] \cdot X \\
& +\Big[u^Y\left(t, X_Y(t), Z_Y(t)\right) - X_Y(t) \cdot u_x^Y\left(t, X_Y(t), Z_Y(t)\right) \\
& \qquad - Z_Y(t) \cdot u_y^Y\left(t, X_Y(t), Z_Y(t)\right)\Big] \cdot Y \\
& +\left[u_y^Y\left(t, X_Y(t), Z_Y(t)\right)\right] \cdot Z
\end{aligned} \tag{8.21}
$$

$$
\begin{aligned}
P_t = & \Big[u^X\left(t, Y_X(t), Z_X(t)\right) - Y_X(t) \cdot u_x^X\left(t, Y_X(t), Z_X(t)\right) \\
& \qquad - Z_X(t) \cdot u_y^X\left(t, Y_X(t), Z_X(t)\right)\Big] \cdot X \\
& +\left[u_x^X\left(t, Y_X(t), Z_X(t)\right)\right] \cdot Y \\
& +\left[u_y^X\left(t, Y_X(t), Z_X(t)\right)\right] \cdot Z
\end{aligned} \tag{8.22}
$$

$$
\begin{aligned}
P_t = & \left[u_x^Z\left(t, X_Z(t), Y_Z(t)\right)\right] \cdot X \\
& +\left[u_y^Z\left(t, X_Z(t), Y_Z(t)\right)\right] \cdot Y \\
& +\Big[u^Z\left(t, X_Z(t), Y_Z(t)\right) - X_Z(t) \cdot u_x^Z\left(t, X_Z(t), Y_Z(t)\right) \\
& \qquad - Y_Z(t) \cdot u_y^Z\left(t, X_Z(t), Y_Z(t)\right)\Big] \cdot Z.
\end{aligned} \tag{8.23}
$$

PROOF The hedging portfolio is given by positions $\Delta^X(t)$, $\Delta^Y(t)$ and $\Delta^Z(t)$ in the underlying assets X, Y and Z. Thus it can be written as

$$P_t = \Delta^X(t) \cdot X + \Delta^Y(t) \cdot Y + \Delta^Z(t) \cdot Z.$$

Since the hedging portfolio is self-financing, it also has to satisfy

$$dP_Y(t) = \Delta^X(t) \cdot dX_Y(t) + \Delta^Z(t) \cdot dZ_Y(t).$$

But we also have

$$
\begin{aligned}
dV_Y(t) &= du^Y\left(t, X_Y(t), Z_Y(t)\right) \\
&= u_x^Y\left(t, X_Y(t), Z_Y(t)\right) dX_Y(t) + u_y\left(t, X_Y(t), Z_Y(t)\right) dZ_Y(t)
\end{aligned}
$$

from Ito's formula after realizing that the dt term is zero. In order to have $P_t = V_t$ at all times, we must have

$$\Delta^X(t) = u_x^Y(t, X_Y(t), Z_Y(t)),$$

and

$$\Delta^Z(t) = u_y^Y(t, X_Y(t), Z_Y(t)).$$

The position in the remaining asset Y is determined from

$$\Delta^Y(t) = P_Y(t) - \Delta^X(t) \cdot X_Y(t) - \Delta^Z(t) \cdot Z_Y(t).$$

This proves the representation of the hedge in terms of the price function u^Y. The remaining expressions are equivalent, which is easy to see from the relationships

$$u^X(t, x, y) = u^Y(t, \tfrac{1}{x}, \tfrac{y}{x}) \cdot x, \qquad u^Z(t, x, y) = u^Y(t, \tfrac{x}{y}, \tfrac{1}{y}) \cdot y.$$

This implies

$$u_x^Y(t, x, y) = u^X(t, \tfrac{1}{x}, \tfrac{y}{x}) - \tfrac{1}{x} \cdot u_x^X(t, \tfrac{1}{x}, \tfrac{y}{x}) - \tfrac{y}{x} \cdot u_y^X(t, \tfrac{1}{x}, \tfrac{y}{x}) = u_x^Z(t, \tfrac{x}{y}, \tfrac{1}{y}),$$

$$u^Y(t, x, y) - x \cdot u_x^Y(t, x, y) - y \cdot u_y^Y(t, x, y) = u_x^X(t, \tfrac{1}{x}, \tfrac{y}{x}) = u_y^Z(t, \tfrac{x}{y}, \tfrac{1}{y}),$$

$$u_y^Y(t, x, y) = u_y^X(t, \tfrac{1}{x}, \tfrac{y}{x}) = u^Z(t, \tfrac{x}{y}, \tfrac{1}{y}) - \tfrac{x}{y} \cdot u_x^Z(t, \tfrac{x}{y}, \tfrac{1}{y}) - \tfrac{1}{y} \cdot u_y^Z(t, \tfrac{x}{y}, \tfrac{1}{y}).$$

□

Example 8.5

Let us determine the hedge of the quanto forward with payoff

$$(X_Y(T) - K) \cdot Z.$$

We have seen that the price of this contract with respect to the reference asset Y is given by

$$u^Y(t, x, y) = \left(x \cdot e^{\rho\sigma_{xy}\sigma_{yz}(T-t)} - K\right) \cdot y,$$

and thus

$$u_x^Y(t, x, y) = e^{\rho\sigma_{xy}\sigma_{yz}(T-t)} \cdot y,$$

and

$$u_y^Y(t, x, y) = x \cdot e^{\rho\sigma_{xy}\sigma_{yz}(T-t)} - K.$$

The hedging portfolio is therefore given by

$$\begin{aligned}
P_t &= \Delta^X(t) \cdot X + \Delta^Y(t) \cdot Y + \Delta^Z(t) \cdot Z \\
&= \left[Z_Y(t) \cdot e^{\rho\sigma_{xy}\sigma_{yz}(T-t)}\right] \cdot X \\
&\qquad - \left[Z_Y(t) \cdot X_Y(t) \cdot e^{\rho\sigma_{xy}\sigma_{yz}(T-t)}\right] \cdot Y + \left[X_Y(t) \cdot e^{\rho\sigma_{xy}\sigma_{yz}(T-t)} - K\right] \cdot Z.
\end{aligned}$$

□

Example 8.6
Similarly we can determine the hedge of the quanto call option with a payoff

$$(X_Y(T) - K)^+ \cdot Z.$$

The price $u^Y(t, x, y)$ of this contract is given in Equation (8.18). Thus we get

$$\begin{aligned} &u_x^Y(t, x, y) = \\ &= e^{\rho\sigma_{xy}\sigma_{yz}(T-t)} \cdot y \cdot N\left(\tfrac{1}{\sigma_{xy}\sqrt{T-t}}\left[\log\left(\tfrac{x\cdot\exp(\rho\sigma_{xy}\sigma_{yz}(T-t))}{K}\right) + \tfrac{1}{2}\sigma_{xy}^2(T-t)\right]\right), \end{aligned}$$

and

$$\begin{aligned} &u_y^Y(t, x, y) = \\ &= x \cdot e^{\rho\sigma_{xy}\sigma_{yz}(T-t)} \cdot N\left(\tfrac{1}{\sigma_{xy}\sqrt{T-t}}\left[\log\left(\tfrac{x\cdot\exp(\rho\sigma_{xy}\sigma_{yz}(T-t))}{K}\right) + \tfrac{1}{2}\sigma_{xy}^2(T-t)\right]\right) \\ &\qquad - K \cdot N\left(\tfrac{1}{\sigma_{xy}\sqrt{T-t}}\left[\log\left(\tfrac{x\cdot\exp(\rho\sigma_{xy}\sigma_{yz}(T-t))}{K}\right) - \tfrac{1}{2}\sigma_{xy}^2(T-t)\right]\right). \end{aligned}$$

The hedging portfolio is therefore given by

$$\begin{aligned} P_t &= \Delta^X(t) \cdot X + \Delta^Y(t) \cdot Y + \Delta^Z(t) \cdot Z \\ &= \Bigg[Z_Y(t) \cdot e^{\rho\sigma_{xy}\sigma_{yz}(T-t)} \times \\ &\qquad \times N\left(\tfrac{1}{\sigma_{xy}\sqrt{T-t}}\left[\log\left(\tfrac{x\cdot\exp(\rho\sigma_{xy}\sigma_{yz}(T-t))}{K}\right) + \tfrac{1}{2}\sigma_{xy}^2(T-t)\right]\right)\Bigg] \cdot X \\ &\quad - \Bigg[Z_Y(t) \cdot X_Y(t) \cdot e^{\rho\sigma_{xy}\sigma_{yz}(T-t)} \times \\ &\qquad \times N\left(\tfrac{1}{\sigma_{xy}\sqrt{T-t}}\left[\log\left(\tfrac{x\cdot\exp(\rho\sigma_{xy}\sigma_{yz}(T-t))}{K}\right) + \tfrac{1}{2}\sigma_{xy}^2(T-t)\right]\right)\Bigg] \cdot Y \\ &\quad + \Bigg[X_Y(t) \cdot e^{\rho\sigma_{xy}\sigma_{yz}(T-t)} \times \\ &\qquad \times N\left(\tfrac{1}{\sigma_{xy}\sqrt{T-t}}\left[\log\left(\tfrac{x\cdot\exp(\rho\sigma_{xy}\sigma_{yz}(T-t))}{K}\right) + \tfrac{1}{2}\sigma_{xy}^2(T-t)\right]\right) \\ &\qquad - K \cdot N\left(\tfrac{1}{\sigma_{xy}\sqrt{T-t}}\left[\log\left(\tfrac{x\cdot\exp(\rho\sigma_{xy}\sigma_{yz}(T-t))}{K}\right) - \tfrac{1}{2}\sigma_{xy}^2(T-t)\right]\right)\Bigg] \cdot Z. \end{aligned}$$

□

References and Further Reading

Options on multiple assets were first priced in Stulz (1982) and later studied by Johnson (1987) and by Boyle et al. (1989). These works generalized the results on the exchange option obtained earlier by Margrabe (1978). Boyle (1988) explored lattice methods for options on multiple assets. Monte Carlo methods were employed by Broadie and Detemple (1997) for American options written on multiple assets. Gerber and Shiu (1996) studied perpetual American options written on two assets. Cherubini et al. (2004) is a monograph on copula methods in finance, which also covers techniques of pricing exotic options. An extensive list of existing option pricing formulas that also covers exotic options appears in Haug (1997).

Exercises

8.1 Show that the price functions in Example 8.3 satisfy the partial differential equations from Theorem 8.1.

8.2 Find the price and the hedging portfolio for a quanto put option with payoff $(K - X_Y(T))^+ \cdot Z$, where the prices of the underlying assets follow geometric Brownian motion models given by Equations (8.8), (8.9), and (8.10).

8.3 **(a)** Find the price and the hedging portfolio of a contract V that pays off the best asset out of X, K_1Y, and K_2Z, or in other words

$$V_T = \max(X_T, K_1Y_T, K_2Z_T)$$

where the prices of the underlying assets follow geometric Brownian motion models given by Equations (8.8), (8.9), and (8.10). As an immediate consequence, find the price and the hedge for the call option on the best of two assets with the payoff

$$(\max(X_T, K_1Y_T) - K_2Z_T)^+.$$

(b) Find the price and the hedging portfolio of a contract that pays off the worst asset

$$\min(X_T, K_1Y_T, K_2Z_T).$$

(c) Find the price and the hedging portfolio of a contract that pays off the "middle" asset, which can be defined as

$$X_T + K_1Y_T + K_2Z_T - \max(X_T, K_1Y_T, K_2Z_T) - \min(X_T, K_1Y_T, K_2Z_T).$$

Chapter 9

Asian Options

Asian options are contracts that depend on underlying assets X and Y and upon the average of the price process $X_Y(t)$. The average price process is captured by a no-arbitrage contract A called the **average asset.** The payoff of the average asset is defined as

$$A_T = \left[\int_0^T X_Y(t)\mu(dt) \right] \cdot Y_T. \tag{9.1}$$

The average asset is a contract that pays off a number of units of an asset Y, where the number of units is the weighted average price of an asset X with respect to the asset Y. The weights are determined by the weighting measure μ which can represent both continuous or discrete averaging. Our definition of the average asset guarantees that its price is always positive, and thus the average asset can be used as a numeraire. The average asset is analogous to the maximal asset M^* that appears in pricing of lookback options. The important difference is that the average asset A turns out to be a no-arbitrage asset in contrast to the maximal asset M^*.

The average asset is typically not traded, but we can still use it as a numeraire in order to derive the pricing equations for Asian options. The pricing techniques for Asian options do not require the existence of the average asset as a traded contract. We will express all hedging positions in terms of assets X and Y only. Moreover as we will show in the following text, the Asian forward can be perfectly replicated by trading in the underlying assets X and Y, and the hedge is model independent. Therefore A itself is a no-arbitrage asset.

We can apply the First Fundamental Theorem of Asset Pricing as long as the assets X and Y are no-arbitrage assets. This is not the case when X is a stock S and Y is dollars \$, when the average asset contract becomes

$$A_T = \left[\int_0^T S_\$(t)\mu(dt) \right] \cdot \$_T.$$

However, we can still rewrite this contract in terms of no-arbitrage assets when the bond price follows a deterministic term structure $B_t^T = e^{-r(T-t)}\$_t$

as

$$A_T = \left[\int_0^T S_{B^T}(t) e^{-r(T-t)} \mu(dt) \right] \cdot B_T^T,$$

which is of the form of (9.1), with the underlying two no-arbitrage assets S and B^T. Note that hedging must be done in no-arbitrage assets exclusively as opposed to arbitrage assets such as currencies. A typical Asian option contract uses equal weights. A continuously sampled average asset pays off

$$A_T = \tfrac{1}{T} \left[\int_0^T S_\$(t) dt \right] \cdot \$_T,$$

which corresponds to an averaging of the form

$$\mu(dt) = \tfrac{1}{T} e^{-r(T-t)} dt,$$

when expressed in terms of S and B^T. A discretely sampled average asset pays off

$$A_T = \tfrac{1}{n} \sum_{k=1}^{n} S_\$(\tfrac{k}{n} T),$$

which corresponds to an averaging of the form

$$\mu(dt) = \tfrac{1}{n} \sum_{k=1}^{n} \delta_{(\frac{k}{n} T)}(t) e^{-r(T-t)} dt$$

when expressed in terms of S and B^T.

Let us define the most general form of an **Asian option.**

DEFINITION 9.1 *An Asian option is a contract that pays off one of the following:*

- $f^Y(X_Y(T), A_Y(T))$ *units of an asset* Y*,*
- $f^X(Y_X(T), A_X(T))$ *units of an asset* X*,*
- $f^A(X_A(T), Y_A(T))$ *units of an asset* A*.*

When the payoff functions are linked by the perspective mapping $f^Y(x, y) = f^X(\frac{1}{x}, \frac{y}{x}) \cdot x = f^A(\frac{x}{y}, \frac{1}{y}) \cdot y$, the three payoffs represent the same contract.

Example 9.1
The Asian call option with a **fixed strike** pays off

$$(A_T - K \cdot Y_T)^+. \tag{9.2}$$

This corresponds to the payoff functions $f^Y(x,y) = (y - K)^+$, $f^X(x,y) = (y - K \cdot x)^+$, or $f^A(x,y) = (1 - K \cdot y)^+$ in the above definition of the Asian option. This means that the payoff can be settled in three equivalent ways:

$$(A_Y(T) - K)^+ \cdot Y = (A_X(T) - K \cdot Y_X(T))^+ \cdot X = (1 - K \cdot Y_A(T))^+ \cdot A.$$

The Asian call option with a **floating strike** pays off

$$(A_T - K \cdot X_T)^+, \tag{9.3}$$

which corresponds to the payoff functions $f^Y(x,y) = (y - K \cdot x)^+$, $f^X(x,y) = (y - K)^+$, or $f^A(x,y) = (1 - K \cdot x)^+$. The payoff can be settled in the following three ways:

$$(A_Y(T) - KX_Y(T))^+ \cdot Y = (A_X(T) - K)^+ \cdot X = (1 - K \cdot X_A(T))^+ \cdot A.$$

Asian options with the fixed or the floating strike are the two most typical Asian option contracts.

It is interesting to note that the prices of the Asian fixed strike and the Asian floating strike options can be written as a Black–Scholes formula. The price of the fixed strike option is simply

$$\mathbb{P}_t^A(A_Y(T) \geq K) - K \cdot \mathbb{P}_t^Y(A_Y(T) \geq K), \tag{9.4}$$

and the price of the floating strike option is

$$\mathbb{P}_t^A(A_X(T) \geq K) - K \cdot \mathbb{P}_t^X(A_X(T) \geq K). \tag{9.5}$$

This follows from the fact that the Asian option can be written as a combination of two Arrow–Debreu securities whose price is given by the above expressions. However, the hard part is that the prices $A_Y(T)$ and $A_X(T)$ do not have a simple analytical distribution as opposed to the case of $X_Y(T)$ which has a known density, and thus determination of the corresponding probabilities is a nontrivial task. Semianalytical representations of these probabilities exist for continuous averaging, but they still require significant computational effort to obtain any numerical result. In our text we present the partial differential equations that correspond to the Asian option pricing problem which applies to both discrete and continuous averaging. These partial differential equations can be solved numerically in a straightforward way.

The foreign exchange market also trades contracts written on the **harmonic average** of the price. The harmonic average is defined as the reciprocal of the arithmetic average of the reciprocals:

$$\frac{1}{\int_0^T \frac{1}{X_Y(t)}\mu(dt)} = \frac{1}{\int_0^T Y_X(t)\mu(dt)}.$$

If we denote by

$$\tilde{A}_T = \left[\int_0^T Y_X(t)\mu(dt) \right] \cdot X_T$$

the average asset where the roles of the assets X and Y are flipped, we can define the harmonic average asset as

$$H_T = \left[\frac{1}{\int_0^T Y_X(t)\mu(dt)} \right] \cdot Y_T = \frac{1}{\tilde{A}_X(T)} \cdot Y_T.$$

Natural contracts to consider are the harmonic Asian option with a fixed strike with payoff

$$(H_T - K \cdot Y_T)^+ = \left(\frac{1}{\tilde{A}_X(T)} \cdot Y_T - K \cdot Y_T \right)^+$$

and the harmonic Asian option with a floating strike with payoff

$$(H_T - K \cdot X_T)^+ = \left(\frac{1}{\tilde{A}_X(T)} \cdot Y_T - K \cdot X_T \right)^+.$$

We can also write the payoffs in terms of the original average asset A_T if we flip the roles of the assets Y and X (it is just a matter of naming the assets). In this case the harmonic Asian option with a fixed strike has payoff

$$\left(\tfrac{1}{A_Y(T)} \cdot X_T - K \cdot X_T \right)^+, \tag{9.6}$$

which corresponds to the payoff functions $f^Y(x,y) = (\frac{x}{y} - K \cdot x)^+$, $f^X(x,y) = (\frac{x}{y} - K)^+$, and $f^A(x,y) = (x \cdot y - K \cdot x)^+$. The harmonic Asian option with a floating strike has payoff

$$\left(\tfrac{1}{A_Y(T)} \cdot X_T - K \cdot Y_T \right)^+, \tag{9.7}$$

which corresponds to the payoff functions $f^Y(x,y) = (\frac{x}{y} - K)^+$, $f^X(x,y) = (\frac{x}{y} - K \cdot x)^+$, and $f^A(x,y) = (x \cdot y - K \cdot y)^+$.

We can also consider more exotic payoffs, such as Asian powers $f^X(x,y) = y^\alpha$. The advantage of this contract is that it admits a closed form solution for integer valued α, and thus it can be used for calibrating numerical schemes. This payoff corresponds to $f^Y(x,y) = y^\alpha \cdot x^{1-\alpha}$, or equivalently to $f^A(x,y) = x^{1-\alpha}$. We can write the payoff as

$$\left(\tfrac{A}{X}\right)_T^\alpha \cdot X = \left(\tfrac{A}{Y}\right)_T^\alpha \cdot \left(\tfrac{X}{Y}\right)_T^{1-\alpha} \cdot Y = \left(\tfrac{X}{A}\right)_T^{1-\alpha} \cdot A.$$

□

Let V denote an Asian option contract. The price of this contract can be expressed in the following ways:

$$V = V_Y(t) \cdot Y = V_X(t) \cdot X = V_A(t) \cdot A.$$

In the Markovian model, we can also write

$$V_t = u^Y(t, X_Y(t), A_Y(t)) \cdot Y = u^X(t, Y_X(T), A_X(T)) \cdot X$$
$$= u^A(t, X_A(T), Y_A(T)) \cdot A,$$

giving us the following relationships between u^Y, u^X, and u^A via the perspective mapping:

$$u^Y(t,x,y) = u^X(t, \tfrac{1}{x}, \tfrac{y}{x}) \cdot x, \quad u^X(t,x,y) = u^Y(t, \tfrac{1}{x}, \tfrac{y}{x}) \cdot x, \tag{9.8}$$

$$u^Y(t,x,y) = u^A(t, \tfrac{x}{y}, \tfrac{1}{y}) \cdot y, \quad u^A(t,x,y) = u^Y(t, \tfrac{x}{y}, \tfrac{1}{y}) \cdot y, \tag{9.9}$$

$$u^X(t,x,y) = u^A(t, \tfrac{1}{y}, \tfrac{x}{y}) \cdot y, \quad u^A(t,x,y) = u^X(t, \tfrac{y}{x}, \tfrac{1}{x}) \cdot x. \tag{9.10}$$

When X and Y are no-arbitrage assets, then A is a no-arbitrage asset (shown below), and from the First Fundamental Theorem of the Asset Pricing we have the following stochastic representations:

$$u^Y(t,x,y) = \mathbb{E}^Y\left[V_Y(T) | X_Y(t) = x, A_Y(t) = y\right]$$
$$= \mathbb{E}^Y\left[f^Y(X_Y(T), A_Y(T)) | X_Y(t) = x, A_Y(t) = y\right], \tag{9.11}$$

$$u^X(t,x,y) = \mathbb{E}^X\left[V_X(T) | Y_X(t) = x, A_X(t) = y\right]$$
$$= \mathbb{E}^X\left[f^X(Y_X(T), A_X(T)) | Y_X(t) = x, A_X(t) = y\right], \tag{9.12}$$

$$u^A(t,x,y) = \mathbb{E}^A\left[V_A(T) | X_A(t) = x, Y_A(t) = y\right]$$
$$= \mathbb{E}^A\left[f^A(X_A(T), Y_A(T)) | X_A(t) = x, Y_A(t) = y\right]. \tag{9.13}$$

Let us show that the average asset A is indeed a no-arbitrage asset.

THEOREM 9.1

Let X and Y be two no-arbitrage assets. Then the replicating portfolio for the average asset contract that pays off

$$A_T = \left[\int_0^T X_Y(y)\mu(dt)\right] \cdot Y_T \tag{9.14}$$

is given by

$$A_t = \left[\int_t^T \mu(ds)\right] \cdot X + \left[\int_0^t X_Y(s)\mu(ds)\right] \cdot Y. \tag{9.15}$$

This result does not depend on the dynamics of the price $X_Y(t)$. In particular,

$$dA_Y(t) = \left[\int_t^T \mu(ds)\right] dX_Y(t). \tag{9.16}$$

PROOF Let $A_t = \bar{\Delta}^X(t)X_t + \bar{\Delta}^Y(t)Y_t$ be the replicating portfolio of the average asset. Then

$$dA_Y(t) = \bar{\Delta}^X(t)dX_Y(t).$$

Using the product rule, this can be rewritten as

$$dA_Y(t) = \bar{\Delta}^X(t)dX_Y(t) = d\left(\bar{\Delta}^X(t) \cdot X_Y(t)\right) - X_Y(t)d\bar{\Delta}^X(t).$$

Integrating this equation, we get

$$A_Y(T) = A_Y(0) + \bar{\Delta}^X(T) \cdot X_Y(T) - \bar{\Delta}^X(0) \cdot X_Y(0) - \int_0^T X_Y(t)d\bar{\Delta}^X(t).$$

Since the terminal position of the average asset is completely invested in the asset Y, and has a zero position in the asset X, we have $\bar{\Delta}^X(T) = 0$. We thus have the following identity:

$$\left(\int_0^T X_Y(t)\mu(dt)\right) = A_Y(0) - \bar{\Delta}^X(0) \cdot X_Y(0) - \int_0^T X_Y(t)d\bar{\Delta}^X(t).$$

The only way to match the payoff is when

$$0 = A_Y(0) - \bar{\Delta}^X(0) \cdot X_Y(0),$$

which is equivalent to

$$A_0 = \bar{\Delta}^X(0)X_0,$$

and

$$\int_0^T X_Y(t)\mu(dt) = -\int_0^T X_Y(t)d\bar{\Delta}^X(t).$$

This implies

$$-d\bar{\Delta}^X(t) = \mu(dt),$$

which is the same as

$$\bar{\Delta}^X(t) = -\int_t^T d\bar{\Delta}^X(s) = \int_t^T \mu(ds).$$

The hedging position $\bar{\Delta}^Y(t)$ in the asset Y follows from the identity

$$A_Y(t) = \bar{\Delta}^X(t)X_Y(t) + \int_0^t X_Y(s)\mu(ds)$$

which concludes the proof. ∎

REMARK 9.1 Note that the hedging position $\bar{\Delta}^X(t)$ in the asset X is deterministic:

$$\boxed{\bar{\Delta}^X(t) = \int_t^T \mu(ds).} \tag{9.17}$$

For instance, when $\mu(dt) = \frac{1}{T}e^{-r(T-t)}dt$, we get

$$\bar{\Delta}^X(t) = \int_t^T \mu(ds) = \int_t^T \tfrac{1}{T}e^{-r(T-s)}ds = \tfrac{1}{rT}\left(1 - e^{-r(T-t)}\right).$$

In the case of uniform weighting $\mu(dt) = \frac{1}{T}dt$, the hedge of the average asset simplifies to

$$\bar{\Delta}^X(t) = \int_t^T \mu(ds) = \int_t^T \tfrac{1}{T}ds = \left(1 - \tfrac{t}{T}\right).$$

For discrete averaging when $\mu(dt) = \frac{1}{n}\sum_{k=1}^n \delta_{(\frac{k}{n}T)}(t)e^{-r(T-t)}dt$, we get

$$\begin{aligned}\bar{\Delta}^X(t) = \int_t^T \mu(ds) &= \tfrac{1}{n}\int_t^T \sum_{k=1}^n \delta_{(\frac{k}{n}T)}(s)e^{-r(T-s)}ds \\ &= \tfrac{1}{n}\sum_{k=\left[\frac{nt}{T}\right]+1}^{n} \exp\left(-r(\tfrac{n-k}{n})T\right),\end{aligned}$$

where $[\cdot]$ denotes the integer part function. This simplifies to

$$\bar{\Delta}^X(t) = 1 - \tfrac{1}{n}\left[n\tfrac{t}{T}\right] \tag{9.18}$$

when the averaging is uniform, i.e. when $\mu(dt) = \frac{1}{n}\sum_{k=1}^n \delta_{(\frac{k}{n}T)}(t)dt$.

∎

The insight of this result is the following: the trader who is replicating the average asset contract starts with a hedging portfolio of $\int_0^T \mu(dt)$ units of X and no units of Y:

$$\bar{\Delta}(0) = (\bar{\Delta}_0^X, \bar{\Delta}_0^Y) = \left(\int_0^T \mu(dt), 0\right).$$

The amount of $\int_0^T \mu(dt)$ units of the asset X is used for replicating the average of the price. The trader then gradually liquidates his position in the asset X, keeping just $\int_t^T \mu(dt)$ fraction of it at time t, and the rest of the portfolio is invested in the asset Y. The position in the asset Y corresponds to the running average $\int_0^t X_Y(s)\mu(ds)$. At the final time T, the hedge becomes

$$\bar{\Delta}(T) = (\bar{\Delta}^X(T), \bar{\Delta}^Y(T)) = \left(0, \int_0^T X_Y(t)\mu(dt)\right),$$

so the asset X is completely unloaded, and the position in the asset Y is the number that corresponds to the average price.

9.1 Pricing in the Geometric Brownian Motion Model

The prices of assets should be martingales under their corresponding numeraire measures. Since we have three underlying assets X, Y and A, we have six price processes to consider: $X_Y(t)$, $A_Y(t)$, $Y_X(t)$, $A_X(t)$, $X_A(t)$, and $Y_A(t)$. The price processes $X_Y(t)$ and $A_Y(t)$ are $\mathbb{P}^Y$ martingales, the price processes $Y_X(t)$ and $A_X(t)$ are $\mathbb{P}^X$ martingales, and the price processes $X_A(t)$ and $Y_A(t)$ are $\mathbb{P}^A$ martingales.

In the geometric Brownian motion model we assume the following price dynamics:

$$\boxed{dX_Y(t) = \sigma X_Y(t)dW^Y(t),} \tag{9.19}$$

and a similar evolution for the inverse price

$$\boxed{dY_X(t) = \sigma Y_X(t)dW^X(t).} \tag{9.20}$$

The evolution of $A_Y(t)$ follows from the hedging formula for the average asset:

$$\boxed{dA_Y(t) = \bar{\Delta}^X(t)dX_Y(t) = \sigma\bar{\Delta}^X(t)X_Y(t)dW^Y(t).} \tag{9.21}$$

Note that this evolution is not Markovian in $A_Y(t)$ since it depends on another process $X_Y(t)$, but it is Markovian in the pair $(X_Y(t), A_Y(t))$. Thus even when the Asian option contract payoff depends only on $A_Y(t)$, the corresponding pricing partial differential equation would depend on both prices.

The evolution of the average asset price under the reference asset X can be expressed as

$$\begin{aligned}
dA_X(t) &= \bar{\Delta}^Y(t)dY_X(t) \\
&= \left[A_Y(t) - \bar{\Delta}^X(t)\cdot X_Y(t)\right] dY_X(t) \\
&= \left[A_Y(t) - \bar{\Delta}^X(t)\cdot X_Y(t)\right] \sigma Y_X(t)dW^X(t) \\
&= \sigma\left[A_X(t) - \bar{\Delta}^X(t)\right] dW^X(t).
\end{aligned}$$

The second equality $\bar{\Delta}^Y(t) = A_Y(t) - \bar{\Delta}^X(t) \cdot X_Y(t)$ follows from the relationship $A_t = \bar{\Delta}^X(t) \cdot X + \bar{\Delta}^Y(t) \cdot Y$. The reason to write the evolution of $A_X(t)$ in terms of $\bar{\Delta}^X(t)$ rather than in terms of $\bar{\Delta}^Y(t)$ is that $\bar{\Delta}^X(t)$ is deterministic, while $\bar{\Delta}^Y(t)$ is stochastic. This means that unlike the price evolution of $A_Y(t)$, the price evolution of $A_X(t)$ is Markovian in just one variable, and thus contracts whose payoff depends only on $A_X(T)$ admit a simpler partial differential equation with one spatial variable. Thus

$$\boxed{dA_X(t) = \sigma\left[A_X(t) - \bar{\Delta}^X(t)\right] dW^X(t).} \tag{9.22}$$

Let us determine the evolution of the remaining prices: $Y_A(t)$, and $X_A(t)$. From Ito's formula we have

$$\begin{aligned} dY_A(t) = dA_Y(t)^{-1} &= -A_Y(t)^{-2} dA_Y(t) + A_Y(t)^{-3} d^2 A_Y(t) \\ &= -Y_A(t)^2 \sigma \bar{\Delta}^X(t) X_Y(t) dW^Y(t) \\ &\qquad + Y_A(t)^3 \sigma^2 \bar{\Delta}^X(t)^2 X_Y(t)^2 dt \\ &= \sigma \bar{\Delta}^X(t) Y_A(t) X_A(t) \Big[- dW^Y(t) + \sigma \bar{\Delta}^X(t) X_A(t) dt \Big]. \end{aligned}$$

According to the First Fundamental Theorem of Asset Pricing, the evolution of $Y_A(t)$ has to be a martingale under the corresponding $\mathbb{P}^A$ measure. Thus we have

$$\boxed{dY_A(t) = \sigma \bar{\Delta}^X(t) Y_A(t) X_A(t) dW^A(t),} \tag{9.23}$$

where $W^A(t)$ is a Brownian motion under $\mathbb{P}^A$ measure. Similarly,

$$\begin{aligned} dX_A(t) &= dA_X(t)^{-1} = -A_X(t)^{-2} dA_X(t) + A_X(t)^{-3} d^2 A_X(t) \\ &= -X_A(t)^2 \sigma \left[A_X(t) - \bar{\Delta}^X(t)\right] dW^X(t) \\ &\qquad + X_A(t)^3 \sigma^2 \left[A_X(t) - \bar{\Delta}^X(t)\right]^2 dt \\ &= \sigma X_A(t) \cdot \left[\bar{\Delta}^X(t) X_A(t) - 1\right] \cdot \Big[dW^X(t) - \sigma \left[1 - \bar{\Delta}^X(t) X_A(t)\right] dt \Big]. \end{aligned}$$

Therefore

$$\boxed{dX_A(t) = \sigma X_A(t) \left[\bar{\Delta}^X(t) X_A(t) - 1\right] dW^A(t),} \tag{9.24}$$

which is a martingale under the $\mathbb{P}^A$ measure.

The price of the Asian option is determined in the next theorem.

THEOREM 9.2
The price function

$$u^Y(t, x, y) = \mathbb{E}^Y\left[f^Y\left(X_Y(T), A_Y(T)\right) | X_Y(t) = x, A_Y(t) = y\right],$$

satisfies partial differential equation

$$u_t^Y(t,x,y) + \tfrac{1}{2}\sigma^2 x^2\Big[u_{xx}^Y(t,x,y) + 2\bar{\Delta}^X(t)u_{xy}^Y(t,x,y) + \bar{\Delta}^X(t)^2 u_{yy}^Y(t,x,y)\Big] = 0 \quad (9.25)$$

with the terminal condition

$$u^Y(T,x,y) = f^Y(x,y). \quad (9.26)$$

The price function

$$u^X(t,x,y) = \mathbb{E}^X\left[f^X\left(Y_X(T), A_X(T)\right)|Y_X(t) = x, A_X(t) = y\right],$$

satisfies partial differential equation

$$u_t^X(t,x,y) + \tfrac{1}{2}\sigma^2\Big[x^2 u_{xx}^X(t,x,y) + 2x(y-\bar{\Delta}^X(t))u_{xy}^X(t,x,y) + (y-\bar{\Delta}^X(t))^2 u_{yy}^X(t,x,y)\Big] = 0, \quad (9.27)$$

with the terminal condition

$$u^X(T,x,y) = f^X(x,y). \quad (9.28)$$

The price function

$$u^A(t,x,y) = \mathbb{E}^A\left[f^A\left(X_A(T), Y_A(T)\right)|X_A(t) = x, Y_A(t) = y\right]$$

satisfies partial differential equation

$$u_t^A(t,x,y) + \tfrac{1}{2}\sigma^2 x^2\Big([x\bar{\Delta}^X(t)-1]^2\cdot u_{xx}^A(t,x,y) + 2y\bar{\Delta}^X(t)[x\bar{\Delta}^X(t)-1]\cdot u_{xy}^A(t,x,y) + y^2(\bar{\Delta}^X(t))^2\cdot u_{yy}^A(t,x,y)\Big) = 0, \quad (9.29)$$

with the terminal condition

$$u^A(T,x,y) = f^A(x,y). \quad (9.30)$$

PROOF The price of the Asian option with respect to the reference asset Y, $u^Y(t, X_Y(t), A_Y(t))$, is a $\mathbb{P}^Y$ martingale, and thus du^Y has a zero dt term. Using Ito's formula, we get

$$\begin{aligned} du^Y &= u_t^Y dt + u_x^Y dX_Y(t) + u_y^Y dA_Y(t) \\ &\quad + \tfrac{1}{2}\left[u_{xx}^Y d^2X_Y(t) + 2u_{xy}^Y dX_Y(t)dA_Y(t) + u_{yy}^Y d^2A_Y(t)\right] \\ &= [u_t^Y + \tfrac{1}{2}\sigma^2x^2(u_{xx}^Y + 2\bar{\Delta}^X(t)u_{xy}^Y + \bar{\Delta}^X(t)^2u_{yy}^Y)]dt \\ &\quad + u_x^Y dX_Y(t) + u_y^Y dA_Y(t). \end{aligned}$$

Since the dt term is zero, we obtain the following partial differential equation:

$$u_t^Y(t,x,y)+\tfrac{1}{2}\sigma^2x^2\left[u_{xx}^Y(t,x,y)+2\bar{\Delta}^X(t)u_{xy}^Y(t,x,y)+\bar{\Delta}^X(t)^2u_{yy}^Y(t,x,y)\right]=0.$$

The terminal condition is given by

$$u^Y(T,x,y)=f^Y(x,y).$$

Similarly, the price of the Asian option with respect to the reference asset X, $u^X(t,Y_X(t),A_X(t))$, is a $\mathbb{P}^X$ martingale, and thus the dt term of du^X is zero. Using the evolution of the price of the average asset under the reference asset X, we get

$$\begin{aligned}
du^X &= u_t^X dt + u_x^X dY_X(t) + u_y^X dA_X(t)\\
&\quad +\tfrac{1}{2}\left[u_{xx}^X d^2Y_X(t) + 2u_{xy}^X dY_X(t)dA_X(t) + u_{yy}^X d^2A_X(t)\right]\\
&= [u_t^X + \tfrac{1}{2}\sigma^2(x^2u_{xx}^X + 2x(y-\bar{\Delta}^X(t))u_{xy}^X + (y-\bar{\Delta}^X(t))^2u_{yy}^X)]dt\\
&\quad + u_x^X dY_X(t) + u_y^X dA_X(t).
\end{aligned}$$

Since the dt term is zero, we have the following partial differential equation

$$\begin{aligned}
u_t^X(t,x,y) &+ \tfrac{1}{2}\sigma^2\Big[x^2u_{xx}^X(t,x,y)\\
&+ 2x(y-\bar{\Delta}^X(t))u_{xy}^X(t,x,y) + (y-\bar{\Delta}^X(t))^2u_{yy}^X(t,x,y)\Big] = 0,
\end{aligned}$$

with the terminal condition

$$u^X(T,x,y)=f^X(x,y).$$

Finally, the price of the Asian option with respect to the reference asset A, $u^A(t,X_A(t),Y_A(t))$, is a $\mathbb{P}^A$ martingale, and thus the dt term of du^A is zero. Using the evolution of the prices of X and Y under the reference asset A, we get

$$\begin{aligned}
du^A &= u_t^A dt + u_x^A dX_A(t) + u_y^A dY_A(t)\\
&\quad +\tfrac{1}{2}\left[u_{xx}^A d^2X_A(t) + 2u_{xy}^A dX_A(t)dY_A(t) + u_{yy}^A d^2Y_A(t)\right]\\
&= \Big[u_t^A + \tfrac{1}{2}\sigma^2x^2[[x\bar{\Delta}^X(t)-1]^2\cdot u_{xx}^A\\
&\quad +2y\bar{\Delta}^X(t)[x\bar{\Delta}^X(t)-1]\cdot u_{xy}^A + y^2\bar{\Delta}^X(t)^2\cdot u_{yy}^A]\Big]dt\\
&\quad + u_x^A dX_A(t) + u_y^A dY_A(t).
\end{aligned}$$

Since the dt term is zero, we have the following partial differential equation

$$\begin{aligned}
u_t^A(t,x,y) &+ \tfrac{1}{2}\sigma^2x^2\Big([x\bar{\Delta}^X(t)-1]^2\cdot u_{xx}^A(t,x,y)\\
&+ 2y\bar{\Delta}^X(t)[x\bar{\Delta}^X(t)-1]\cdot u_{xy}^A(t,x,y) + y^2\bar{\Delta}^X(t)^2\cdot u_{yy}^A(t,x,y)\Big) = 0,
\end{aligned}$$

with the terminal condition

$$u^A(T,x,y) = f^A(x,y). \tag{9.31}$$

□

9.2 Hedging of Asian Options

Since Asian options depend on three assets: X, Y, and the Asian forward A, the hedge should take positions in all these assets. The hedging portfolio should be of the form

$$P_t = \Delta^X(t) \cdot X + \Delta^Y(t) \cdot Y + \Delta^A(t) \cdot A. \tag{9.32}$$

However, the average asset itself can be hedged by assets X and Y:

$$A_t = \bar{\Delta}^X(t) \cdot X + \bar{\Delta}^Y(t) \cdot Y, \tag{9.33}$$

and thus the Asian option hedge can be reduced to positions in just two assets, X and Y:

$$P_t = [\Delta^X(t) + \Delta^A(t) \cdot \bar{\Delta}^X(t)] \cdot X + [\Delta^Y(t) + \Delta^A(t) \cdot \bar{\Delta}^Y(t)] \cdot Y. \tag{9.34}$$

The hedging position in the underlying assets X and Y has two components: one part ($\Delta^X(t)$ or $\Delta^Y(t)$) represents the usual delta sensitivity of the Asian option price with respect to the price of the underlying asset, and the other part represents the delta sensitivity of the Asian option price with respect to the average asset price ($\Delta^A(t)$), multiplied by the hedge of the average asset in terms of the assets X and Y ($\bar{\Delta}^X(t)$, or $\bar{\Delta}^Y(t)$). This feature is rather unique among contingent claims. The exact forms of the hedging portfolio are given in the following theorem. Recall that

$$\bar{\Delta}^X(t) = \int_t^T \mu(ds),$$

and

$$\bar{\Delta}^Y(t) = \int_0^t X_Y(s)\mu(ds).$$

THEOREM 9.3

The hedging portfolio P_t of the Asian option admits each of the following *equivalent represenations:*

$$P_t = \left[u_x^Y(t, X_Y(t), A_Y(t)) + \bar{\Delta}^X(t) \cdot u_y^Y(t, X_Y(t), A_Y(t))\right] \cdot X$$
$$+ \left[u^Y(t, X_Y(t), A_Y(t)) - X_Y(t) \cdot u_x^Y(t, X_Y(t), A_Y(t)) + \left(\bar{\Delta}^Y(t) - A_Y(t)\right) \cdot u_y^Y(t, X_Y(t), A_Y(t))\right] \cdot Y, \quad (9.35)$$

$$P_t = \left[u^X(t, Y_X(t), A_X(t)) - Y_X(t) \cdot u_x^X(t, Y_X(t), A_X(t)) + \left(\bar{\Delta}^X(t) - A_X(t)\right) \cdot u_y^X(t, Y_X(t), A_X(t))\right] \cdot X$$
$$+ \left[u_x^X(t, Y_X(t), A_X(t)) + \bar{\Delta}^Y(t) \cdot u_y^X(t, Y_X(t), A_X(t))\right] \cdot Y, \quad (9.36)$$

$$P_t = \left[\left[u^A(t, X_A(t), Y_A(t)) - u_y^A(t, X_A(t), Y_A(t)) \cdot Y_A(t)\right] \cdot \bar{\Delta}^X(t) + u_x^A(t, X_A(t), Y_A(t)) \cdot \left[1 - \bar{\Delta}^X(t) X_A(t)\right]\right] \cdot X$$
$$+ \left[\left[u^A(t, X_A(t), Y_A(t)) - u_x^A(t, X_A(t), Y_A(t)) \cdot X_A(t)\right] \cdot \bar{\Delta}^Y(t) + u_y^A(t, X_A(t), Y_A(t)) \cdot \left[1 - \bar{\Delta}^Y(t) Y_A(t)\right]\right] \cdot Y. \quad (9.37)$$

PROOF Let us find a hedge for the Asian option of the form

$$P_t = \Delta^X(t) \cdot X + \Delta^Y(t) \cdot Y.$$

Using the fact that the process $u^Y(t, X_Y(t), A_Y(t))$ has a zero dt term, we get

$$\begin{aligned} du^Y &= u_x^Y \cdot dX_Y(t) + u_y^Y \cdot dA_Y(t) \\ &= \left(u_x^Y + \bar{\Delta}^X(t) u_y^Y\right) \cdot dX_Y(t). \end{aligned}$$

Thus the hedging position in the asset X is given by the formula

$$\Delta^X(t, X_Y(t), A_Y(t)) = u_x^Y(t, X_Y(t), A_Y(t)) + \bar{\Delta}^X(t) u_y^Y(t, X_Y(t), A_Y(t)). \quad (9.38)$$

Similarly, using the evolution of $u^X(t, Y_X(t), A_X(t))$

$$\begin{aligned} du^X &= u_x^X \cdot dY_X(t) + u_y^X \cdot dA_X(t) \\ &= \left(u_x^X + \bar{\Delta}^Y(t) \cdot u_y^X\right) \cdot dY_X(t), \end{aligned}$$

we get the following representation of the hedging position in the asset Y:

$$\Delta^Y(t, Y_X(t), A_X(t)) = u_x^X(t, Y_X(t), A_X(t)) + \bar{\Delta}^Y(t) u_y^X(t, Y_X(t), A_X(t)). \tag{9.39}$$

Therefore the hedging portfolio takes the following form

$$\begin{aligned} P_t = &\left[u_x^Y(t, X_Y(t), A_Y(t)) + \bar{\Delta}^X(t) u_y^Y(t, X_Y(t), A_Y(t))\right] \cdot X \\ &+ \left[u_x^X(t, Y_X(t), A_X(t)) + \bar{\Delta}^Y(t) u_y^X(t, Y_X(t), A_X(t))\right] \cdot Y. \end{aligned} \tag{9.40}$$

We can also rewrite the above representation of the hedging portfolio using the function u^Y of the function u^X only. From

$$u^X(t, x, y) = u^Y(t, \tfrac{1}{x}, \tfrac{y}{x}) \cdot x,$$

we get

$$u_x^X(t, x, y) = u^Y(t, \tfrac{1}{x}, \tfrac{y}{x}) - \tfrac{1}{x} \cdot u_x^Y(t, \tfrac{1}{x}, \tfrac{y}{x}) - \tfrac{y}{x} \cdot u_y^Y(t, \tfrac{1}{x}, \tfrac{y}{x}),$$

and

$$u_y^X(t, x, y) = u_y^Y(t, \tfrac{1}{x}, \tfrac{y}{x}).$$

Substituting into (9.40), we get

$$\begin{aligned} P_t = &\left[u_x^Y(t, X_Y(t), A_Y(t)) + \bar{\Delta}^X(t) \cdot u_y^Y(t, X_Y(t), A_Y(t))\right] \cdot X \\ &+ \Big[u^Y(t, X_Y(t), A_Y(t)) - X_Y(t) \cdot u_x^Y(t, X_Y(t), A_Y(t)) \\ &\qquad + \left(\bar{\Delta}^Y(t) - A_Y(t)\right) \cdot u_y^Y(t, X_Y(t), A_Y(t))\Big] \cdot Y. \end{aligned}$$

Similarly, from

$$u^Y(t, x, y) = u^X(t, \tfrac{1}{x}, \tfrac{y}{x}) \cdot x,$$

we get

$$u_x^Y(t, x, y) = u^X(t, \tfrac{1}{x}, \tfrac{y}{x}) - \tfrac{1}{x} \cdot u_x^X(t, \tfrac{1}{x}, \tfrac{y}{x}) - \tfrac{y}{x} \cdot u_y^X(t, \tfrac{1}{x}, \tfrac{y}{x}),$$

and

$$u_y^Y(t, x, y) = u_y^X(t, \tfrac{1}{x}, \tfrac{y}{x}).$$

Substituting to (9.40), we get

$$P_t = \Big[u^X(t, Y_X(t), A_X(t)) - Y_X(t) \cdot u_x^X(t, Y_X(t), A_X(t))$$

$$+ \left(\bar{\Delta}_t^X - A_X(t)\right) \cdot u_y^X(t, Y_X(t), A_X(t))\Big] \cdot X$$

$$+ \left[u_x^X(t, Y_X(t), A_X(t)) + \bar{\Delta}^Y(t) \cdot u_y^X(t, Y_X(t), A_X(t))\right] \cdot Y.$$

Finally, from

$$du^A = u_x^A dX_A(t) + u_y^A dY_A(t),$$

we get a hedging portfolio representation of the form

$$P_t = u^A \cdot A_t = u_x^A \cdot X_t + u_y^A \cdot Y_t + \left[u^A - X_A(t) \cdot u_x^A - Y_A(t) \cdot u_y^A\right] \cdot A_t.$$

Using the fact that $A_t = \bar{\Delta}^X(t) \cdot X + \bar{\Delta}^Y(t) \cdot Y$, we conclude that

$$P_t = \Big[\left[u^A(t, X_A(t), Y_A(t)) - u_y^A(t, X_A(t), Y_A(t)) \cdot Y_A(t)\right] \cdot \bar{\Delta}^X(t)$$

$$+ u_x^A(t, X_A(t), Y_A(t)) \cdot \left[1 - \bar{\Delta}^X(t) X_A(t)\right]\Big] \cdot X$$

$$+ \Big[\left[u^A(t, X_A(t), Y_A(t)) - u_x^A(t, X_A(t), Y_A(t)) \cdot X_A(t)\right] \cdot \bar{\Delta}^Y(t)$$

$$+ u_y^A(t, X_A(t), Y_A(t)) \cdot \left[1 - \bar{\Delta}^Y(t) Y_A(t)\right]\Big] \cdot Y.$$

□

9.3 Reduction of the Pricing Equations

When the Asian option contract depends only on the assets A and X, such as in the case of an Asian call option with a floating strike that has a payoff $(A_T - K \cdot X_T)^+$, the option pricing problem depends only on the price process $A_X(t)$, and thus the corresponding partial differential equations depend only on one spatial variable. In this case the pricing equation (9.27) does not depend on the variable x that represents the price $Y_X(t)$ that is irrelevant to this problem, and thus it reduces to the partial differential equation

$$\boxed{u_t^X(t,y) + \tfrac{1}{2}\sigma^2(y - \bar{\Delta}^X(t))^2 u_{yy}^X(t,y) = 0,} \tag{9.41}$$

with the terminal condition

$$u^X(T,y) = f^X(y), \tag{9.42}$$

where

$$u^X(t,y) = \mathbb{E}^X[f^X(A_X(T))|A_X(t) = y].$$

We keep the notation y (as opposed to x) for the only spatial variable in order to be consistent with the pricing problem (9.27). Similarly, when A is a reference asset and the payoff depends only on $X_A(T)$, the pricing equation (9.29) does not depend on the variable y, and the partial differential equation simplifies to

$$\boxed{u_t^A(t,x) + \tfrac{1}{2}\sigma^2 x^2[x\bar{\Delta}^X(t) - 1]^2 \cdot u_{xx}^A(t,x) = 0,} \tag{9.43}$$

with the terminal condition

$$u^A(T,x) = f^A(x), \tag{9.44}$$

where

$$u^A(t,x) = \mathbb{E}^A[f^A(X_A(T))|X_A(t) = x].$$

The formulas for the hedging portfolio given in Equations (9.36) and (9.37) also simplify to

$$\begin{aligned} P_t = &\left[u^X\left(t, A_X(t)\right) + \left(\bar{\Delta}^X(t) - A_X(t)\right) \cdot u_y^X\left(t, A_X(t)\right)\right] \cdot X \\ &+ \left[\bar{\Delta}^Y(t) \cdot u_y^X\left(t, A_X(t)\right)\right] \cdot Y, \end{aligned} \tag{9.45}$$

and

$$\begin{aligned} P_t = &\left[\bar{\Delta}^X(t) \cdot u^A\left(t, X_A(t)\right) + u_x^A\left(t, X_A(t)\right) \cdot \left[1 - \bar{\Delta}^X(t) X_A(t)\right]\right] \cdot X \\ &+ \left[\bar{\Delta}^Y(t) \cdot \left[u^A\left(t, X_A(t)\right) - u_x^A\left(t, X_A(t)\right) \cdot X_A(t)\right]\right] \cdot Y. \end{aligned} \tag{9.46}$$

The pricing equation (9.25) does not reduce in this case, and it is strictly suboptimal to employ it for pricing Asian options that do not depend on the asset Y.

When the contract depends on the assets A and Y only, such as in the case of the Asian call option with a fixed strike that has a payoff $(A_T - K \cdot Y_T)^+$, the reduction of the pricing equations is possible only in special cases, not in general. The reason is that the evolution of the price process $A_Y(t)$ depends on both prices $A_Y(t)$ and $X_Y(t)$ (in contrast to the evolution of the price $A_X(t)$ that depends only on itself), and thus the partial differential equation (9.25) cannot be reduced to only one spatial variable. However, when the payoff of the contract is only a function of the asset F known as the **Asian forward** defined as

$$F_T = A_T - K_1 Y_T, \tag{9.47}$$

a reduction of the pricing problem similar to Equation (9.41) is possible when the asset X is taken as a numeraire. Consider a contract that pays off $f^X(F_X(T))$ units of an asset X, where K_1 in (9.47) is a constant. When the payoff function is given by $f^X(x) = (x - K_2)^+$, the contract that corresponds to it is

$$[F_X(T) - K_2]^+ \cdot X = (A_X(T) - K_1 Y_X(T) - K_2)^+ \cdot X = (A_T - K_1 Y_T - K_2 X_T)^+,$$

which covers both the floating strike option when $K_1 = 0$, and the fixed strike option when $K_2 = 0$.

Let us define

$$u^X(t,x) = \mathbb{E}^X[f^X(F_X(T))|F_X(t) = x].$$

In order to get the partial differential equation for u^X, we need to determine $dF_X(t)$. Note that

$$\begin{aligned}
dF_X(t) &= d\left[A_X(t) - K_1 Y_X(t)\right] \\
&= \left[\bar{\Delta}^Y(t) - K_1\right] \cdot dY_X(t) \\
&= \left[A_Y(t) - \bar{\Delta}^X(t) \cdot X_Y(t) - K_1\right] \cdot dY_X(t) \\
&= \left[A_Y(t) - \bar{\Delta}^X(t) \cdot X_Y(t) - K_1\right] \sigma Y_X(t) dW^X(t) \\
&= \sigma\left[\left[A_X(t) - K_1 Y_X(t)\right] - \bar{\Delta}^X(t)\right] dW^X(t) \\
&= \sigma\left[F_X(t) - \bar{\Delta}^X(t)\right] dW^X(t).
\end{aligned}$$

Therefore

$$dF_X(t) = \sigma\left[F_X(t) - \bar{\Delta}^X(t)\right] dW^X(t),$$

which is identical to an evolution of the average asset A. Therefore the pricing partial differential equation takes the same form as (9.41):

$$u_t^X(t,x) + \tfrac{1}{2}\sigma^2(x - \bar{\Delta}^X(t))^2 u_{xx}^X(t,x) = 0 \tag{9.48}$$

with the terminal condition

$$u^X(T,x) = f^X(x). \tag{9.49}$$

Thus we can also efficiently solve the Asian call option with the fixed strike using the above partial differential equation. Note the important difference from Equation (9.41). In the previous case, the basic price process was $A_X(t)$, the price of the average asset A in terms of the reference asset X. The partial differential equation (9.48) applies to the price process $F_X(t)$, the price of the Asian forward F in terms of the reference asset X. The corresponding spatial variables are shifted by the factor $K_1Y_X(t)$ as

$$A_X(t) - F_X(t) = A_X(t) - A_X(t) + K_1Y_X(t) = K_1Y_X(t).$$

Note that while $A_X(t)$ is always positive, $F_X(t)$ can become zero or even become negative, and thus the Asian forward F cannot be used as a reference asset for the purposes of pricing. Thus in contrast to the case of the average asset A, there is no partial differential equation where F serves as a reference asset.

The hedging portfolio agrees with (9.45), but the value of $A_X(t)$ is replaced by $F_X(t)$:

$$P_t = \left[u^X(t, F_X(t)) + (\bar{\Delta}^X(t) - F_X(t)) \cdot u_x^X(t, F_X(t))\right] \cdot X + \left[\bar{\Delta}^Y(t) \cdot u_x^X(t, F_X(t))\right] \cdot Y. \tag{9.50}$$

References and Further Reading

The approach to Asian options presented in this text extends previous works of Vecer (2001, 2002) which used the average asset as the natural asset for pricing. The characterization of the Asian option price with partial differential equations was known even earlier; see for instance Rogers and Shi (1995), but the asset that was considered for pricing was the running average which is an arbitrage asset, and the corresponding partial differential equation had extra terms that appear in the connection with the time value of the running average. Use of the average asset for pricing Asian options is not limited to the geometric Brownian motion. It is possible to generalize this approach to other martingale models of the price as shown by Fouque and Han (2003) for stochastic volatility or for models with jumps as shown in Vecer and Xu (2004), and later by Bayraktar and Xing (2010). Hoogland and Neumann

(2000) and Henderson and Wojakowski (2002) pointed out the symmetries between the fixed and the floating strike Asian options. Other relevant papers include Geman and Yor (1993), Curran (1994), Linetsky (2004), Dufresne (2000), D'Halluin et al. (2005), Milevsky and Posner (1998), or Nielsen and Sandmann (2003).

Exercises

9.1 Show that 1, x, and y are solutions of the partial differential equations for the Asian option prices u^Y, u^X, and u^A. This is useful in calibrating numerical schemes. What are the contracts corresponding to these solutions?

9.2 Find the price of the contract that pays off

$$V_T = [A_X(T)]^2 \cdot X = \frac{[A_Y(T)]^2}{X_Y(T)} \cdot Y,$$

where A is the average asset.
Hint: The price function $u^X(t,x) = V_X(t)$ satisfies the partial differential equation

$$u_t^X(t,y) + \tfrac{1}{2}\sigma^2(y - \bar{\Delta}^X(t))^2 u_{yy}^X(t,y) = 0,$$

with the terminal condition

$$u^X(T,y) = y^2.$$

Consider a solution of the form

$$u^X(t,y) = a_2(t)y^2 + a_1(t)y + a_0(t).$$

Plug it into the partial differential equation for $u^X(t,x)$, and find the functions $a_2(t)$, $a_1(t)$, and $a_0(t)$ by solving the resulting ordinary differential equations.

Chapter 10

Jump Models

This chapter studies jump models of a price evolution. A martingale in continuous time can be written as a sum of a diffusion martingale and a pure jump martingale. Therefore a jump evolution is the second way of describing dynamics of the price. Since real markets quote and trade only at discrete levels and at discrete times, it may not be obvious if the true underlying price process is continuous or if it has jumps, as long as the jump sizes are relatively small and frequent.

The most basic process with jumps is a Poisson process $N(t)$. The Poisson process makes a jump in the time interval $[t, t + \Delta t]$ with probability $\lambda \Delta t$, where λ is the intensity of the process. The Poisson process itself is nondecreasing, and thus it is not a martingale. However, the compensated Poisson process, $N(t) - \lambda t$, is a martingale, and thus can serve as the basic model of market noise with jumps. In analogy to the geometric Brownian motion model of the price which is driven by Brownian motion, we consider a geometric Poisson process model of the price, which is driven by a compensated Poisson process.

To price European options, we need to know the evolution of both $X_Y(t)$ and $Y_X(t)$ in order to determine both martingale measures $\mathbb{P}^Y$ and $\mathbb{P}^X$. It is possible to preserve the symmetry of the evolution of the prices with the exception that the jump preserves the direction: when $X_Y(t)$ jumps up, $Y_X(t)$ jumps down, and vice versa. The jump $N(t)$ belongs to the pair of X and Y; it cannot be individualized to one asset in contrast to the geometric Brownian motion model, where the noise factor W^Y is associated with the asset Y, and the noise factor W^X is associated with the asset X. In the case of Poisson evolution, it is the intensity λ of the Poisson process that is associated with the particular asset. Under the $\mathbb{P}^Y$ measure, the process $N(t) - \lambda^Y t$ is a martingale, while under the $\mathbb{P}^X$ measure, the process $N(t) - \lambda^X t$ is a martingale. We will show that the values of λ^Y and λ^X are linked by the relationship $\lambda^X = e^\gamma \lambda^Y$, where γ is the size of the jump of $\log(X_Y(t))$.

The price of a European option can be computed via the Black–Scholes formula. Equivalently, the price satisfies a difference differential equation when the price follows a geometric Poisson process. The geometric Poisson process model of the price represents a complete market, and thus it is possible to construct a perfect hedge for a European contract. It turns out that the hedging position in the asset X does not agree with $\mathbb{P}_t^X(X_Y(T) \geq K)$ in contrast to the geometric Brownian motion model; it is slightly smaller when the price process $X_Y(t)$ has negative jumps. We determine the explicit form of the hedging position, and study the difference with $\mathbb{P}_t^X(X_Y(T) \geq K)$.

A more general model of the evolution of the price with jumps considers jumps of various sizes and intensities. When the price process is driven by a noise with independent and identically distributed increments, we call this noise a Lévy process. The prices of European contingent claims are characterized by an integro-differential equation. Models with multiple jump sizes are incomplete, and no perfect hedging is possible in this situation.

10.1 Poisson Process

The Poisson process $N(t)$ is a continuous time Markov process that takes integer values and satisfies the following:

$$\mathbb{P}(N(t+\Delta t) = N(t)) = 1 - \lambda\Delta t + o(\Delta t), \tag{10.1}$$

$$\mathbb{P}(N(t+\Delta t) = N(t)+1) = \lambda\Delta t + o(\Delta t), \tag{10.2}$$

$$\mathbb{P}(N(t+\Delta t) = N(t)+2) = o(\Delta t), \tag{10.3}$$

where $o(\Delta t)$ represents some function that is much smaller than Δt for small Δt, meaning

$$\lim_{\Delta t \to 0} \frac{o(\Delta t)}{\Delta t} = 0.$$

The parameter λ is known as the **intensity** of the Poisson process. Note that the increment of the Poisson process $N(t) - N(s)$ is independent of the information set $\mathcal{F}_s$.

One can think of a Poisson process as a limit of a binomial distribution. We can write

$$N(t) = \sum_{i=1}^{n} \left[N\left(\tfrac{i}{n}\cdot t\right) - N\left(\tfrac{i-1}{n}\cdot t\right)\right].$$

Each Poisson increment $N\left(\frac{i}{n}\cdot t\right) - N\left(\frac{i-1}{n}\cdot t\right)$ has approximately Bernoulli distribution, meaning that it takes only two values 0 or 1 with corresponding probabilities $1 - \lambda\frac{t}{n}$ and $\lambda\frac{t}{n}$. Thus $N(t)$ has approximately a binomial distribution with

$$\mathbb{P}(N(t) = k) \approx \binom{n}{k}\left(\frac{\lambda t}{n}\right)^k \left(1 - \frac{\lambda t}{n}\right)^{n-k}.$$

As $n \to \infty$, we have

$$\lim_{n\to\infty} \binom{n}{k} n^{-k} = \lim_{n\to\infty} \frac{n(n-1)\dots(n-k+1)}{k!n^k} = \frac{1}{k!},$$

and

$$\lim_{n\to\infty} \left(1 - \frac{\lambda t}{n}\right)^{n-k} = \lim_{n\to\infty} \left(1 - \frac{\lambda t}{n}\right)^{n} \lim_{n\to\infty} \left(1 - \frac{\lambda t}{n}\right)^{-k} = e^{-\lambda t}.$$

Hence,

$$\mathbb{P}(N(t) = k) = e^{-\lambda t} \frac{(\lambda t)^k}{k!}.$$

In other words, $N(t)$ is a Poisson random variable with parameter λt.

Let τ be the time of the first jump of the Poisson process. Note that

$$\mathbb{P}(\tau \le t) = \mathbb{P}(N(t) \ge 1) = 1 - \mathbb{P}(N(t) = 0) = 1 - e^{-\lambda t}.$$

The first equality follows from the fact that if there is a jump before time t, the value of the Poisson process must be greater than or equal to 1 at time t. The density of the time of the first jump is given by

$$f(t) = \frac{\partial}{\partial t}\mathbb{P}(\tau \le t) = \lambda e^{-\lambda t},$$

which corresponds to an exponential random variable with parameter λ.

For modeling price processes we need a martingale market noise that would ensure that the resulting portfolios are arbitrage free. In pricing models with continuous paths, the role of the market noise was played by Brownian motion, where all prices were represented as stochastic integrals with respect to Brownian motion. In the case of jump models, a Poisson process $N(t)$ itself is nondecreasing, and thus it is not a martingale. However, **a compensated Poisson process,** $N(t) - \lambda t$, is a martingale. This is easy to see from

$$\begin{aligned}\mathbb{E}_s[N(t) - \lambda t] &= \mathbb{E}_s[(N(t) - N(s)) + N(s) - \lambda t] \\ &= \mathbb{E}[N(t) - N(s)] + N(s) - \lambda t = \lambda(t-s) + N(s) - \lambda t = N_s - \lambda s.\end{aligned}$$

The compensated Poisson process $N(t) - \lambda t$ is a jump analog of Brownian motion $W(t)$ in diffusion models. It can be regarded as model of a market noise. Figure 10.1 shows a sample path of a compensated Poisson process. Note that the path is discontinuous at the time of each jump; the process jumps up by 1.

The basic idea is that all price processes that are driven by the compensated Poisson process market noise can be represented as stochastic integrals with

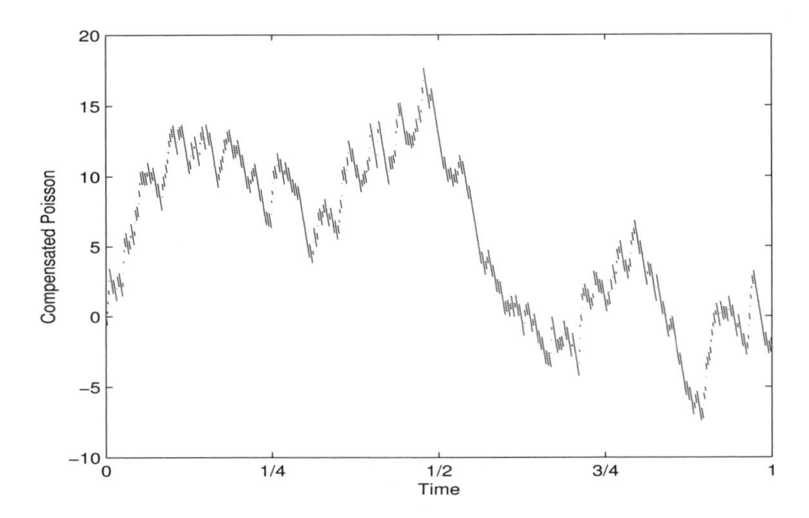

FIGURE 10.1: A sample path of a compensated Poisson process with $\lambda^Y = 300$, $T = 1$.

respect to that noise. The resulting price process should be a martingale in order not to have any arbitrage opportunity. As the following example suggests, one has to exclude integrands that have prior information about the jumps, and thus can create an opportunity for a risk-free profit. Note the following relationship holds

$$\int_0^t \Delta N(s) d(N(s) - \lambda s) = \int_0^t dN(s) = N(t),$$

where $\Delta N(t) = N(t) - N(t-) = 1$ at the time of the jump and zero otherwise. This shows that the stochastic integrals with respect to a martingale may not end up being a martingale. This results in an arbitrage opportunity, which would make this model undesirable. However, taking a position $\Delta N(t)$ is problematic at the time of the jump. This would require that the investor set his position in the underlying market noise $N(t) - \lambda t$ to one at the exact time of the jump, which is not possible since the jump happens unexpectedly. In practice, there is a delay between first observing the jump and being able to switch the position in the underlying asset. Thus we must incorporate this delay in our analysis. The delay itself may be infinitesimally small as long as the model preserves the correct order of the two events: the jump precedes the switch of the position.

The integrands that preserve the precedence of the jump before switching the position in the underlying asset are known as **predictable processes.** In particular, processes that are continuous from the left are suitable integrands as the change in the trading position does not come unexpectedly in this case. This is not the case for a Poisson process $N(t)$ which is not left-continuous, and the jump comes unexpectedly. A function h is called left-continuous if

$$h(t) = \lim_{s \uparrow t} h(s).$$

In order to use functions of the Poisson process $N(t)$ as a legitimate position in the underlying asset, the process must be slightly "delayed" in order to make sure that the jump has been observed first. The corresponding delayed process is given by

$$N(t-) = \lim_{s \uparrow t} N(s), \tag{10.4}$$

which gives the pre-jump value of the Poisson process $N(t)$.

10.2 Geometric Poisson Process

Let $\mathcal{X}(t)$ be a general market noise, such as a Brownian motion $W(t)$, or a compensated Poisson process $N(t) - \lambda t$. The market noise can take negative values, and as such, it does not serve as a good model for a possible price evolution. One can consider the so-called **stochastic exponential** $\mathcal{E}(t)$ of the process $\mathcal{X}(t)$ instead which takes positive values. The stochastic exponential of $\mathcal{X}(t)$ is defined by

$$d\mathcal{E}(t) = \mathcal{E}(t-)d\mathcal{X}(t) \tag{10.5}$$

with $\mathcal{E}(0) = 1$. When $\mathcal{X}(t)$ is a linear function $\mathcal{X}(t) = at$, the stochastic exponential reduces to an ordinary exponential $\mathcal{E}(t) = e^{at}$. The linear function at does not correspond to a market noise, but the stochastic exponential is still well defined, and it shows the relationship between an ordinary and a stochastic exponential. When $\mathcal{X}(t)$ is a Brownian motion scaled by a factor σ, so that $\mathcal{X}(t) = \sigma W(t)$, the corresponding stochastic exponential satisfies

$$d\mathcal{E}(t) = \mathcal{E}(t)d(\sigma \cdot W(t)) = \sigma\mathcal{E}(t)dW(t),$$

and its solution is a geometric Brownian motion $\mathcal{E}(t) = \exp(\sigma W(t) - \frac{1}{2}\sigma^2 t)$.

Let us determine the exact formula for a **geometric Poisson process** that has the following dynamics

$$dX_Y(t) = (e^\gamma - 1) \cdot X_Y(t-)d(N(t) - \lambda^Y t). \tag{10.6}$$

This corresponds to the driving noise $(e^\gamma - 1) \cdot (N(t) - \lambda^Y t)$. The factor $(e^\gamma - 1)$ plays a similar role as the volatility σ in the geometric Brownian motion model. It is more convenient to use $(e^\gamma - 1)$ as opposed to simply using γ as it leads to a symmetric solution for the inverse price $Y_X(t)$. Note that using a Taylor series $e^\gamma - 1 \approx \gamma$.

The stochastic differential equation (10.6) is analogous to the stochastic differential equation for geometric Brownian motion

$$dX_Y(t) = \sigma X_Y(t) dW^Y(t).$$

However, there is a major difference between the diffusion and jump models. In diffusion models, the noise factors come with each individual reference asset. For instance in the geometric Brownian motion model, the asset Y has its own noise factor W^Y, and the asset X has its own noise factor W^X. In even more complex models, such as in the case of power options, each power option R^α comes with its own noise factor $W^{(\alpha)}$. Although the driving noises (such as W^Y and W^X) are perfectly correlated in the diffusion models, one can still assign a proprietary noise to each asset.

This is not the case for jump models. The jump $N(t)$ belongs to the pair of assets X and Y, and it cannot be individualized to just one asset. When the price $X_Y(t)$ jumps up, the inverse price $Y_X(t)$ jumps down, and vice versa. The individual part of the noise is the compensation factor λ^Y; different assets have different compensators. The driving process, $N(t) - \lambda^Y t$, is a martingale under the $\mathbb{P}^Y$ measure that corresponds to the reference asset Y. This means that the Poisson process $N(t)$ has intensity λ^Y, or in other words,

$$\mathbb{P}^Y(N(t) = k) = \exp(-\lambda^Y t) \cdot \frac{(\lambda^Y t)^k}{k!}.$$

The individual assets may disagree on the distribution of jumps. The reader should keep in mind that this does not indicate how likely it is that we will get a particular number of jumps $N(t)$, but how costly it is to deliver the underlying asset on such an outcome.

The geometric Poisson process itself is an example of a complete model of the price process in which the contingent claims can be perfectly replicated by trading in the underlying assets. When $\gamma > 0$, the price process jumps up at each jump time, but it is compensated by a deterministic drift factor in the exponential function $-(e^\gamma - 1)\lambda^Y t < 0$. On the other hand when $\gamma < 0$, the price process jumps down at each jump time, but it is deterministically increasing between jumps with drift factor $-(e^\gamma - 1)\lambda^Y t > 0$.

If there were no jump, the price process would follow a simple deterministic evolution

$$dX_Y(t) = -(e^\gamma - 1) \cdot X_Y(t-)\lambda^Y dt,$$

that admits an exponential solution

$$X_Y(t) = X_Y(0) \exp\left(-(e^\gamma - 1)\lambda^Y t\right). \tag{10.7}$$

If there were only jumps, the price process would follow

$$dX_Y(t) = (e^\gamma - 1) \cdot X_Y(t-)dN(t),$$

meaning that at the time of the jump, the price process adds $(e^\gamma - 1) \cdot X_Y(t-)$ to its pre-jump value $X_Y(t-)$ so that

$$X_Y(t) = (e^\gamma - 1) \cdot X_Y(t-) + X_Y(t-) = e^\gamma \cdot X_Y(t-),$$

and

$$\Delta X_Y(t) = X_Y(t) - X_Y(t-) = (e^\gamma - 1) \cdot X_Y(t-).$$

The price process jumps by a factor of e^γ at the time of the jump. In conclusion, we have

$$X_Y(t) = X_Y(0)e^{\gamma N(t)}. \tag{10.8}$$

Combining the results from (10.7) and (10.8), we conclude that the geometric Poisson process is given by

$$\boxed{X_Y(t) = X_Y(0)\exp\left(\gamma \cdot N(t) - (e^\gamma - 1)\lambda^Y t\right).} \tag{10.9}$$

REMARK 10.1 The stochastic exponential for a general process $\mathcal{X}(t)$ is given by the Doléans-Dade formula

$$\mathcal{E}(t) = \exp\left(\mathcal{X}(t) - \tfrac{1}{2}[\mathcal{X},\mathcal{X}]^c(t)\right) \prod_{s \le t}(1 + \Delta\mathcal{X}(s))\exp(-\Delta\mathcal{X}(s)), \tag{10.10}$$

where $[\mathcal{X},\mathcal{X}](t)$ is the quadratic variation of $\mathcal{X}(t)$ defined by

$$[\mathcal{X},\mathcal{X}](t) = [\mathcal{X}(t)]^2 - 2\int_0^t \mathcal{X}(s-)d\mathcal{X}(s),$$

and where $[\mathcal{X},\mathcal{X}]^c(t)$ is the continuous part of $[\mathcal{X},\mathcal{X}](t)$.

For instance, when $\mathcal{X}(t) = \sigma W(t)$, the process $\mathcal{X}(t)$ has no jumps, and thus the part in (10.10) in the product is simply equal to one. Quadratic variation of $\mathcal{X}(t)$ (Brownian motion scaled by a factor σ) is given by

$$\begin{aligned}[\sigma W, \sigma W](t) &= [\sigma W(t)]^2 - 2\int_0^t [\sigma W(s)]d[\sigma W(s)] \\ &= [\sigma W(t)]^2 - ([\sigma W(t)]^2 - \sigma^2 t) = \sigma^2 t.\end{aligned}$$

Thus we have $\mathcal{E}(t) = \exp\left(\sigma W(t) - \frac{1}{2}\sigma^2 t\right)$.

Similarly, when $\mathcal{X}(t) = (e^\gamma - 1)(N(t) - \lambda t)$, the continuous part of the quadratic variation $[\mathcal{X},\mathcal{X}]^c(t)$ is zero (only jumps contribute to the quadratic variation in this case). At the time of the jump, $\Delta\mathcal{X}(t) = \mathcal{X}(t) - \mathcal{X}(t-) = e^\gamma - 1$. Thus we have

$$\begin{aligned}\mathcal{E}(t) &= \exp\left(\mathcal{X}(t) - \tfrac{1}{2}[\mathcal{X},\mathcal{X}]^c(t)\right)\prod_{s\le t}(1+\Delta\mathcal{X}(s))\exp(-\Delta\mathcal{X}(s))\\ &= \exp\left((e^\gamma-1)(N(t)-\lambda t)\right)\prod_{s\le t}(1+(e^\gamma-1))\cdot\exp(-(e^\gamma-1))\\ &= \exp\left((e^\gamma-1)(N(t)-\lambda t)\right)\cdot e^{\gamma N(t)}\cdot\exp(-(e^\gamma-1)N(t))\\ &= \exp\left(\gamma\cdot N(t) - (e^\gamma-1)\lambda t\right),\end{aligned}$$

which agrees with the formula for the geometric Poisson process. ∎

For pricing European options written on assets X and Y, we also need to determine the probability measure $\mathbb{P}^X$. We can deduce this probability measure from the dynamics of the inverse price, $Y_X(t)$. We can apply the **Ito's formula for jump processes**

$$\begin{aligned}f(X(t)) = f(X(0)) &+ \int_0^t f'(X(s-))dX(s)\\ &+ \sum_{0\le s\le t}\left[f(X(s)) - f(X(s-)) - f'(X(s-))\Delta X(s)\right]\end{aligned}\tag{10.11}$$

for the choice of the function $f(x) = \frac{1}{x}$ applied to the process $X_Y(t)$. Using the fact that $f'(x) = -\frac{1}{x^2}$, we get

$$\begin{aligned}\frac{1}{X_Y(t)} &= \frac{1}{X_Y(0)} + \int_0^t -\frac{1}{X_Y(s-)^2}\cdot(e^\gamma-1)\cdot X_Y(s-)d(N(s)-\lambda^Y s)\\ &\quad + \sum_{0\le s\le t}\left[\frac{1}{X_Y(s)} - \frac{1}{X_Y(s-)} + \frac{1}{X_Y(s-)^2}\Delta X_Y(s)\right]\\ &= Y_X(0) - \int_0^t (e^\gamma-1)Y_X(s-)d(N(s)-\lambda^Y s)\\ &\quad + \int_0^t \left[\tfrac{1}{e^\gamma} - 1 + (e^\gamma-1)\right]Y_X(s-)dN(s)\\ &= Y_X(0) + \int_0^t \left[e^{-\gamma}-1\right]Y_X(s-)dN(s) + \int_0^t (1-e^\gamma)Y_X(s-)d(-\lambda^Y s).\end{aligned}$$

Rewriting these dynamics in the differential form, we get

$$\begin{aligned}dY_X(t) &= \left[e^{-\gamma}-1\right]Y_X(t-)dN(t) - e^\gamma\cdot\left[e^{-\gamma}-1\right]Y_X(t-)\lambda^Y dt\\ &= \left[e^{-\gamma}-1\right]Y_X(t-)d(N(t) - e^\gamma\lambda^Y t).\end{aligned}$$

The inverse price $Y_X(t)$ should have a martingale evolution under the probability measure that corresponds to the reference asset X. This is possible only when the driving process $N(t) - e^{\gamma}\lambda^Y t$ itself is a martingale under the $\mathbb{P}^X$ measure. This means that the Poisson process $N(t)$ has intensity $\lambda^X = e^{\gamma}\lambda^Y$ under the $\mathbb{P}^X$ measure:

$$\mathbb{P}^X(N(t) = k) = \exp(-\lambda^X t) \cdot \frac{(\lambda^X t)^k}{k!},$$

where

$$\boxed{\lambda^X = e^{\gamma}\lambda^Y.} \tag{10.12}$$

We conclude that the dynamics of the inverse price $Y_X(t)$ satisfy

$$dY_X(t) = \left[e^{-\gamma} - 1\right] Y_X(t-)d(N(t) - \lambda^X t), \tag{10.13}$$

which admits the following solution:

$$\boxed{Y_X(t) = Y_X(0) \exp\left(-\gamma N(t) - (e^{-\gamma} - 1)\lambda^X t\right).} \tag{10.14}$$

Note that at the time of the jump we have

$$Y_X(t) = \left(e^{-\gamma} - 1\right) \cdot Y_X(t-) + Y_X(t-) = e^{-\gamma} \cdot Y_X(t-),$$

and

$$\Delta Y_X(t) = Y_X(t) - Y_X(t-) = \left(e^{-\gamma} - 1\right) \cdot Y_X(t-).$$

The roles of X and Y in (10.6) and (10.13) are exchangeable when we substitute γ for $-\gamma$, and λ^Y for $\lambda^X = e^{\gamma}\lambda^Y$.

Using the above arguments, we have shown the following mathematical result:

THEOREM 10.1 Poisson Change of Measure
Let $N(t)$ be a Poisson process with intensity λ^Y under the $\mathbb{P}^Y$ measure. Then $N(t)$ is a Poisson process with intensity $\lambda^X = e^{\gamma}\lambda^Y$ under the $\mathbb{P}^X$ measure, where

$$\mathbb{Z}(t) = \frac{d\mathbb{P}^X}{d\mathbb{P}^Y} = \frac{X_Y(t)}{X_Y(0)} = \exp\left(\gamma \cdot N(t) - \lambda^Y \left(e^{\gamma} - 1\right) t\right).$$

Example 10.1
Let A be the event $\{N(T) = k\}$. Using the Radon–Nikodým derivative, we can write

$$\begin{aligned}
\mathbb{P}^X(N(T) = k) = \mathbb{P}^X(A) &= \int_A \mathbb{Z}(T, \omega) d\mathbb{P}^Y(\omega) \\
&= \exp\left(\gamma \cdot k - \lambda^Y \left(e^{\gamma} - 1\right) T\right) \cdot \exp(-\lambda^Y T) \cdot \frac{(\lambda^Y T)^k}{k!} \\
&= \exp(-e^{\gamma}\lambda^Y T) \cdot \frac{(e^{\gamma}\lambda^Y T)^k}{k!} \\
&= \exp(-\lambda^X T) \cdot \frac{(\lambda^X T)^k}{k!},
\end{aligned}$$

which is a Poisson distribution with parameter $\lambda^X T$. □

10.3 Pricing Equations

Let V be a contract that pays off $f^Y(X_Y(T))$ units of an asset Y at time T. The price of the contract V with respect to the reference asset Y is given by

$$V_Y(t) = \mathbb{E}_t^Y\left[f^Y(X_Y(T))\right]. \tag{10.15}$$

The conditional expectation on the right hand side of the above equation which gives the price of the contract with respect to Y, is a $\mathbb{P}^Y$ martingale, and its value depends only on the price of $X_Y(t)$. Thus we can write

$$u^Y(t,x) = \mathbb{E}^Y\left[f^Y(X_Y(T))\,|X_Y(t) = x\right]. \tag{10.16}$$

We can also compute the price of this contract with respect to a reference asset X as

$$V_X(t) = \mathbb{E}_t^X\left[f^X(Y_X(T))\right], \tag{10.17}$$

where f^X is a payoff function in terms of the asset X. The functions f^Y and f^X are related by a perspective mapping $f^X(x) = f^Y(\frac{1}{x})\cdot x$. The price function $u^X(t,x)$ is defined

$$u^X(t,x) = \mathbb{E}^X\left[f^X(Y_X(T))\,|Y_X(t) = x\right]. \tag{10.18}$$

As in the geometric Brownian motion model, the price functions $u^Y(t,x)$ and $u^X(t,x)$ in the jump model are solutions to an equation, in this case of a difference-differential type.

THEOREM 10.2
The price function $u^Y(t,x) = \mathbb{E}^Y\left[f^Y(X_Y(T))\,|X_Y(t) = x\right]$ satisfies

$$\boxed{u_t^Y(t,x) + \lambda^Y\left[u^Y(t,e^\gamma x) - u^Y(t,x) - (e^\gamma - 1)x u_x^Y(t,x)\right] = 0.} \tag{10.19}$$

with the terminal condition

$$u^Y(T,x) = f^Y(x). \tag{10.20}$$

The hedging portfolio is given by

$$P_t = \left[\frac{u^Y(t,e^\gamma X_Y(t)) - u^Y(t,X_Y(t))}{(e^\gamma - 1)X_Y(t)}\right]\cdot X + \left[\frac{e^\gamma\cdot u^Y(t,X_Y(t)) - u^Y(t,e^\gamma X_Y(t))}{e^\gamma - 1}\right]\cdot Y. \tag{10.21}$$

The price function $u^X(t,x) = \mathbb{E}^X\left[f^X\left(Y_X(T)\right)|Y_X(t)=x\right]$ *satisfies*

$$\boxed{u_t^X(t,x) + \lambda^X\left[u^X\left(t,e^{-\gamma}x\right) - u^X(t,x) - \left(e^{-\gamma}-1\right)xu_x^X(t,x)\right] = 0.} \tag{10.22}$$

with the terminal condition

$$u^X(T,x) = f^X(x). \tag{10.23}$$

The hedging portfolio is given by

$$P_t = \left[\frac{e^{-\gamma}\cdot u^X(t,Y_X(t)) - u^X(t,e^{-\gamma}Y_X(t))}{e^{-\gamma}-1}\right]\cdot X + \left[\frac{u^X\left(t,e^{-\gamma}Y_X(t)\right) - u^X\left(t,Y_X(t)\right)}{\left(e^{-\gamma}-1\right)Y_X(t)}\right]\cdot Y. \tag{10.24}$$

PROOF According to Ito's formula for jump processes, we have

$$\begin{aligned}
u^Y\left(t,X_Y(t)\right) &= u^Y\left(0,X_Y(0)\right) + \int_0^t u_t^Y\left(s,X_Y(s-)\right)ds \\
&\quad + \int_0^t u_x^Y\left(s,X_Y(s-)\right)dX_Y(s) \\
&\quad + \sum_{0\le s\le t}\Big[u^Y\left(s,X_Y(s)\right) - u^Y\left(s,X_Y(s-)\right) \\
&\quad - u_x^Y\left(s,X_Y(s-)\right)\Delta X_Y(s)\Big] \\
&= u^Y\left(0,X_Y(0)\right) + \int_0^t u_t^Y\left(s,X_Y(s-)\right)ds \\
&\quad + \int_0^t u_x^Y\left(s,X_Y(s-)\right)(e^\gamma-1)X_Y(s-)d(N(s)-\lambda^Y s) \\
&\quad + \int_0^t \Big[u^Y\left(s,e^\gamma X_Y(s-)\right) - u^Y\left(s,X_Y(s-)\right) \\
&\qquad - u_x^Y\left(s,X_Y(s-)\right)(e^\gamma-1)X_Y(s-)\Big]dN(s) \\
&= u^Y\left(0,X_Y(0)\right) \\
&\quad + \int_0^t \Big[u_t^Y\left(s,X_Y(s-)\right) - \lambda^Y(e^\gamma-1)X_Y(s-)u_x^Y\left(s,X_Y(s-)\right) \\
&\quad + \lambda^Y\left(u^Y\left(s,e^\gamma X_Y(s-)\right) - u^Y\left(s,X_Y(s-)\right)\right)\Big]ds \\
&\quad + \int_0^t \left(u^Y\left(s,e^\gamma X_Y(s-)\right) - u^Y\left(s,X_Y(s-)\right)\right)d(N(s)-\lambda^Y s).
\end{aligned}$$

We have subtracted the

$$\left[u^Y\left(s,e^\gamma X_Y(s-)\right) - u^Y\left(s,X_Y(s-)\right)\right]d(\lambda^Y s)$$

term from $dN(s)$ in order to obtain the martingale part of the equality and added it back to the corresponding ds term. Since $u^Y(t, X_Y(t))$ is a $\mathbb{P}^Y$-martingale, the corresponding ds term that is not a part of the martingale $N(s) - \lambda^Y s$ must vanish. Thus we must have

$$u_t^Y(t,x) + \lambda^Y \left[u^Y(t, e^\gamma x) - u^Y(t,x) - (e^\gamma - 1)x u_x^Y(t,x)\right] = 0.$$

The terminal condition is given by

$$u^Y(T,x) = f^Y(x).$$

The contract follows a martingale evolution given by

$$du^Y(t, X_Y(t)) = \left[u^Y(t, e^\gamma X_Y(t-)) - u^Y(t, X_Y(t-))\right] d(N(t) - \lambda^Y t). \tag{10.25}$$

If we consider a replicating portfolio for this contract of the form

$$P_t = \Delta^X(t) \cdot X + \Delta^Y(t) \cdot Y,$$

where $\Delta^X(t)$ is the number of units of an asset X, and $\Delta^Y(t)$ is the number of units of an asset Y, we also have

$$dP_Y(t) = \Delta^X(t) dX_Y(t) = \Delta^X(t)(e^\gamma - 1)X_Y(t-)d(N(t) - \lambda^Y t). \tag{10.26}$$

Should the price and the dynamics of the replicating portfolio P be identical to the price and the dynamics of the contract V, we must also have

$$dP_Y(t) = dV_Y(t) = du^Y(t, X_Y(t)).$$

Comparing (10.25) with (10.26), we get the following representation for the hedge:

$$\Delta^X(t) = \frac{u^Y(t, e^\gamma X_Y(t-)) - u^Y(t, X_Y(t-))}{(e^\gamma - 1)X_Y(t-)}.$$

This hedging position should be held at all times, whether they are jump times or not, and thus we can also write

$$\Delta^X(t) = \frac{u^Y(t, e^\gamma X_Y(t)) - u^Y(t, X_Y(t))}{(e^\gamma - 1)X_Y(t)}.$$

The hedging position $\Delta^Y(t)$ in the asset Y is given by

$$\begin{aligned}
\Delta^Y(t) &= u^Y(t, X_Y(t)) - \Delta^X(t) \cdot X_Y(t) \\
&= u^Y(t, X_Y(t)) - \frac{u^Y(t, e^\gamma X_Y(t)) - u^Y(t, X_Y(t))}{(e^\gamma - 1)X_Y(t)} \cdot X_Y(t) \\
&= \frac{e^\gamma \cdot u^Y(t, X_Y(t)) - u^Y(t, e^\gamma X_Y(t))}{e^\gamma - 1}.
\end{aligned}$$

This concludes the proof for the price function u^Y. The result for the price function u^X is proved in a similar way. □

10.4 European Call Option in Geometric Poisson Model

The price and the hedging formulas can be further simplified for a European call option with payoff $(X_T - K \cdot Y_T)^+$. The price of an asset X in terms of a reference asset Y is given by

$$\begin{aligned} X_Y(T) &= X_Y(t) \exp\left(\gamma \cdot N(T-t) - (e^\gamma - 1)\lambda^Y (T-t)\right) \\ &= X_Y(t) \exp\left(\gamma \cdot N(T-t) + (e^{-\gamma} - 1)\lambda^X (T-t)\right), \end{aligned}$$

where $\lambda^X = e^\gamma \lambda^Y$ in the geometric Poisson model. The price of a European call option V is given by the Black–Scholes formula

$$V_t = \mathbb{P}_t^X (X_Y(T) \geq K) \cdot X - K \cdot \mathbb{P}_t^Y (X_Y(T) \geq K) \cdot Y.$$

The goal of this subsection is to determine the probabilities $\mathbb{P}_t^X(X_Y(T) \geq K)$ and $\mathbb{P}_t^Y(X_Y(T) \geq K)$. The event

$$X_Y(T) \geq K \tag{10.27}$$

is equivalent to

$$N(T-t) \leq \tfrac{1}{\gamma} \cdot \left[-\log\left(\tfrac{1}{K} \cdot X_Y(t)\right) + (e^\gamma - 1)\lambda^Y (T-t)\right] \tag{10.28}$$

when $\gamma < 0$, and to

$$N(T-t) \geq \tfrac{1}{\gamma} \cdot \left[-\log\left(\tfrac{1}{K} \cdot X_Y(t)\right) + (e^\gamma - 1)\lambda^Y (T-t)\right] \tag{10.29}$$

when $\gamma > 0$. When $\gamma < 0$, we have

$$\begin{aligned} &\mathbb{P}_t^X (X_Y(T) \geq K) \\ &\quad = \mathcal{P}\left(\tfrac{1}{\gamma} \cdot \left[-\log\left(\tfrac{1}{K} \cdot X_Y(t)\right) + (e^\gamma - 1)\lambda^Y (T-t)\right]; \lambda^X (T-t)\right), \end{aligned} \tag{10.30}$$

and

$$\begin{aligned} &\mathbb{P}_t^Y (X_Y(T) \geq K) \\ &\quad = \mathcal{P}\left(\tfrac{1}{\gamma} \cdot \left[-\log\left(\tfrac{1}{K} \cdot X_Y(t)\right) + (e^\gamma - 1)\lambda^Y (T-t)\right]; \lambda^Y (T-t)\right), \end{aligned} \tag{10.31}$$

where $\mathcal{P}(x; \lambda)$ is the cumulative distribution function of the Poisson random variable with intensity λ defined as

$$\mathcal{P}(x; \lambda) = \sum_{k=0}^{\lfloor x \rfloor} e^{-\lambda} \cdot \frac{\lambda^k}{k!}.$$

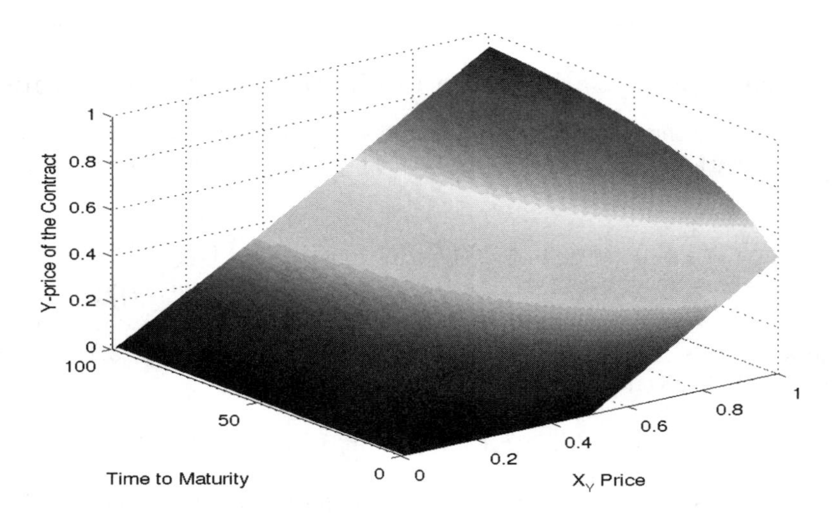

FIGURE 10.2: The price of a call option V_Y with payoff $(X_T - K \cdot Y_T)^+$ as a function of X_Y and time to maturity. The parameters are $K = \frac{1}{2}$, $\gamma = -0.02$, $\lambda^Y = 300$.

The price of the European call option in the case of $\gamma < 0$ is given by the formula

$$V_Y(t) = X_Y(t) \cdot \mathcal{P}\left(d; \lambda^X(T-t)\right) - K \cdot \mathcal{P}\left(d; \lambda^Y(T-t)\right), \tag{10.32}$$

where

$$\boxed{d = \frac{1}{\gamma} \cdot \left[\log\left(\frac{K}{X_Y(t)}\right) + (e^\gamma - 1)\lambda^Y(T-t)\right].} \tag{10.33}$$

Similarly when $\gamma > 0$, the price of the European call option is given by

$$V_Y(t) = X_Y(t) \cdot (1 - \mathcal{P}\left(d; \lambda^X(T-t)\right)) - K \cdot (1 - \mathcal{P}\left(d; \lambda^Y(T-t)\right)). \tag{10.34}$$

The price function $u^Y(t,x)$ is plotted in Figure 10.2.

Let us determine the hedging portfolio for the European call option. According to the Theorem 10.2, the hedging formulas are given by

$$\begin{aligned}
\Delta^X(t) &= \frac{u^Y(t, e^\gamma X_Y(t)) - u^Y(t, X_Y(t))}{(e^\gamma - 1)X_Y(t)} \\
&= \frac{e^\gamma X_Y(t) \cdot \mathbb{P}^X\Big(X_Y(T) \geq K | e^\gamma X_Y(t)\Big) - K \cdot \mathbb{P}^Y\Big(X_Y(T) \geq K | e^\gamma X_Y(t)\Big)}{(e^\gamma - 1)X_Y(t)} \\
&\quad - \frac{X_Y(t) \cdot \mathbb{P}^X\Big(X_Y(T) \geq K | X_Y(t)\Big) - K \cdot \mathbb{P}^Y\Big(X_Y(T) \geq K | X_Y(t)\Big)}{(e^\gamma - 1)X_Y(t)} \\
&= \frac{e^\gamma \cdot \mathbb{P}^X\Big(X_Y(T) \geq K | e^\gamma X_Y(t)\Big) - \mathbb{P}^X\Big(X_Y(T) \geq K | X_Y(t)\Big)}{(e^\gamma - 1)} \\
&\quad - K \cdot \frac{\mathbb{P}^Y\Big(X_Y(T) \geq K | e^\gamma X_Y(t)\Big) - \mathbb{P}^Y\Big(X_Y(T) \geq K | X_Y(t)\Big)}{(e^\gamma - 1)X_Y(t)}
\end{aligned}$$

for the case of the hedging position in the asset X, and as

$$\begin{aligned}
\Delta^Y(t) &= \frac{e^\gamma \cdot u^Y(t, X_Y(t)) - u^Y(t, e^\gamma X_Y(t))}{e^\gamma - 1} \\
&= \frac{e^\gamma X_Y(t) \cdot \mathbb{P}^X\Big(X_Y(T) \geq K | X_Y(t)\Big) - K e^\gamma \cdot \mathbb{P}^Y\Big(X_Y(T) \geq K | X_Y(t)\Big)}{e^\gamma - 1} \\
&\quad - \frac{e^\gamma X_Y(t) \cdot \mathbb{P}^X\Big(X_Y(T) \geq K | e^\gamma X_Y(t)\Big) - K \cdot \mathbb{P}^Y\Big(X_Y(T) \geq K | e^\gamma X_Y(t)\Big)}{e^\gamma - 1} \\
&= e^\gamma X_Y(t) \cdot \frac{\mathbb{P}^X\Big(X_Y(T) \geq K | X_Y(t)\Big) - \mathbb{P}^X\Big(X_Y(T) \geq K | e^\gamma X_Y(t)\Big)}{e^\gamma - 1} \\
&\quad - K \cdot \frac{e^\gamma \mathbb{P}^Y\Big(X_Y(T) \geq K | X_Y(t)\Big) - \mathbb{P}^Y\Big(X_Y(T) \geq K | e^\gamma X_Y(t)\Big)}{e^\gamma - 1}
\end{aligned}$$

for the case of the hedging position in the asset Y.

The formulas for $\Delta^X(t)$ and $\Delta^Y(t)$ can be further simplified if we express them in terms of a cumulative distribution function of the Poisson random variable. Consider the case when $\gamma < 0$. Then we have

$$\mathbb{P}\Big(X_Y(T) \geq K | X_Y(t)\Big) = \mathcal{P}(d; \lambda),$$

and

$$\mathbb{P}\Big(X_Y(T) \geq K | e^\gamma X_Y(t)\Big) = \mathcal{P}(d - 1; \lambda).$$

Therefore we can write

$$\begin{aligned}
\Delta^X(t) &= \frac{e^\gamma \cdot \mathbb{P}^X\Big(X_Y(T) \geq K | e^\gamma X_Y(t)\Big) - \mathbb{P}^X\Big(X_Y(T) \geq K | X_Y(t)\Big)}{(e^\gamma - 1)} \\
&\qquad - K \cdot \frac{\mathbb{P}^Y\Big(X_Y(T) \geq K | e^\gamma X_Y(t)\Big) - \mathbb{P}^Y\Big(X_Y(T) \geq K | X_Y(t)\Big)}{(e^\gamma - 1) X_Y(t)} \\
&= \frac{e^\gamma \cdot \mathcal{P}(d-1; \lambda^X(T-t)) - \mathcal{P}(d; \lambda^X(T-t))}{(e^\gamma - 1)} \\
&\qquad - K \cdot \frac{\mathcal{P}(d-1; \lambda^Y(T-t)) - \mathcal{P}(d; \lambda^Y(T-t))}{(e^\gamma - 1) X_Y(t)} \\
&= \mathcal{P}(d; \lambda^X(T-t)) \\
&\qquad + \frac{e^\gamma}{e^\gamma - 1} \cdot \left(\mathcal{P}(d-1; \lambda^X(T-t)) - \mathcal{P}(d; \lambda^X(T-t))\right) \\
&\qquad - \frac{K}{(e^\gamma - 1) X_Y(t)} \cdot \left(\mathcal{P}(d-1; \lambda^Y(T-t)) - \mathcal{P}(d; \lambda^Y(T-t))\right) \\
&= \mathcal{P}(d; \lambda^X(T-t)) - \frac{e^\gamma}{e^\gamma - 1} \cdot e^{-\lambda^X(T-t)} \cdot \frac{(\lambda^X(T-t))^{\lfloor d \rfloor}}{\lfloor d \rfloor !} \\
&\qquad + \frac{K}{(e^\gamma - 1) X_Y(t)} \cdot e^{-\lambda^Y(T-t)} \cdot \frac{(\lambda^Y(T-t))^{\lfloor d \rfloor}}{\lfloor d \rfloor !} \\
&= \mathcal{P}(d; \lambda^X(T-t)) \\
&\qquad - \frac{(\lambda^Y(T-t))^{\lfloor d \rfloor}}{\lfloor d \rfloor !} \cdot e^{-\lambda^Y(T-t)} \cdot \frac{K}{X_Y(t)(1 - e^\gamma)} \cdot \Big(1 - e^{\gamma(\lfloor d \rfloor + 1 - d)}\Big).
\end{aligned}$$

In contrast to the geometric Brownian motion model, the hedging position in the asset X in the geometric Poisson process model, $\Delta^X(t)$, is not equal to $\mathbb{P}_t^X(X_Y(T) \geq K) = \mathcal{P}(d; \lambda^X(T-t))$. While $\Delta^X(t)$ is a smooth function of the price of the underlying asset $X_Y(t)$, $\mathbb{P}_t^X(X_Y(T) \geq K)$ is a piecewise constant function. Figure 10.3 illustrates this situation.

The difference between $\mathbb{P}^X(X_Y(T) \geq K)$ and $\Delta^X(t)$ is equal to

$$\frac{(\lambda^Y(T-t))^{\lfloor d \rfloor}}{\lfloor d \rfloor !} \cdot e^{-\lambda^Y(T-t)} \cdot \frac{K}{X_Y(t)(1 - e^\gamma)} \cdot \Big(1 - e^{\gamma(\lfloor d \rfloor + 1 - d)}\Big).$$

Since $0 < (\lfloor d \rfloor + 1 - d) \leq 1$, the factor $(1 - e^{\gamma(\lfloor d \rfloor + 1 - d)})$ is at least zero and at most $1 - e^\gamma$. Therefore

$$\begin{aligned}
0 &< \frac{(\lambda^Y(T-t))^{\lfloor d \rfloor}}{\lfloor d \rfloor !} \cdot e^{-\lambda^Y(T-t)} \cdot \frac{K}{X_Y(t)(1 - e^\gamma)} \cdot \Big(1 - e^{\gamma(\lfloor d \rfloor + 1 - d)}\Big) \\
&\leq \frac{(\lambda^Y(T-t))^{\frac{1}{\gamma} \cdot \left[\log\left(\frac{K}{X_Y(t)}\right) + (e^\gamma - 1)\lambda^Y(T-t)\right]}}{\Gamma\left(\frac{1}{\gamma} \cdot \left[\log\left(\frac{K}{X_Y(t)}\right) + (e^\gamma - 1)\lambda^Y(T-t)\right] + 1\right)} \cdot e^{-\lambda^Y(T-t)} \cdot \frac{K}{X_Y(t)}.
\end{aligned}$$

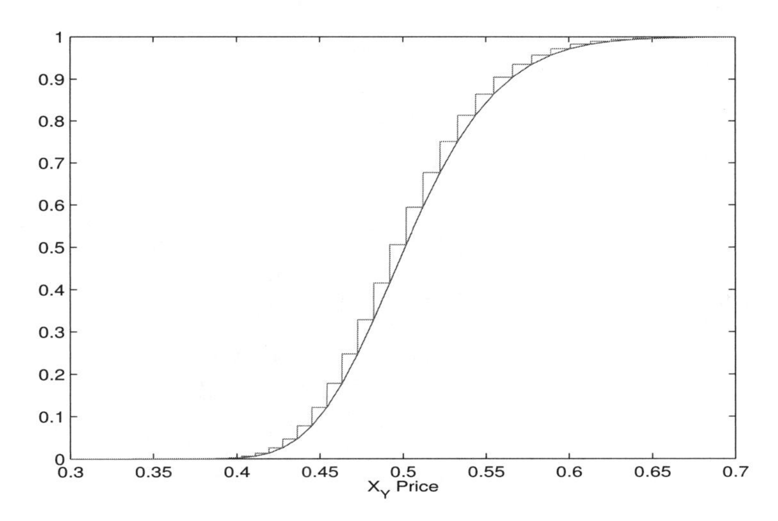

FIGURE 10.3: Comparison of Δ^X (smooth function) and $\mathbb{P}^X(X_Y(T) \geq K)$ (step function) for parameters $\gamma = -0.02$, $\lambda^Y = 20$, and $T = 1$.

Figure 10.4 illustrates the difference between $\mathbb{P}^X(X_Y(T) \geq K)$ and $\Delta^X(t)$. The difference can be anywhere between zero and the upper bound derived in the previous computation. The exact value depends on the difference $\lfloor d \rfloor + 1 - d$.

The hedging position $\Delta^Y(t)$ is given by

$$\begin{aligned}\Delta^Y(t) &= V_Y(t) - \Delta^X(t) X_Y(t) \\ &= -K \cdot \mathcal{P}(d, \lambda^Y(T-t)) \\ &\quad + \frac{(\lambda^Y(T-t))^{\lfloor d \rfloor}}{\lfloor d \rfloor !} \cdot e^{-\lambda^Y(T-t)} \cdot \frac{K}{(1-e^{\gamma})} \cdot \left(1 - e^{\gamma(\lfloor d \rfloor + 1 - d)}\right).\end{aligned}$$

Figures 10.5 and 10.6 show the hedging positions in the assets X and Y as a function of the price X_Y and time to maturity. The hedging positions in the geometric Poisson process model are similar to the hedging positions in the geometric Brownian motion model. Figure 10.7 shows a sample path of a geometric Poisson process together with the price of the European option with payoff $(X_T - K \cdot Y_T)^+$. Figure 10.8 shows the corresponding hedging positions in the assets X and Y. Since the option expires worthless, the terminal hedging positions are zero. Note that the prices and the hedging positions exhibit jumps. Figure 10.9 shows a different sample path of a geometric Poisson process, but this time the European option expires in the money. Thus the terminal hedging positions are given by $\Delta^X(T) = 1$, and $\Delta^Y = -K$ as seen from Figure 10.10.

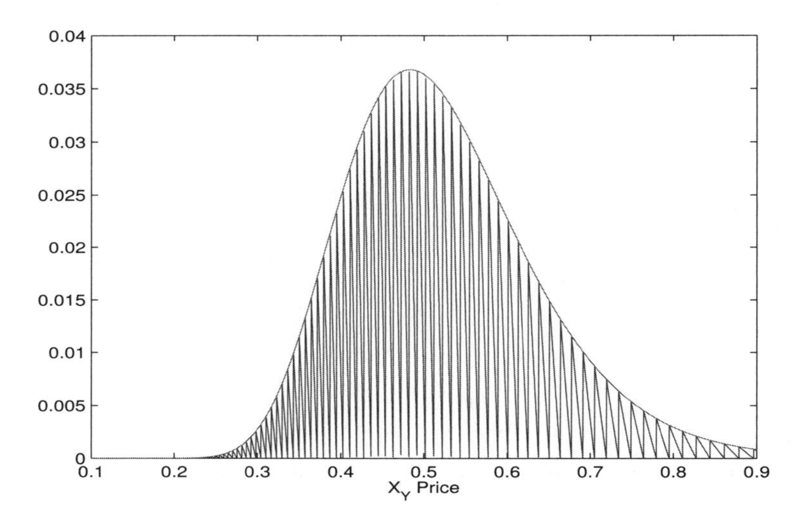

FIGURE 10.4: Difference between $\mathbb{P}^X(X_Y(T) \geq K)$ and Δ^X for parameters $\gamma = -0.02$, $\lambda^Y = 120$, and $T = 1$.

10.5 Lévy Models with Multiple Jump Sizes

The above results easily generalize to the case of jumps with different sizes and intensities. Let us assume that the jump process has independent and stationary increments, and let $\mu(dx, dt)$ denote a random measure associated with jumps of the price process. The random measure has the following interpretation. The quantity $\int_0^t \int_A \mu(dx, ds)$ represents the number of jumps of sizes in the set A that happened in the time interval $[0, t]$. Let us denote

$$\nu(t, A) = \mathbb{E} \int_0^t \int_A \mu(dx, ds), \tag{10.35}$$

which is the expected number of jumps of sizes in the set A. This is known as a **compensator.** The process

$$\int_0^t \int_{-\infty}^{\infty} (\mu(dx, dt) - \nu(dx, dt)) \tag{10.36}$$

is a martingale, and it can serve as a model of the market noise. Since the process has independent increments, it is time homogeneous, and we can also

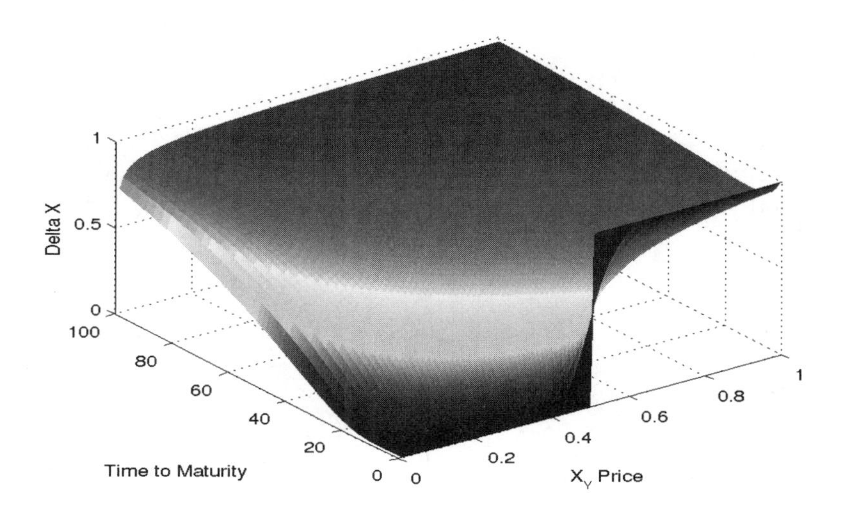

FIGURE 10.5: The hedging position in the asset X of a call option V with payoff $(X_T - K \cdot Y_T)^+$ as a function of X_Y and time to maturity. The parameters are $K = \frac{1}{2}$, $\gamma = -0.02$, $\lambda^Y = 300$.

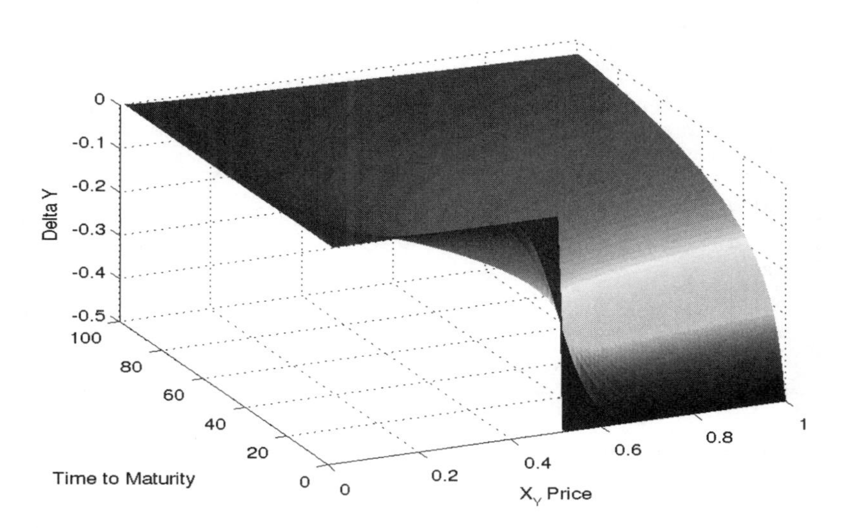

FIGURE 10.6: The hedging position in the asset Y of a call option V with payoff $(X_T - K \cdot Y_T)^+$ as a function of X_Y and time to maturity. The parameters are $K = \frac{1}{2}$, $\gamma = -0.02$, $\lambda^Y = 300$.

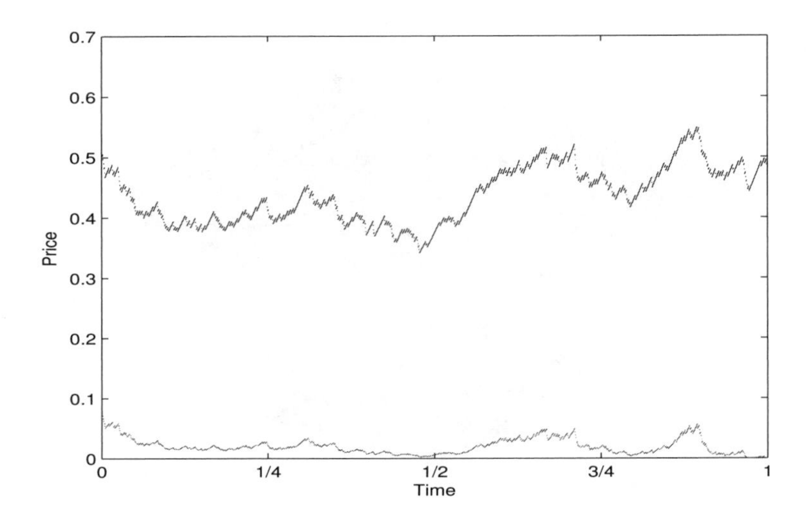

FIGURE 10.7: A sample path of the price of an asset X_Y (top) and the price of a corresponding call option V_Y (bottom) with payoff $(X_T - K \cdot Y_T)^+$ as a function of time. The parameters are $K = \frac{1}{2}$, $\gamma = -0.02$, $\lambda^Y = 300$, $T = 1$.

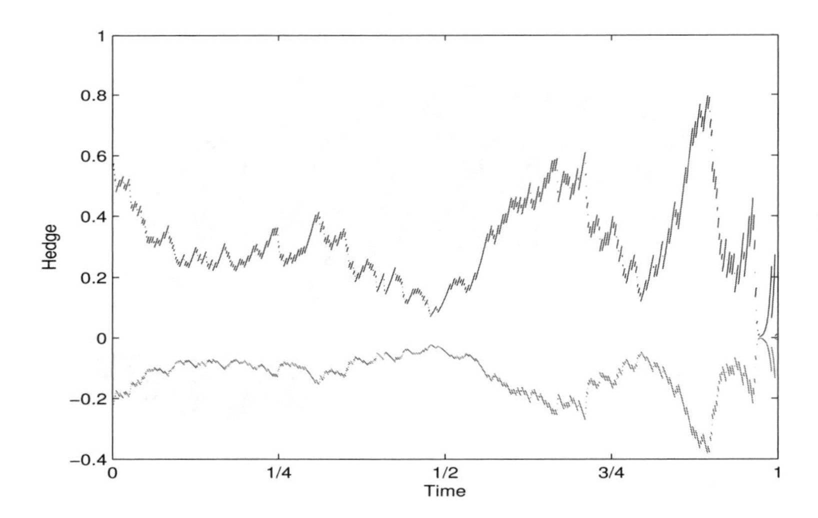

FIGURE 10.8: A sample path of the hedging position $\Delta^X(t)$ and $\Delta^Y(t)$ that corresponds to the previous example as a function of time. The parameters are $K = \frac{1}{2}$, $\gamma = -0.02$, $\lambda^Y = 300$, $T = 1$. The option expires out of the money, and thus the terminal positions are zero.

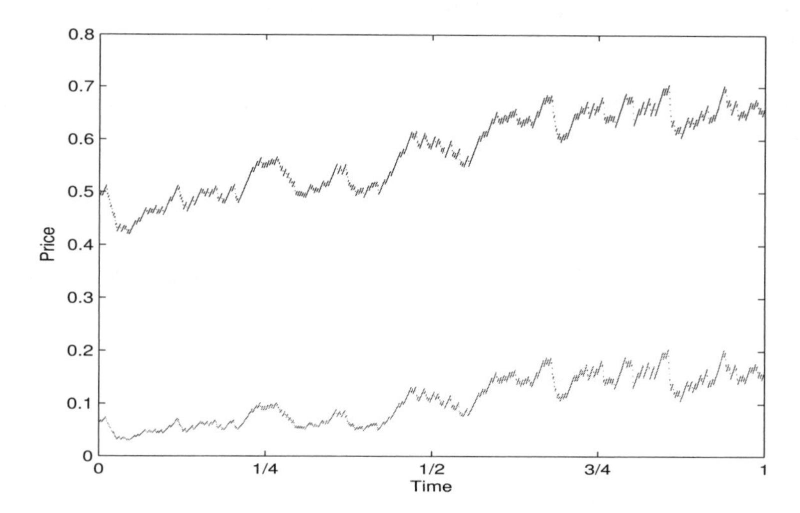

FIGURE 10.9: A sample path of the price of an asset X_Y (top) and the price of a corresponding call option V_Y (bottom) with payoff $(X_T - K \cdot Y_T)^+$ as a function of time. The parameters are $K = \frac{1}{2}$, $\gamma = -0.02$, $\lambda^Y = 300$, $T = 1$.

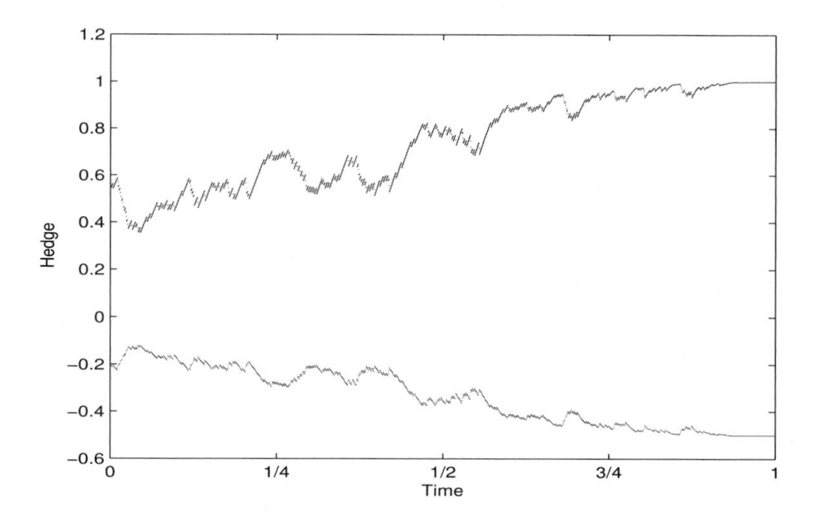

FIGURE 10.10: A sample path of the hedging position $\Delta^X(t)$ and $\Delta^Y(t)$ that corresponds to the previous example as a function of time. The parameters are $K = \frac{1}{2}$, $\gamma = -0.02$, $\lambda^Y = 300$, $T = 1$. The option expires in the money, and thus the terminal positions are fully invested in the underlying assets.

write $\nu(dx, dt) = \nu(dx)dt$, so the dynamics of the jump process are determined by the sizes of the jumps, not by time.

We have previously considered market noise driven by Brownian motion or a Poisson process. The market noise in (10.36) is a generalization of a compensated Poisson process, which corresponds to the choice of

$$\nu(dx, dt) = \lambda\delta_{(1)}(x)dx,$$

where $\delta_{(c)}$ is the Dirac delta function (A.1). A compensated compound Poisson process corresponds to

$$\nu(dx, dt) = \sum_{i=1}^{n} \lambda^i \delta_{(c^i)}(x)dx,$$

which means that the process makes jumps of sizes c^i with intensities λ^i, $i = 1, \ldots, n$. This is just a sum of n compensated Poisson processes with the corresponding jump sizes and intensities. However, the jump process can take a more general form; it can even exhibit infinitely large activity for small enough jumps, meaning that $\nu(x)$ may converge to infinity for $x \to 0$. It is possible that $\int_{-\infty}^{\infty} \nu(dx) = \infty$, which represents the case of infinite activity of jumps, but we must assume that $\int_{-\infty}^{\infty}(|x|^2 \wedge 1)\nu(dx) < \infty$ in order to prevent explosions in the market noise. An example of a compensator that may have infinite jump activity is for instance

$$\nu(x) = \begin{cases} C\frac{\exp(-G|x|)}{|x|^{1+Y}}, & x < 0 \\ C\frac{\exp(-M|x|)}{|x|^{1+Y}}, & x > 0, \end{cases}$$

which corresponds to a CGMY model.

Let us assume that the price process is driven by the jump process with a random measure $\mu(dx, dt)$ and a compensator $\nu(dx)$. Then both price processes $X_Y(t)$ and $Y_X(t)$ are driven by the same jumps, but if one price jumps up, the inverse price jumps down accordingly. Thus the jump measure $\mu(dx, dt)$ driving the price processes differs only in the sign of the jump: $\mu^Y(dx, dt) = \mu^X(-dx, dt)$. The main difference is that the reference assets Y and X place different intensity on the jumps, so ν^Y represents the compensator associated with the reference asset Y, while ν^X represents the compensator associated with the reference asset X. Let us find the relationship between ν^Y and ν^X.

The price process $X_Y(t)$ can be written as

$$dX_Y(t) = \int_{-\infty}^{\infty} (e^x - 1) \cdot X_Y(t-)\Big(\mu^Y(dx, dt) - \nu^Y(dx)dt\Big), \tag{10.37}$$

which is a generalization of Equation (10.6). The integral is over different jump sizes. From Ito's formula, the inverse price process $Y_X(t)$ satisfies

$$dY_X(t) = \int_{-\infty}^{\infty} (e^{-x} - 1) \cdot Y_X(t-)\Big(\mu^Y(dx, dt) - e^x \nu^Y(dx)dt\Big), \tag{10.38}$$

or

$$dY_X(t) = \int_{-\infty}^{\infty} (e^{x} - 1) \cdot Y_X(t-)\Big(\mu^X(dx, dt) - e^{-x} \nu^Y(-dx)dt\Big) \tag{10.39}$$

after switching the sign. From the symmetry between X and Y, this equation can be written as

$$dY_X(t) = \int_{-\infty}^{\infty} (e^{x} - 1) \cdot Y_X(t-)\Big(\mu^X(dx, dt) - \nu^X(dx)dt\Big). \tag{10.40}$$

Comparing the two above equations, we conclude that the relationship of the Lévy measures ν^Y and ν^X is given by

$$\boxed{\nu^X(x) = e^{-x}\nu^Y(-x).} \tag{10.41}$$

Let us consider a price process $X_Y(t)$ of the form

$$dX_Y(t) = \int_{-\infty}^{\infty} (e^{x} - 1) \cdot X_Y(t-)\Big(\mu^Y(dx, dt) - \nu^Y(dx)dt\Big). \tag{10.42}$$

The corresponding inverse price process satisfies

$$dY_X(t) = \int_{-\infty}^{\infty} (e^{x} - 1) \cdot Y_X(t-)\Big(\mu^X(dx, dt) - \nu^X(dx)dt\Big), \tag{10.43}$$

where $\nu^X(x) = e^{-x}\nu^Y(-x)$.

THEOREM 10.3

The price function $u^Y(t,x) = \mathbb{E}^Y[V_Y(T)|X_Y(t) = x]$ satisfies integro-differential equation

$$\boxed{u_t^Y(t,x) + \int_{-\infty}^{\infty} \nu^Y(dy)\left[u^Y(t, e^y x) - u^Y(t,x) - (e^y - 1)x u_x^Y(t,x)\right] = 0} \tag{10.44}$$

with the terminal condition

$$u^Y(T,x) = f^Y(x), \tag{10.45}$$

and the price function $u^X(t,x) = \mathbb{E}^X[V_X(T)|Y_X(t) = x]$ satisfies integro-differential equation

$$\boxed{u_t^X(t,x) + \int_{-\infty}^{\infty} \nu^X(dy)\left[u^X(t, e^y x) - u^X(t,x) - (e^y - 1)\, x u_x^X(t,x)\right] = 0} \tag{10.46}$$

with the terminal condition

$$u^X(T,x) = f^X(x). \tag{10.47}$$

PROOF Let us show the partial integro-differential equation for u^Y. According to Ito's formula for the jump processes, we have

$$\begin{aligned}
u^Y(t, X_Y(t)) &= u^Y(0, X_Y(0)) + \int_0^t u_t^Y(s, X_Y(s-))\,ds \\
&\quad + \int_0^t u_x^Y(s, X_Y(s-))\,dX_Y(s) \\
&\quad + \sum_{0\le s\le t}\Big[u^Y(s, X_Y(s)) - u^Y(s, X_Y(s-)) \\
&\qquad - u_x^Y(s, X_Y(s-))\,\Delta X_Y(s)\Big] \\
&= u^Y(0, X_Y(0)) + \int_0^t u_t^Y(s, X_Y(s-))\,ds \\
&\quad + \int_0^t \int_{-\infty}^{\infty} u_x^Y(s, X_Y(s-))\,(e^y - 1)X_Y(s-) \\
&\qquad \cdot\Big(\mu^Y(dy, ds) - \nu^Y(dy)ds\Big) \\
&\quad + \int_0^t \int_{-\infty}^{\infty} \big[u^Y(s, e^y X_Y(s-)) - u^Y(s, X_Y(s-)) \\
&\qquad - u_x^Y(s, X_Y(s-))\,(e^y - 1)X_Y(s-)\big]\,\mu^Y(dy, ds) \\
&= u^Y(0, X_Y(0)) + \int_0^t u_t^Y(s, X_Y(s-))\,ds \\
&\quad + \int_0^t \int_{-\infty}^{\infty} \nu^Y(dy)\,\big[-(e^y - 1)X_Y(s-)u_x^Y(s, X_Y(s-)) \\
&\quad + \big(u^Y(s, e^y X_Y(s-)) - u^Y(s, X_Y(s-))\big)\big]\,ds \\
&\quad + \int_0^t \int_{-\infty}^{\infty} \big(u^Y(s, e^y X_Y(s-)) - u^Y(s, X_Y(s-))\big) \\
&\qquad \cdot\Big(\mu^Y(dy, ds) - \nu^Y(dy)ds\Big).
\end{aligned}$$

We have subtracted

$$\int_{-\infty}^{\infty} \big[u^Y(s, e^y X_Y(s-)) - u^Y(s, X_Y(s-))\big]\,\nu^Y(dy)ds$$

from the $\mu^Y(dy, ds)$ term in order to obtain the martingale part of the equality, and added it back to the corresponding ds term. Since $u(t, X_Y(t))$ is a $\mathbb{P}^Y$-martingale, the corresponding ds term that is not a part of the martingale $\mu^Y(dy, ds) - \nu^Y(dy)ds$ must vanish. Thus we have the following identity:

$$u_t^Y(t,x) + \int_{-\infty}^{\infty} \nu^Y(dy) \left[u^Y(t, e^y x) - u^Y(t,x) - (e^y - 1)x u_x^Y(t,x)\right] = 0 \tag{10.48}$$

The terminal condition is given by

$$u^Y(T,x) = f^Y(x). \tag{10.49}$$

□

The Lévy Jump Model is in general incomplete with the exception of a single jump size model.

References and Further Reading

Jump models of the price date back to Merton (1976) who assumed normal distribution of the jump sizes. Kou (2002) suggested exponential distribution of the jump sizes, leading to a better fit with empirical data. Carr et al. (2002) suggested an even more general class of jump models that also covers exponential distribution of the jump sizes, which also covers processes with finite and infinite activity. Symmetry analysis of the price processes driven by jumps appears in Jamshidian (2007) and in Fajardo and Mordecki (2006). A standard reference for stochastic calculus with jump processes is Protter (2005). The reader can also refer to the monographs on financial modeling with jump processes by Schoutens (2003), and Cont and Tankov (2003). Cont and Voltchkova (2005b,a) study numerical implementations of partial integro-differential equations that arise in pricing of financial contracts. Carr et al. (2001) discuss hedging in incomplete markets.

Exercises

10.1 Show that the following processes are martingales.

(a) $(N(t) - \lambda t)^2 - \lambda t$.

(b) $\exp\left(\gamma N(t) - (e^\gamma - 1)\lambda t\right)$.

10.2 Assume that the price process $X_Y(t)$ follows

$$dX_Y(t) = (e^{\gamma} - 1) \cdot X_Y(t-)d(N(t) - \lambda^Y t).$$

Use Ito's formula for the jump processes to determine

$$d\log(X_Y(t)).$$

10.3 Use Ito's formula for the jump processes to determine

$$d[X_Y(t)]^{\alpha},$$

where $X_Y(t)$ is a geometric Poisson process.

10.4 Consider a geometric Poisson model for two no-arbitrage assets X and Y, where the price follows

$$dX_Y(t) = (e^{\gamma} - 1) \cdot X_Y(t-)d(N(t) - \lambda^Y t).$$

(a) Compute the price of a contract that pays off $V_T = \mathbb{I}(N(T) = 0) \cdot Y_T$ (in terms of a reference asset Y, and the probability measure $\mathbb{P}^Y$ determined by λ^Y). This contract gives its holder the asset Y if there is no jump in the price at time T.

(b) Double check the result from (a) by computing the price of $V_X(T)$ using the reference asset X. Use the relationship between λ^Y and λ^X.

(c) Compute the hedge of the contract: $\Delta^X(t)$ and $\Delta^Y(t)$.

(d) Compute the price of a contract that pays off $V_T = \mathbb{I}(N(T) = k) \cdot Y_T$ for general $k \in \mathbb{N}$ together with $\Delta^X(t)$ and $\Delta^Y(t)$.

10.5 Consider the geometric Poisson model with the price evolution

$$dX_Y(t) = (e^{\gamma} - 1)X_Y(t-)d(N(t) - \lambda^Y t).$$

(a) Find the price of a contract that pays off a unit of an asset X when $N(T) = k$.

(b) Find the price of a contract that pays off a unit of an asset Y when $N(T) = k$.

(c) Which of the two contracts is more valuable? Find a condition for k for which the contract that pays off the asset X is more valuable (inequality in terms of γ, λ^Y, T, and $X_Y(0)$).

10.6 Show that $u^Y(t, x) = 1$ and $u^Y(t, x) = x$ are solutions of the integro-differential equation (10.44). An analogous result is valid for the integro-differential equation (10.46). What are the financial contracts that correspond to these solutions? How would you hedge these contracts?

10.7 Consider the average asset A defined in (9.1).

(a) Find

$$dA_Y(t)$$

in the geometric Poisson process model.

(b) Let $u^Y(t, X_Y(t), A_Y(t))$ be the price of the Asian option with payoff $f^Y(X_Y(T), A_Y(T))$. Using the Ito's formula for jumps (with 2 spatial variables), find the integro-differential equation for $u^Y(t, x, y)$. Show that $u^Y(t, x, y) = 1$, $u^Y(t, x, y) = x$, and $u^Y(t, x, y) = y$ are solutions of this equation. Also check that when the contract does not depend on A, the price function $u^Y(t, x, y)$ does not depend on the variable x, and the integro-differential equation simplifies to (10.44).

(c) Find

$$dA_X(t)$$

and the integro-differential equation for $u^X(t, x, y)$. Show that $u^X(t, x, y) = 1$, $u^X(t, x, y) = x$, and $u^X(t, x, y) = y$ are solutions of this equation. Note that when the payoff does not depend on the asset Y, the integro-differential equation does not depend on the variable x, and the equation simplifies.

The price processes Y_A and X_A do not preserve the stationarity property, and thus the equation for u^A cannot be written in terms of a compensator ν^A as the compensator does not exist.

Appendix A

Elements of Probability Theory

A.1 Probability, Random Variables

This section summarizes some basic concepts in probability theory. **Probability** is a triplet $(\Omega, \mathcal{F}, \mathbb{P})$, where Ω is a set of outcomes (also known as sample space), $\mathcal{F}$ is a set of events, and $\mathbb{P}$ is the probability measure. We denote by ω individual outcomes from the set Ω. The set of events $\mathcal{F}$ includes combinations of outcomes, and thus each event from $\mathcal{F}$ is a subset of Ω. Moreover, the set of events $\mathcal{F}$ includes the set of all outcomes Ω, and is closed under complements and countable unions. In mathematical notation

$$A \in \mathcal{F} \Rightarrow (\Omega \setminus A) \in \mathcal{F}, \qquad A_i \in \mathcal{F} \Rightarrow \bigcup_{i=1}^{\infty} A_i \in \mathcal{F}.$$

A probability measure assigns a value on the interval $[0, 1]$ to events in $\mathcal{F}$ with the following restrictions: the probability of the set of all outcomes Ω is one, and when the sets A_i are disjoint ($A_i \cap A_j = \emptyset,\ i \neq j$), probability of $\bigcup_i A_i$ is the sum of probabilities of individual events:

$$\mathbb{P}(\Omega) = 1, \qquad \mathbb{P}(\bigcup_{i=1}^{\infty} A_i) = \sum_{i=1}^{\infty} \mathbb{P}(A_i).$$

Example A.1
Consider a coin toss. The two possible outcomes are "Head" and "Tail," so $\Omega = \{\text{Head, Tail}\}$. The set of events is $\mathcal{F} = \{\emptyset, \{\text{Head}\}, \{\text{Tail}\}, \Omega\}$. The probability measure $\mathbb{P}$ is determined by the value of $\mathbb{P}(\{\text{Head}\}) = p \in [0, 1]$, because $\mathbb{P}(\{\text{Tail}\}) + \mathbb{P}(\{\text{Head}\}) = \mathbb{P}(\Omega) = 1$, and thus $\mathbb{P}(\{\text{Tail}\}) = 1 - p$. Note that we may have different probability measures determined by the value p, that comes with the same set of outcomes Ω and the same set of events $\mathcal{F}$. □

What is the purpose of assigning probability to events $\mathcal{F}$ rather than assigning the probability directly to individual outcomes ω? We have seen in Example A.1 that the probability of the individual outcomes determines the probability of all events, but this does not work in general. One example of why it is not sufficient to assign probability just to individual outcomes is a situation when the set of outcomes is uncountable (such as the real line). In this case the probability of each individual outcome may be zero, and it would be impossible to reconstruct the probability of events from outcomes with zero probability. For that one has to start with events that contain more than a critical fraction of outcomes, such as intervals in the case of continuous distributions on the real line.

Note that the set of outcomes may have nonnumerical values, such as in the case of the coin toss when the set of outcomes is $\Omega = \{\text{Head}, \text{Tail}\}$. When the outcomes are assigned a number X, then we say X is a **random variable.** Formally, a random variable is a mapping

$$X : \Omega \to \mathbb{R}$$

with the property that for each Borel set B,

$$\{X \in B\} = \{\omega \in \Omega; X(\omega) \in B\}$$

is an event from the set of events $\mathcal{F}$. This assures that the probability of the event $\{X \in B\}$ is well defined. Every Borel set can be obtained from closed intervals $[a, b]$ by taking complements, or countable unions of intersections of these sets.

We call

$$F(x) = \mathbb{P}(X \leq x),$$

a cumulative distribution function. The cumulative distribution function already determines $\mathbb{P}(X \in B)$ for all Borel sets, in which case we talk about the distribution of a random variable. The derivative f of the cumulative distribution function, if it exists, is known as a **density**

$$f(x)dx = dF(x).$$

In the case when the cumulative distribution function has jumps (in the case of discrete distributions), the density function does not exist in the classical sense as a function. However, it exists in terms of linear functionals, measures that operate on a space of functions. For instance the Dirac delta $\delta_{(c)}(x)$ is defined by the following relationship

$$\int_{-\infty}^{\infty} g(x)\delta_{(c)}(x)dx = g(c). \tag{A.1}$$

The Dirac delta itself is not a function, but one can view it as a graph that has zero value outside c, but when evaluated at c it behaves like an impulse of a size 1. Indeed, we have

$$\int_{c-\epsilon}^{c+\epsilon} \delta_{(c)}(x)dx = 1$$

for any $\epsilon > 0$.

The expectation of a random variable is defined as

$$\mathbb{E}X = \int_\Omega X(\omega)d\mathbb{P}(\omega) = \int_{-\infty}^{\infty} xdF(x) = \int_{-\infty}^{\infty} xf(x)dx.$$

Example A.2 Uniform Distribution
The discrete uniform distribution on the interval $[0,1]$ is determined by the following probabilities

$$\mathbb{P}(X = \tfrac{k}{n}) = \tfrac{1}{n+1}, \quad k = 0, 1, \ldots, n.$$

This can be viewed as a density using the Dirac deltas:

$$f(x) = \tfrac{1}{n+1} \cdot \delta_{(k/n)}(x).$$

The expectation of this random variable is given by

$$\mathbb{E}X = \int_\Omega X(\omega)d\mathbb{P}(\omega) = \int_{-\infty}^{\infty} x \cdot f(x)dx = \sum_{k=0}^{n} \tfrac{k}{n} \cdot \tfrac{1}{n+1} = \tfrac{1}{2}.$$

The continuous uniform distribution has density

$$f(x) = \mathbb{I}([0,1])(x);$$

i.e., it is equal to 1 on the interval $[0,1]$, and 0 everywhere else. The expectation is $\frac{1}{2}$, which is the same as in the discrete case. □

Example A.3 Binomial Distribution
The binomial distribution $B(n,p)$ is given by the following probabilities:

$$\mathbb{P}(X = k) = \binom{n}{k} p^k \cdot (1-p)^{n-k}, \quad \text{for } k = 0, 1, \ldots, n.$$

In terms of the density function, this can be expressed as

$$f(x) = \binom{n}{k} p^k \cdot (1-p)^{n-k} \delta_{(k)}(x).$$

Its expectation is given by

$$\mathbb{E}X = \int_\Omega X(\omega)d\mathbb{P}(\omega) = \int_{-\infty}^{\infty} x \cdot f(x)dx = \sum_{k=0}^{n} k \cdot \binom{n}{k} p^k \cdot (1-p)^{n-k} = n \cdot p.$$

□

Example A.4 Poisson Distribution
The Poisson distribution is given by the following probabilities:

$$\mathbb{P}(X = k) = e^{-\lambda} \cdot \frac{\lambda^k}{k!},$$

where $\lambda > 0$ is a parameter. We can write the density of Poisson distribution as

$$f(x) = e^{-\lambda} \cdot \frac{\lambda^k}{k!} \delta_{(k)}(x).$$

Its expectation is given by

$$\mathbb{E}X = \int_{\Omega} X(\omega) d\mathbb{P}(\omega) = \int_{-\infty}^{\infty} x \cdot f(x) dx = \sum_{k=0}^{\infty} k \cdot e^{-\lambda} \cdot \frac{\lambda^k}{k!} = \lambda.$$

□

Example A.5 Normal Distribution
The normal distribution $N(\mu, \sigma^2)$ has the following density:

$$f(x) = \frac{1}{\sigma\sqrt{2\pi}} \cdot \exp\left(-\frac{(x-\mu)^2}{2\sigma^2}\right).$$

Its expectation is given by

$$\begin{aligned} \mathbb{E}X = \int_{\Omega} X(\omega) d\mathbb{P}(\omega) &= \int_{-\infty}^{\infty} x \cdot f(x) dx \\ &= \int_{-\infty}^{\infty} x \cdot \frac{1}{\sigma\sqrt{2\pi}} \cdot \exp\left(-\frac{(x-\mu)^2}{2\sigma^2}\right) = \mu. \end{aligned}$$

□

Example A.6 Exponential Distribution
The exponential distribution has a density given by

$$f(x) = \lambda e^{-\lambda x} \mathbb{I}_{(x>0)} dx.$$

Its expectation is equal to

$$\mathbb{E}X = \int_{\Omega} X(\omega) d\mathbb{P}(\omega) = \int_{-\infty}^{\infty} x \cdot f(x) dx = \int_{0}^{\infty} x \cdot \lambda e^{-\lambda x} dx = \frac{1}{\lambda}.$$

□

Example A.7 Gamma Distribution
The gamma distribution is given by the density

$$f(x) = \frac{x^{\alpha-1}e^{-\frac{x}{\beta}}}{\Gamma(\alpha)\beta^{\alpha}}\mathbb{I}_{(x>0)}dx.$$

Its expectation is equal to

$$\mathbb{E}X = \int_{\Omega} X(\omega)d\mathbb{P}(\omega) = \int_{-\infty}^{\infty} x \cdot f(x)dx = \int_{0}^{\infty} x \cdot \frac{x^{\alpha-1}e^{-\frac{x}{\beta}}}{\Gamma(\alpha)\beta^{\alpha}}dx = \alpha \cdot \beta.$$

□

A.2 Conditional Expectation

A σ-algebra is a set of events that is closed under complements, countable unions, and intersections. We denote by $\sigma(A_s)$ the smallest σ-algebra that contains all sets A_s. Let X be a random variable. By information of a random variable $\sigma(X)$ we mean the σ-algebra generated by the sets

$$\{\omega \in \Omega : X(\omega) \in B\}$$

for any Borel set B. Borel sets on a real line are open intervals, or sets that are created by taking countable unions or intersections of the previously obtained sets, starting from the open intervals.

DEFINITION A.1 Conditional Expectation
The conditional expectation of a random variable X with respect to the information set $\mathcal{F}$, $\mathbb{E}[X|\mathcal{F}]$, is a random variable that satisfies the following properties:

1. $\sigma(\mathbb{E}[X|\mathcal{F}]) \subseteq \mathcal{F}$,
2. *for each* $A \in \mathcal{F}$, $\mathbb{E}[\mathbb{E}[X|\mathcal{F}] \cdot \mathbb{I}_A] = \mathbb{E}[X \cdot \mathbb{I}_A]$.

Property 1 of the conditional expectation means that the information of the conditional expectation as a random variable is not larger than the information set $\mathcal{F}$. Property 2 ensures that the conditional expectation has the same average as the original random variable on sets that belong to $\mathcal{F}$.

Example A.8
Consider a set of outcomes that corresponds to 2 coin tosses: $\Omega = \{HH, HT, TH, TT\}$. Let us assume that the coin toss is fair, and all the out-

comes have the same probability $\frac{1}{4}$: $\mathbb{P}(HH) = \mathbb{P}(HT) = \mathbb{P}(TH) = \mathbb{P}(TT) = \frac{1}{4}$. Define a random variable X with the following property:

$$\text{X} = (\text{1st coin resulted in H}) \times 1 + (\text{2nd coin toss resulted in H}) \times 2,$$

or in other words,

$$X(HH) = 3, \quad X(HT) = 1, \quad X(TH) = 2, \quad X(TT) = 0.$$

Let us consider three times: $t = 0$ before any coin toss, $t = 1$ after the first coin toss but before the second coin toss, and $t = 2$ after the second coin toss. At time $t = 0$, there is no information available, and thus the corresponding information set is trivially $\mathcal{F}_0 = \{\emptyset, \Omega\}$. At time $t = 1$, it is possible to distinguish between events $\{HH, HT\}$ and $\{TH, TT\}$, but it is not possible to distinguish between $\{HH\}$ and $\{HT\}$, or between $\{TH\}$ and $\{TT\}$ since the second coin toss has not yet been observed. Thus the information set $\mathcal{F}_1$ contains events $\mathcal{F}_1 = \{\emptyset, \Omega, \{HH, HT\}, \{TH, TT\}\}$. At time $t = 2$, we have a complete information set that can distinguish the events $\{HH\}$, $\{HT\}$, $\{TH\}$, $\{TT\}$. Thus the information set is given by $\mathcal{F}_2 = \sigma(\{HH\}, \{HT\}, \{TH\}, \{TT\})$.

Let us determine the conditional expectations $\mathbb{E}[X|\mathcal{F}_0](\omega)$, $\mathbb{E}[X|\mathcal{F}_1](\omega)$, $\mathbb{E}[X|\mathcal{F}_2](\omega)$. Let us start with $\mathbb{E}[X|\mathcal{F}_0](\omega)$. From Property 1 of the conditional expectation, we must have

$$\sigma(\mathbb{E}[X|\mathcal{F}_0]) \subseteq \mathcal{F}_0 = \{\emptyset, \Omega\}.$$

This means $\mathbb{E}[X|\mathcal{F}_0](\omega)$ must be a constant for all $\omega \in \Omega$. From Property 2 of the conditional expectation, using $A = \Omega$ and $\mathbb{I}_\Omega = 1$, we have

$$\mathbb{E}[\mathbb{E}[X|\mathcal{F}_0]] = \mathbb{E}[X].$$

Since $\mathbb{E}[X|\mathcal{F}_0]$ must be a constant, we conclude that

$$\mathbb{E}[X|\mathcal{F}_0] = \mathbb{E}[X] = (3 \cdot \tfrac{1}{4} + 1 \cdot \tfrac{1}{4} + 2 \cdot \tfrac{1}{4} + 0 \cdot \tfrac{1}{4}) = \tfrac{3}{2}.$$

Conditional expectation $\mathbb{E}[X|\mathcal{F}_1](\omega)$ must satisfy Property 1, namely

$$\sigma(\mathbb{E}[X|\mathcal{F}_1]) \subseteq \mathcal{F}_1 = \{\emptyset, \Omega, \{HH, HT\}, \{TH, TT\}\}.$$

The information set is richer than in the case of $\mathcal{F}_0$, which means that the conditional expectation can distinguish between events $\{HH, HT\}$ and $\{TH, TT\}$, so its value may differ on them. However, the conditional expectation does not distinguish events $\{HH\}$ and $\{HT\}$, or $\{TH\}$ and $\{TT\}$, meaning that

$$\mathbb{E}[X|\mathcal{F}_1](HH) = \mathbb{E}[X|\mathcal{F}_1](HT) = C_1; \quad \mathbb{E}[X|\mathcal{F}_1](TH) = \mathbb{E}[X|\mathcal{F}_1](TT) = C_2.$$

This ensures Property 1. Let us compute C_1 and C_2 using Property 2. Let us consider $A = \{HH, HT\}$. We get

$$\begin{aligned}\mathbb{E}[\mathbb{E}[X|\mathcal{F}_1] \cdot \mathbb{I}_A] = \mathbb{E}[X|\mathcal{F}_1](HH) \cdot 1 \cdot \tfrac{1}{4} + \mathbb{E}[X|\mathcal{F}_1](HT) \cdot 1 \cdot \tfrac{1}{4} \\ + \mathbb{E}[X|\mathcal{F}_1](TH) \cdot 0 \cdot \tfrac{1}{4} + \mathbb{E}[X|\mathcal{F}_1](TT) \cdot 0 \cdot \tfrac{1}{4} = \tfrac{C_1}{2}.\end{aligned}$$

$$\begin{aligned}\mathbb{E}[X \cdot \mathbb{I}_A] = X(HH) \cdot 1 \cdot \tfrac{1}{4} + X(HT) \cdot 1 \cdot \tfrac{1}{4} + X(TH) \cdot 0 \cdot \tfrac{1}{4} + X(TT) \cdot 0 \cdot \tfrac{1}{4} \\ = 3 \cdot 1 \cdot \tfrac{1}{4} + 1 \cdot 1 \cdot \tfrac{1}{4} + 2 \cdot 0 \cdot \tfrac{1}{4} + 0 \cdot 0 \cdot \tfrac{1}{4} = 1.\end{aligned}$$

Since

$$\mathbb{E}[\mathbb{E}[X|\mathcal{F}_1] \cdot \mathbb{I}_A] = \mathbb{E}[X \cdot \mathbb{I}_A],$$

we must have $\frac{C_1}{2} = 1$, or in other words, $C_1 = 2$. We can similarly determine C_2 by considering $A = \{TH, TT\}$ and using Property 2:

$$\begin{aligned}\mathbb{E}[\mathbb{E}[X|\mathcal{F}_1] \cdot \mathbb{I}_A] = \mathbb{E}[X|\mathcal{F}_1](HH) \cdot 0 \cdot \tfrac{1}{4} + \mathbb{E}[X|\mathcal{F}_1](HT) \cdot 0 \cdot \tfrac{1}{4} \\ + \mathbb{E}[X|\mathcal{F}_1](TH) \cdot 1 \cdot \tfrac{1}{4} + \mathbb{E}[X|\mathcal{F}_1](TT) \cdot 1 \cdot \tfrac{1}{4} = \tfrac{C_2}{2}.\end{aligned}$$

$$\begin{aligned}\mathbb{E}[X \cdot \mathbb{I}_A] = X(HH) \cdot 0 \cdot \tfrac{1}{4} + X(HT) \cdot 0 \cdot \tfrac{1}{4} + X(TH) \cdot 1 \cdot \tfrac{1}{4} + X(TT) \cdot 1 \cdot \tfrac{1}{4} \\ = 3 \cdot 0 \cdot \tfrac{1}{4} + 1 \cdot 0 \cdot \tfrac{1}{4} + 2 \cdot 1 \cdot \tfrac{1}{4} + 0 \cdot 1 \cdot \tfrac{1}{4} = \tfrac{1}{2}.\end{aligned}$$

Since

$$\mathbb{E}[\mathbb{E}[X|\mathcal{F}_1] \cdot \mathbb{I}_A] = \mathbb{E}[X \cdot \mathbb{I}_A],$$

we must have $\frac{C_2}{2} = \frac{1}{2}$, or in other words, $C_2 = 1$. The values of $\mathbb{E}[X|\mathcal{F}_1]$ are listed in Table A.1.

As for $\mathbb{E}[X|\mathcal{F}_2]$, from Property 1 we get

$$\sigma(\mathbb{E}[X|\mathcal{F}_2]) \subseteq \mathcal{F}_2 = \sigma(\{HH\}, \{HT\}, \{TH\}, \{TT\}),$$

meaning that the conditional expectation can distinguish between all outcomes ω from Ω. We can write

$$\mathbb{E}[\mathbb{E}[X|\mathcal{F}_2] \cdot \mathbb{I}_{(\omega)}] = \mathbb{E}[X|\mathcal{F}_2](\omega) \cdot \mathbb{P}(\omega),$$

and

$$\mathbb{E}[X \cdot \mathbb{I}_{(\omega)}] = X(\omega) \cdot \mathbb{P}(\omega).$$

From Property 2, we must have

$$\mathbb{E}[X|\mathcal{F}_2](\omega) \cdot \mathbb{P}(\omega) = X(\omega) \cdot \mathbb{P}(\omega),$$

or in other words

$$\mathbb{E}[X|\mathcal{F}_2] = X,$$

meaning that the conditional expectation is the original random variable X itself.

$\square$

TABLE A.1: Conditional expectation.

ω	$X(\omega)$	$\mathbb{E}[X\vert\mathcal{F}_0](\omega)$	$\mathbb{E}[X\vert\mathcal{F}_1](\omega)$	$\mathbb{E}[X\vert\mathcal{F}_2](\omega)$
HH	3	$\frac{3}{2}$	2	3
HT	1	$\frac{3}{2}$	2	1
TH	2	$\frac{3}{2}$	1	2
TT	0	$\frac{3}{2}$	1	0

A.2.1 Some Properties of Conditional Expectation

Let us mention some important properties of the conditional expectation.
Linearity:

$$\mathbb{E}[aX + bY|\mathcal{F}] = a\mathbb{E}[X|\mathcal{F}] + b\mathbb{E}[Y|\mathcal{F}].$$

Independence: If X is integrable and independent of $\mathcal{F}$:

$$\mathbb{E}[X|\mathcal{F}] = \mathbb{E}[X].$$

Taking out what is known: If $\sigma(Y) \subseteq \mathcal{F}$

$$\mathbb{E}[X \cdot Y|\mathcal{F}] = Y \cdot \mathbb{E}[X|\mathcal{F}].$$

The special situation when $X = 1$ reads as

$$\mathbb{E}[Y|\mathcal{F}] = Y.$$

Tower property: If $\mathcal{G} \subseteq \mathcal{F}$

$$\mathbb{E}[\mathbb{E}[X|\mathcal{F}]|\mathcal{G}] = \mathbb{E}[X|\mathcal{G}].$$

Projection: The random variables $X - \mathbb{E}[X|\mathcal{F}]$ and $\mathbb{E}[X|\mathcal{F}]$ have zero correlation:

$$\mathbb{E}\left[[X - \mathbb{E}[X|\mathcal{F}]] \cdot \mathbb{E}[X|\mathcal{F}]\right] = 0.$$

A.3 Martingales

A process $\mathcal{M}(t)$ is called a **martingale** if

$$\mathbb{E}[\mathcal{M}(t)|\mathcal{F}_s] = \mathcal{M}(s). \tag{A.2}$$

Let us introduce the following notation. When the filtration $\mathcal{F}_t$ represents information up to time t, we will simply write

$$\mathbb{E}_t[V] := \mathbb{E}[V|\mathcal{F}_t].$$

The martingale property can be rewritten as

$$\mathbb{E}_s[\mathcal{M}(t)] = \mathcal{M}(s).$$

Let us show several processes that are martingales. Let $X(i)$ be a random variable

$$X(i) = \begin{cases} 1 & p = \frac{1}{2}, \\ -1 & 1 - p = \frac{1}{2}. \end{cases}$$

Then $S(n) = \sum_{i=1}^{n} X(i)$ is a martingale, and $S^2(n) - n$ is a martingale. In order to prove the martingale property, it is enough to show that

$$\mathbb{E}_n[\mathcal{M}(n+1)] = \mathcal{M}(n).$$

In particular,

$$\mathbb{E}_n[S(n+1)] = \mathbb{E}_n[S(n) + X(n+1)] = S(n) + \mathbb{E}[X(n+1)] = S(n).$$

The second equality follows from the fact that $S(n)$ is known at time n, and $X(n+1)$ is independent of $\mathcal{F}_n$, and thus the conditional expectation becomes a simple expectation. As for the second statement, we have

$$\begin{aligned} \mathbb{E}_n[S^2(n+1) - (n+1)] &= \mathbb{E}_n[(S(n) + X(n+1))^2] - (n+1) \\ &= \mathbb{E}_n[S^2(n) + 2S(n) \cdot X(n+1) + X^2(n+1)] - (n+1) \\ &= S^2(n) + 2S(n) \cdot \mathbb{E}[X(n+1)] + \mathbb{E}[X^2(n+1)] - (n+1) \\ &= S^2(n) + 1 - n - 1 = S^2(n) - n. \end{aligned}$$

The process $S(n)$ is known as a **discrete random walk.** One can think about this process as a basic market noise which comes from playing a fair game with outcomes 1 and -1. At each time step, the player bets one dollar and has a fifty percent chance of winning an extra dollar, and a fifty percent chance of losing the stake. The process $S(n)$ measures the player's winning or losing balance after n steps.

Martingales represent fair games. The First Fundamental Theorem of Asset Pricing (Chapter 1) states that prices must be martingales (under a certain measure associated with the reference asset) in order to prevent an opportunity for risk-free profit. Intuitively, the market cannot be beaten by playing such a game. In particular, the expected value of the martingale stays the same, so we have

$$\mathbb{E}[S(n)] = S(0) = 0.$$

A natural question is whether one can beat the market (or the coin tossing game in this case) by taking a more sophisticated strategy. The player's controls are twofold: the size of the bet, and when to quit playing the game. Let us first check the varying size of the bet. The player bets $\Delta(n)$ at time n, so the resulting profit or loss P from betting satisfies

$$P(n) = \sum_{i=1}^{n} \Delta(i) \cdot X(i).$$

The process $P(n)$ is still a martingale

$$\begin{aligned}\mathbb{E}_n[P(n+1)] &= \mathbb{E}_n[P(n) + \Delta(n+1) \cdot X(n+1)] \\ &= P(n) + \Delta(n+1) \cdot \mathbb{E}[X(n+1)] = P(n).\end{aligned}$$

The crucial part is that the size of the bet, $\Delta(n+1)$, is known at time n (before the outcome of the bet $X(n+1)$), so that it can be taken out of the conditional expectation. This is always the case in real markets or in casino games (think about roulette). If $\Delta(n+1)$ could be set after observing the outcome of $X(n+1)$, this would lead to a possibility of a risk-free profit. One can take for instance $\Delta(n+1) = X(n+1)$, in which case $P(n) = \sum_{i=1}^{n} \Delta(i) \cdot X(i) = \sum_{i=1}^{n} X^2(i) = n$.

Example A.9 **Doubling strategy**

A particular case of a betting strategy is a doubling strategy. The player's initial bet is $\Delta(1) = 1$. If he wins, he would collect one dollar (two dollars won minus one dollar bet), and quit the game. If he loses, he would double his previous bet ($\Delta(2) = 2$). If he wins in the second round, he would collect one dollar (four dollars won minus three dollars bet). If he loses again, he would double his bets, and continue in this procedure. After n steps, the probability that he won by that time is $1 - \frac{1}{2^n}$. The winning is always one dollar:

$$2^n - (1 + 2 + \cdots + 2^{n-1}) = 1.$$

The chance that he has not won by time n is $\frac{1}{2^n}$, in which case the total loss is equal to

$$-(1 + 2 + \cdots + 2^{n-1}) = -(2^n - 1).$$

Thus we have

$$P(n) = \begin{cases} 1 & p = 1 - \frac{1}{2^n}, \\ -(2^n - 1) & 1 - p = \frac{1}{2^n}. \end{cases}$$

While $\lim_{n\to\infty} \mathbb{P}(P(n) = 1) = 1$, we still have $\mathbb{E}[P(n)] = 0$. The probability of winning is indeed large and converges to one, but one still does not beat the expected value of the game. The reason is the unlikely event that one would not win a single time in n trials, but the loss corresponding to this case is astronomical. This strategy would eventually lead to a bankruptcy. ☐

The second control over the game from the perspective of a player is to choose a time when to stop. Let us introduce the following concept.

DEFINITION A.2 *A* **stopping time** τ *is a random time with the property that*

$$\{\tau \leq t\} \in \mathcal{F}_t. \tag{A.3}$$

The condition $\{\tau \leq t\} \in \mathcal{F}_t$ means that at every moment t, one is able to tell if the event $\{\tau \leq t\}$ has happened or not. Being unsure about this event is not an option. The simplest example of a stopping time is when $\tau = N$, where N is a constant. Practical examples of stopping times include the time when a price of a stock crosses a certain level, time of the arrival of a train, or the time of getting married. These examples illustrate that one can tell if the event happened or not. For instance, one can always tell in principle if he or she is married or not married, although in this case one may not be certain about it, like after a good party in Las Vegas. This was the case of a basketball player Dennis Rodman who got married to Carmen Electra in the state of intoxication, and had no recollection of the event (this marriage was later annulled). This is also a good illustration of asymmetrical information since the time of his marriage was not a stopping time with respect to Rodman's information set, but it was a stopping time with respect to the information set of the state of Nevada (they could still tell if he was married or not). A simpler example of a time that is not a stopping time is for instance a half-life of the event. The information about the exact time of a half-life will be known only after observing the event in the future, and one is not sure if the half-life time already happened or not.

Stopping times play an important role in finance since one can construct contracts that pay off at such stopping times. For instance, barrier options become dead or alive depending on the price hitting a certain level, where the time of hitting is a stopping time. Contracts based on the stopping times are at least in legal terms well defined, as opposed to contracts where one is uncertain if the contractual party already qualifies for a payoff or not.

Let us return to the question if one can beat the casino (at least in a fair game) by adopting a strategy that involves a stopping time. Take for instance the first time τ when one is winning a dollar, defined by

$$\tau = \min\{n \geq 0 : S(n) = 1\}.$$

Clearly, $P(\tau) = 1$, and we have that

$$1 = \mathbb{E}[S(\tau)] \neq S(0) = 0.$$

Moreover, $\mathbb{P}(\tau < \infty) = 1$, which is not obvious, but still true. This means that one can potentially beat a fair game by waiting for the first time when his profit from the game is positive. However, this win does not come easily; it turns out that $\mathbb{E}[\tau] = \infty$, so the expected profit per expected time to reach it is still zero. The random variable τ is finite, but it has an infinite expectation.

In more practical situations, such as in the case of a bounded stopping time, one cannot beat a fair game. For this result to hold, the value of the game cannot take large values on small probability events. More precisely, the following theorem holds.

THEOREM A.1 Optional Sampling Theorem

Let $\mathcal{M}$ be a martingale, and τ be a stopping time with $\mathbb{P}(\tau < \infty) = 1$. If $\mathbb{E}[|\mathcal{M}(\tau)|] < \infty$, and

$$\lim_{n\to\infty} \mathbb{E}[|\mathcal{M}(n)|\mathbb{I}(\tau > n)] = 0, \tag{A.4}$$

then

$$\mathbb{E}[\mathcal{M}(\tau)] = \mathbb{E}[\mathcal{M}(0)]. \tag{A.5}$$

The condition (A.4) ensures that the martingale does not dissipate in space. For instance, in the case of a discrete random walk $S(n)$, $\mathbb{P}(\tau > n) \approx \sqrt{n}$, and $\mathbb{E}[|S(n)|\mathbb{I}(\tau > n)]$ does not go to zero.

Example A.10 Gambler's ruin problem

Suppose that a player starts with a dollars ($S(0) = a$), and plays a fair game so that his wealth follows a discrete random walk. He stops in two situations: he either loses all his money, so that $S(n) = 0$, or he wins enough to quit, so that $S(n) = N$. The stopping time is given by

$$\tau = \min\{n \geq 0 : S(n) = 0 \text{ or } S(n) = N\}.$$

Since $S(n)$ is bounded (it stays between 0 and N), the conditions of the Optional Sampling Theorem are satisfied, and we have

$$\mathbb{E}[S(\tau)] = \mathbb{E}[S(0)] = a.$$

In this case,

$$\mathbb{E}[S(\tau)] = N \cdot \mathbb{P}(S(\tau) = N),$$

and therefore

$$\mathbb{P}(S(\tau) = N) = \tfrac{a}{N}.$$

□

A.4 Brownian Motion

This section introduces a Brownian motion (also known as a Wiener Process) and shows its basic properties.

DEFINITION A.3 Brownian Motion
Brownian motion $W(t)$ is a process with the following properties:

1. *$W(t) - W(s)$ is independent of $\mathcal{F}_s$, $s < t$,*
2. *$W(t) - W(s)$ has a normal $N(0, t-s)$ distribution,*
3. *$W(t)$ is a continuous process.*

It is not obvious that a process with these properties exists, but the answer is indeed affirmative. Brownian motion can be constructed as a limit of a discrete random walk in the following sense. Let $S(n)$ be a discrete random walk, and define a process $W^{(n)}$ by

$$W^{(n)}(t) = \frac{1}{\sqrt{n}} S(nt)$$

for t of the form $\frac{k}{n}$, where k is an integer. If t is not in this form, we define $W^{(n)}(t)$ as a linear interpolation between its values at the nearest points $\frac{k}{n}$. When $t > s$ are equal to $\frac{k}{n}$ for integer k, we have that $W^{(n)}(t) - W^{(n)}(s)$ is independent of the information set $\mathcal{F}_s$. Furthermore,

$$\mathbb{E}[W^{(n)}(t) - W^{(n)}(s)] = \mathbb{E}\left[\frac{1}{\sqrt{n}} S(nt) - \frac{1}{\sqrt{n}} S(ns)\right] = 0,$$

and

$$\begin{aligned}
\text{Var}\left(W^{(n)}(t) - W^{(n)}(s)\right) &= \mathbb{E}[W^{(n)}(t) - W^{(n)}(s)]^2 \\
&= \mathbb{E}\left[\frac{1}{\sqrt{n}} S(nt) - \frac{1}{\sqrt{n}} S(ns)\right]^2 = \mathbb{E}\left[\frac{1}{\sqrt{n}} \cdot \sum_{k=ns}^{nt} X(k)\right]^2 \\
&= \frac{1}{n} \sum_{k=ns}^{nt} \mathbb{E}\left[X(k)^2\right] = \frac{1}{n} \sum_{k=ns}^{nt} 1 = \frac{1}{n} \cdot [nt - ns] = t - s.
\end{aligned}$$

When $n \to \infty$, the increments remain independent of $\mathcal{F}_s$, they will be normally distributed because of the Central Limit Theorem with the mean 0 and the variance $t - s$, and the limiting process will be continuous.

REMARK A.1 Martingales of Brownian motion

The following processes are martingales: $W(t)$, $W(t)^2 - t$, $e^{\sigma W(t) - \frac{1}{2}\sigma^2 t}$. This can be seen from the following relationships.

$$\mathbb{E}_s[W(t)] = \mathbb{E}_s[(W(t) - W(s)) + W(s)] = \mathbb{E}_s[W(t) - W(s)] + W(s) = W(s).$$

$$\begin{aligned}
\mathbb{E}_s[W(t)^2 - t] &= \mathbb{E}_s[(W(t) - W(s) + W(s))^2 - t] \\
&= \mathbb{E}_s[(W(t) - W(s))^2 - 2 \cdot W(s) \cdot (W(t) - W(s)) + W(s)^2 - t] \\
&= \mathbb{E}[(W(t) - W(s))^2] - 2 \cdot W(s) \cdot \mathbb{E}[W(t) - W(s)] + W(s)^2 - t \\
&= (t - s) - 2 \cdot W(s) \cdot 0 + W(s)^2 - t = W(s)^2 - s.
\end{aligned}$$

$$\begin{aligned}
\mathbb{E}_s\left[e^{\sigma W(t) - \frac{1}{2}\sigma^2 t}\right] &= \mathbb{E}_s\left[e^{\sigma((W(t) - W(s)) + W(s)) - \frac{1}{2}\sigma^2 t}\right] \\
&= e^{\sigma W(s) - \frac{1}{2}\sigma^2 t} \cdot \mathbb{E}\left[e^{\sigma(W(t) - W(s))}\right] \\
&= e^{\sigma W(s) - \frac{1}{2}\sigma^2 t} \cdot e^{\frac{1}{2}\sigma^2 (t-s)} = e^{\sigma W(s) - \frac{1}{2}\sigma^2 s}.
\end{aligned}$$

□

REMARK A.2 Hitting time of Brownian motion

Let us denote by τ_a the first time when a Brownian motion reaches a level $a > 0$:

$$\tau_a = \inf\{t \geq 0 : W(t) = a\}.$$

Note that we can write

$$\begin{aligned}
\mathbb{P}(\tau_a \leq t) &= \mathbb{P}(\tau_a \leq t, W(t) \geq a) + \mathbb{P}(\tau_a \leq t, W(t) < a) \\
&= 2\mathbb{P}(\tau_a \leq t, W(t) \geq a) = 2\mathbb{P}(W(t) \geq a) \\
&= \int_a^\infty \frac{1}{\sqrt{2\pi t}} \cdot e^{-\frac{y^2}{2t}} dy.
\end{aligned}$$

The event $W(t) \geq a$ already implies the event $\tau_a \leq t$. The equality $\mathbb{P}(\tau_a \leq t, W(t) \geq a) = \mathbb{P}(\tau_a \leq t, W(t) < a)$ follows from the so-called **reflection principle.** Once Brownian motion W reaches a, it is equally likely that the process will end up above the level a at a fixed future time T, or end up below a because of the symmetry of distribution of Brownian motion. We can obtain the density of the hitting time by taking a derivative of $\mathbb{P}(\tau_a \leq t)$ with respect to t

$$f(t) = \frac{\partial \mathbb{P}(\tau_a \leq t)}{\partial t} = \frac{a}{t\sqrt{2\pi t}} \cdot e^{-\frac{a^2}{2t}}.$$

□

REMARK A.3 Quadratic variation of Brownian motion
The **quadratic variation** of a process $\mathcal{X}$ up to time T is defined as

$$[\mathcal{X}, \mathcal{X}](T) = \lim_{\|\Pi\| \to 0} \sum_{i=1}^{n} [\mathcal{X}(t_i) - \mathcal{X}(t_{i-1})]^2, \tag{A.6}$$

where $\Pi = \{t_0 = 0, t_1, \ldots, t_n = T\}$ is a partition of the interval $[0, T]$, and $\|\Pi\| = \max_{1 \le i \le n} |t_i - t_{i-1}|$. A Brownian motion has the quadratic variation equal to T

$$\boxed{[W, W](T) = T.} \tag{A.7}$$

This is a remarkable result since each of the factors $(W(t_i) - W(t_{i-1})^2$ in the definition of quadratic variation is random, and thus $[W, W](T)$ is also a random variable. But the resulting value of the quadratic variation, T, has zero variance. It means that each path of a Brownian motion W has the same quadratic variation.

We can prove the result by showing that $\mathbb{E}[W, W](T) = T$, and $\mathrm{Var}([W, W](T)) = 0$. Define

$$Q_\Pi = \sum_{i=1}^{n} [W(t_i) - W(t_{i-1})]^2$$

the sampled quadratic variation that corresponds to the partition Π so that $\lim_{\|\Pi\| \to 0} Q_\Pi = [W, W](T)$. First,

$$\mathbb{E}[Q_\Pi] = \mathbb{E}\left[\sum_{i=1}^{n} [W(t_i) - W(t_{i-1})]^2\right] = \sum_{i=1}^{n} [t_i - t_{i-1}] = T.$$

From the independence of Brownian increments,

$$\mathrm{Var}(Q_\Pi) = \sum_{i=1}^{n} \mathrm{Var}\left([W(t_i) - W(t_{i-1})]^2\right),$$

so the variance of Q_Π is the sum of the variances of the squared increments of a Brownian motion. Since

$$\begin{aligned}
\mathrm{Var}\left([W(t_i) - W(t_{i-1})]^2\right) &= \mathbb{E}\left[\left([W(t_i) - W(t_{i-1})]^2 - (t_i - t_{i-1})\right)^2\right] \\
= \mathbb{E}\left[W(t_i) - W(t_{i-1})\right]^4 &- 2(t_i - t_{i-1})\mathbb{E}\left[W(t_i) - W(t_{i-1})\right]^2 + (t_i - t_{i-1})^2 \\
&= 3(t_i - t_{i-1})^2 - 2(t_i - t_{i-1})^2 + (t_i - t_{i-1})^2 = 2(t_i - t_{i-1})^2,
\end{aligned}$$

one finds

$$\mathrm{Var}(Q_\Pi) = \sum_{i=1}^{n} 2(t_i - t_{i-1})^2 \le 2 \sum_{i=1}^{n} \|\Pi\| \cdot (t_i - t_{i-1}) = 2\|\Pi\| \cdot T \to 0.$$

The relationship (A.7) can be also rewritten in infinitesimal form as

$$\boxed{(dW(t))^2 = dt.} \tag{A.8}$$

□

Note that the quadratic variation of a linear function t is zero, and the cross variation of W and t is also zero. To show these results, note that

$$[T,T](T) = \lim_{\|\Pi\|\to 0} \sum_{i=1}^{n} (t_i - t_{i-1})^2 \leq \lim_{\|\Pi\|\to 0} \|\Pi\| \cdot T = 0,$$

and

$$\begin{aligned}
[W,T](T) &= \lim_{\|\Pi\|\to 0} \sum_{i=1}^{n} (W(t_i) - W(t_{i-1}))(t_i - t_{i-1}) \\
&\leq \lim_{\|\Pi\|\to 0} \max_{1\leq i\leq n} |W(t_i) - W(t_{i-1})| \sum_{i=1}^{n} (t_i - t_{i-1}) = 0.
\end{aligned}$$

The term $\max_{1\leq i\leq n} |W(t_i) - W(t_{i-1})|$ converges to zero because of the continuity of a Brownian motion W. The two relationships can be written in the infinitesimal form as

$$\boxed{dt \cdot dt = 0,} \qquad \boxed{dW(t) \cdot dt = 0.} \tag{A.9}$$

The quadratic variation of a function f that has a continuous derivative is zero. From the mean value theorem, there exists $t_i^* \in [t_{i-1}, t_i]$ such that

$$f(t_i) - f(t_{i-1}) = f'(t_i^*) \cdot (t_i - t_{i-1}).$$

Therefore

$$\begin{aligned}
[f,f](T) &= \lim_{\|\Pi\|\to 0} \sum_{i=1}^{n} (f(t_i) - f(t_{i-1}))^2 \\
&= \lim_{\|\Pi\|\to 0} \sum_{i=1}^{n} (f'(t_i^*))^2 \cdot (t_i - t_{i-1})^2 \\
&\leq \lim_{\|\Pi\|\to 0} \|\Pi\| \cdot \sum_{i=1}^{n} (f'(t_i^*))^2 \cdot (t_i - t_{i-1}) \\
&= \lim_{\|\Pi\|\to 0} \|\Pi\| \cdot \int_0^T (f'(t))^2 dt = 0.
\end{aligned}$$

The fact that the quadratic variation of a Brownian motion is equal to T, and not to zero, implies that a Brownian motion does not have a continuous derivative. It turns out that it does not have any derivative at any point, so each Brownian path is “edgy.”

A.5 Stochastic Integration

In analogy to a discrete random walk, Brownian motion can represent the profit or loss resulting from a fair game played in continuous time. As such, it is a good model for market noise. In the discrete case, we considered the situation when the player controls the size of the bet $\Delta(n)$ at time n, and his change of wealth can be represented as

$$P(n) = \sum_{i=1}^{n} \Delta(i) \cdot X(i).$$

In the discrete case, the random variable $X(i)$ which takes values 1 or -1 represents the market noise. In the continuous case, the increment of the Brownian motion $W(t_i) - W(t_{i-1})$ represents the market noise. If one also controls the size of the position in the underlying noise, the total profit or loss resulting from this betting strategy, or trading in a market noise, is given by

$$I(T) = \sum_{i=1}^{n} \Delta(t_{i-1}) \cdot (W(t_i) - W(t_{i-1})). \tag{A.10}$$

Note that the position in the underlying noise $\Delta(t_{i-1})$ should be set before having any information about the increment $W(t_i) - W(t_{i-1})$, so the position Δ must be set at or before time t_{i-1}. Otherwise one would be able to make a risk-free profit. Mathematically one can also consider other times $t_i^* \in [t_{i-1}, t_i]$ and define $\bar{I}(T) = \sum_{i=1}^{n} \Delta(t_i^*) \cdot (W(t_i) - W(t_{i-1}))$, but only (A.10) makes financial sense. If we allow for continuous changes of the position $\Delta(t)$ the above sum will in the limit become an integral

$$I(T) = \int_0^T \Delta(t) dW(t). \tag{A.11}$$

We can also rewrite the above relationship in a differential form as

$$dI(t) = \Delta(t) dW(t). \tag{A.12}$$

It turns out that in contrast to an ordinary integral $\int_0^T \Delta(t)dt$, it does matter which time $t_i^* \in [t_{i-1}, t_i]$ is taken in the evaluation of the integrand $\Delta(t_i^*)$. Different choices of t_i^* lead to different stochastic integrals. The reason is that Brownian motion is not differentiable, so we cannot write $dW(t) = W'(t)dt$, as there is no such thing as $W'(t)$. Furthermore,

$$\int_0^T |dW(t)| = \lim_{n \to \infty} \sum_{i=1}^{n} |W(t_i) - W(t_{i-1})| = \infty,$$

so Brownian motion takes an infinite path between any two time points. Therefore it is not surprising that the choice of time t_i^* in the evaluation of the integrand $\Delta(t_i^*)$ makes a difference. When the integrand is evaluated at the left end of the interval t_{i-1}, it is called **Ito's integral.** It is the only stochastic integral based on Brownian motion that makes financial sense.

Let us list some important properties of the Ito's integral. We will prove the listed properties for the integrands that are constant on a finite number of intervals. The properties are valid for general integrands by considering an approximating sequence of the constant integrands, and passing to the limit.

Linearity:

$$\int_0^T (aX(t) + bY(t))dW(t) = a\int_0^T X(t)dW(t) + b\int_0^T Y(t)dW(t).$$

Martingale property:
$I(T)$ is a martingale.

$$\begin{aligned}
\mathbb{E}_t[I(T)] &= \mathbb{E}_t\left[I(t) + \sum_{i:t\leq t_{i-1}} \Delta(t_{i-1})\cdot(W(t_i) - W(t_{i-1}))\right] \\
&= I(t) + \sum_{i:t\leq t_{i-1}} \mathbb{E}_t\left[\Delta(t_{i-1})\cdot(W(t_i) - W(t_{i-1}))\right] \\
&= I(t) + \sum_{i:t\leq t_{i-1}} \mathbb{E}_t\mathbb{E}_{t_{i-1}}\left[\Delta(t_{i-1})\cdot(W(t_i) - W(t_{i-1}))\right] \\
&= I(t) + \sum_{i:t\leq t_{i-1}} \mathbb{E}_t\left[\Delta(t_{i-1})\cdot\mathbb{E}_{t_{i-1}}[W(t_i) - W(t_{i-1})]\right] \\
&= I(t).
\end{aligned}$$

Ito's isometry:

$$\mathrm{Var}(I(T)) = \mathbb{E}[I(T)]^2 = \mathbb{E}\left[\int_0^T \Delta(t)^2 dt\right]. \tag{A.13}$$

This follows from

$$
\begin{aligned}
\mathbb{E}[I(T)]^2 &= \mathbb{E}\left[\sum_{i=1}^{n} \Delta(t_{i-1}) \cdot (W(t_i) - W(t_{i-1}))\right]^2 \\
&= \mathbb{E}\sum_{i=1}^{n} \Delta(t_{i-1})^2 \cdot (W(t_i) - W(t_{i-1}))^2 \\
&+ 2\mathbb{E} \sum_{1\le i<j\le n} \Delta(t_{i-1}) \cdot \Delta(t_{j-1}) \cdot (W(t_i) - W(t_{i-1})) \cdot (W(t_j) - W(t_{j-1})) \\
&= \sum_{i=1}^{n} \mathbb{E}\mathbb{E}_{t_{i-1}} \left[\Delta(t_{i-1})^2 \cdot (W(t_i) - W(t_{i-1}))^2\right] \\
&+ 2 \sum_{1\le i<j\le n} \mathbb{E}[\Delta(t_{i-1}) \cdot \Delta(t_{j-1}) \cdot (W(t_i) - W(t_{i-1}))] \cdot \mathbb{E}[W(t_j) - W(t_{j-1})] \\
&= \sum_{i=1}^{n} \mathbb{E}\left[\Delta(t_{i-1})^2 \cdot \mathbb{E}_{t_{i-1}} \left[(W(t_i) - W(t_{i-1}))^2\right]\right] \\
&= \sum_{i=1}^{n} \mathbb{E}\left[\Delta(t_{i-1})^2 \cdot (t_i - t_{i-1})\right] = \mathbb{E}\left[\int_0^T \Delta(t)^2 dt\right].
\end{aligned}
$$

A.6 Stochastic Calculus

Stochastic calculus explains the evolution of functions of diffusion processes. Let $f(t,x)$ be a function for which the partial derivatives f_t, f_x, and f_{xx} exist. From Taylor's expansion we can write

$$
\begin{aligned}
f(t+dt, W(t+dt)) - f(t, W(t)) = \; & f_t(t, W(t))dt \\
& + f_x(t, W(t)) \cdot (W(t+dt) - W(t)) \\
& + \tfrac{1}{2} f_{xx}(t, W(t)) \cdot (W(t+dt) - W(t))^2 \\
& + \text{higher order terms.}
\end{aligned}
$$

We have already seen that $(dW(t))^2 = dt$, and the higher order terms do not contribute in the limit to the above expression as $dt \cdot dW(t) = 0$, and $(dt)^2 = 0$. Therefore we have **Ito's formula**

$$
\boxed{df(t, W(t)) = \left[f_t(t, W(t)) + \tfrac{1}{2} f_{xx}(t, W(t))\right] dt + f_x(t, W(t))dW(t).} \qquad \text{(A.14)}
$$

Example A.11
Consider the process $W(t)^2 - t$, which corresponds to the choice of $f(t,x) = x^2 - t$. The partial derivatives that appear in the Ito's formula are given by

$f_t(t,x) = -1$, $f_x(t,x) = 2x$, and $f_{xx}(t,x) = 2$. Therefore we get

$$d(W(t)^2 - t) = [-1 + \tfrac{1}{2} \cdot 2]dt + 2W(t)dW(t) = 2W(t)dW(t).$$

Thus the following relationship holds

$$\int_0^T W(t)dW(t) = \tfrac{1}{2}[W(T)^2 - T].$$

Since stochastic integrals are martingales, this is another proof of the martingale property of the process $W(t)^2 - t$. □

Example A.12

Consider a geometric Brownian motion $S(t) = S(0) \cdot \exp(\sigma W(t) - \frac{1}{2}\sigma^2 t)$, which corresponds to the choice of $f(t,x) = S(0) \cdot \exp(\sigma x - \frac{1}{2}\sigma^2 t)$. The partial derivatives of f are given by $f_t(t,x) = -\frac{1}{2}\sigma^2 f(t,x)$, $f_x(t,x) = \sigma f(t,x)$, and $f_{xx}(t,x) = \sigma^2 f(t,x)$. Therefore we have

$$\begin{aligned} d\left[S(0) \cdot \exp(\sigma W(t) - \tfrac{1}{2}\sigma^2 t)\right] &= [-\tfrac{1}{2}\sigma^2 f(t,W(t)) + \tfrac{1}{2}\sigma^2 f(t,W(t))]dt \\ &\quad + \sigma f(t,W(t))dW(t) \\ &= \sigma \cdot S(0) \cdot \exp(\sigma W(t) - \tfrac{1}{2}\sigma^2 t)dW(t), \end{aligned}$$

or in other words,

$$dS(t) = \sigma S(t)dW(t).$$

Thus we can write

$$S(T) = S(0) + \int_0^T \sigma S(t)dW(t),$$

showing that the geometric Brownian motion $S(t)$ can be represented as a stochastic integral, so in particular, it is a martingale. □

Ito's formula can be generalized for functions of a general diffusion process as follows.

THEOREM A.2 Ito's formula for a general diffusion process

Let $X(t)$ be a diffusion process with dynamics

$$dX(t) = a(t,X(t))dt + b(t,X(t))dW(t). \tag{A.15}$$

Then

$$\begin{aligned} df(t,X(t)) = \Big[&f_t(t,X(t)) + a(t,X(t)) \cdot f_x(t,X(t)) \\ &+ \tfrac{1}{2}b^2(t,X(t)) \cdot f_{xx}(t,X(t))\Big]dt + b(t,X(t)) \cdot f_x(t,X(t))dW(t). \end{aligned} \tag{A.16}$$

PROOF We have

$$\begin{aligned} df(t,X(t)) &= f_t(t,X(t))dt + f_x(t,X(t))dX(t) + \tfrac{1}{2}f_{xx}(t,X(t))(dX(t))^2 \\ &= f_t(t,X(t))dt + f_x(t,X(t))\cdot(a(t,X(t))dt + b(t,X(t))dW(t)) \\ &\quad +\tfrac{1}{2}f_{xx}(t,X(t))\cdot(a(t,X(t))dt + b(t,X(t))dW(t))^2 \\ &= \Big[f_t(t,X(t)) + a(t,X(t))\cdot f_x(t,X(t)) \\ &\quad +\tfrac{1}{2}b^2(t,X(t))\cdot f_{xx}(t,X(t))\Big]dt + b(t,X(t))\cdot f_x(t,X(t))dW(t). \end{aligned}$$

□

REMARK A.4 Product rule
When $X(t)$ and $Y(t)$ are two diffusions, we have

$$d(X(t)\cdot Y(t)) = X(t)dY(t) + Y(t)dX(t) + dX(t)dY(t).$$

One can show this result by applying the arguments presented above to the function $f(x,y) = x\cdot y$. □

A.7 Connections with Partial Differential Equations

According to the First Fundamental Theorem of Asset Pricing, the prices of no-arbitrage assets are martingales under the probability measure corresponding to the reference asset. The Martingale Representation Theorem states that every martingale with continuous path adapted to a filtration generated by Brownian motion is in fact a diffusion, and it solves the following stochastic differential equation

$$d\mathcal{M}(t) = \phi(t)dW(t) \tag{A.17}$$

for some $\phi(t)$ that is adapted to $\mathcal{F}_t^W$.

THEOREM A.3 Martingale Representation Theorem
A martingale $\mathcal{M}(t)$ with continuous paths adapted to a filtration $\mathcal{F}_t^W$ generated by a Brownian motion W admits the following representation

$$\mathcal{M}^c(t) = \mathcal{M}^c(0) + \int_0^t \phi(s)dW(s), \tag{A.18}$$

where $\phi(t)$ is adapted to $\mathcal{F}_t^W$.

In particular, a martingale has a zero "dt" term. Let us first identify a class of martingales that can be written simply as a function f of time t and

Brownian motion $W(t)$, so the martingale is in the form $\mathcal{M}(t) = f(t, W(t))$. According to the Ito's formula,

$$d\mathcal{M}(t) = df(t, W(t)) = [f_t(t, W(t)) + \tfrac{1}{2} f_{xx}(t, W(t))]dt + f_x(t, W(t))dW(t).$$

But since the corresponding "*dt*" term must be zero, the function f must satisfy the following partial differential equation

$$f_t(t, x) + \tfrac{1}{2} f_{xx}(t, x) = 0. \tag{A.19}$$

Example A.13
The functions $f(t, x) = 1$, $f(t, x) = x$, $f(t, x) = x^2 - t$, or $f(t, x) = \exp(\sigma x - \frac{1}{2}\sigma^2 t)$ satisfy the partial differential equation (A.19). Therefore a constant 1, Brownian motion $W(t)$, process $W(t)^2 - t$, or geometric Brownian motion $\exp(\sigma W(t) - \frac{1}{2}\sigma^2 t)$ are examples of martingales that can be expressed as a function of time t and of a Brownian motion $W(t)$. □

We can obtain a similar connection between functions of a general diffusion process and partial differential equations. Let $X(t)$ be a diffusion process

$$dX(t) = a(t, X(t))dt + b(t, X(t))dW(t).$$

Consider a function $f(t, X(t))$ that depends on the time t and the value of the diffusion process $X(t)$. According to Ito's formula we have

$$\begin{aligned} df(t, X(t)) = \Big[& f_t(t, X(t)) + a(t, X(t)) \cdot f_x(t, X(t)) \\ & + \tfrac{1}{2} b^2(t, X(t)) \cdot f_{xx}(t, X(t)) \Big] dt + b(t, X(t)) \cdot f_x(t, X(t))dW(t). \end{aligned}$$

Since $f(t, X(t))$ should be a martingale, the "*dt*" term must be zero, leading to the partial differential equation

$$f_t(t, x) + a(t, x) \cdot f_x(t, x) + \tfrac{1}{2} b^2(t, x) \cdot f_{xx}(t, x) = 0. \tag{A.20}$$

Note that the equation (A.19) is a special case of the above partial differential equation for the choice of $a(t, x) = 0$, and $b(t, x) = 1$.

Example A.14
When $a(t, x) = 0$ and $b(t, x) = \sigma x$, we get

$$dX(t) = \sigma X(t)dW(t),$$

so this case corresponds to a geometric Brownian motion. Martingales that are functions of a geometric Brownian motion thus satisfy

$$f_t(t, x) + \tfrac{1}{2}\sigma^2 x^2 f_{xx}(t, x) = 0.$$

□

An alternative characterization of a martingale that is a function of a diffusion is via conditional expectation. Let $\mathcal{M}(T) = g(X(T))$ be the final value of the martingale $\mathcal{M}$ expressed as a function of the value of $X(T)$. We can define

$$f(t, x) = \mathbb{E}[g(X(T))|X(t) = x].$$

Note that $f(t, X(t))$ is a martingale, which easily follows from the tower property

$$f(s, X(s)) = \mathbb{E}[g(X(T))|X(s)] = \mathbb{E}[\mathbb{E}[g(X(T))|X(t)]|X(s)] = \mathbb{E}_s[f(t, X(t))].$$

Since $f(t, X(t))$ is a martingale, the function $f(t, x)$ must satisfy partial differential equation (A.20). We have proved the following theorem.

THEOREM A.4 Feynman–Kac Theorem
A function $f(t, x)$ defined by

$$f(t, x) = \mathbb{E}[g(X(T))|X(t) = x],$$

where

$$dX(t) = a(t, X(t))dt + b(t, X(t))dW(t)$$

satisfies partial differential equation

$$f_t(t, x) + a(t, x) \cdot f_x(t, x) + \tfrac{1}{2}b^2(t, x) \cdot f_{xx}(t, x) = 0$$

with the terminal condition

$$f(T, x) = g(x).$$

Example A.15
Consider the partial differential equation

$$f_t(t, x) + \mu x \cdot f_x(t, x) + \tfrac{1}{2}\sigma^2 x^2 \cdot f_{xx}(t, x) = 0$$

with the terminal condition

$$f(T, x) = \log(x).$$

According to the Feynman–Kac theorem, this partial differential equation admits a solution with stochastic representation

$$f(t, x) = \mathbb{E}[\log(X(T))|X(t) = x],$$

where

$$dX(t) = \mu X(t)dt + \sigma X(t)dW(t).$$

The process $X(t)$ is a geometric Brownian motion with drift μ that admits a closed form solution

$$X(T) = X(t) \cdot \exp((\mu - \tfrac{1}{2}\sigma^2)(T-t) + \sigma W(T-t)),$$

and thus

$$\begin{aligned} f(t,x) &= \mathbb{E}[\log(X(T))|X(t)=x] \\ &= \mathbb{E}[\log(X(t) \cdot \exp((\mu - \tfrac{1}{2}\sigma^2)(T-t) + \sigma W(T-t)))|X(t)=x] \\ &= \log(x) + \mathbb{E}[(\mu - \tfrac{1}{2}\sigma^2)(T-t) + \sigma W(T-t)] \\ &= \log(x) + (\mu - \tfrac{1}{2}\sigma^2)(T-t). \end{aligned}$$

One can check that $f(t,x) = \log(x) + (\mu - \frac{1}{2}\sigma^2)(T-t)$ is indeed the solution of the partial differential equation. ∎

References and Further Reading

A more detailed treatment of the concepts of conditional expectation and martingales can be found in Williams (1991). The books by Lawler (2006) and Mikosch (1999) serve as introductory references for martingales, Markov chains, and stochastic calculus. The reader interested in a more advanced theory of stochastic calculus should refer to Karatzas and Shreve (1991), Oksendal (2007), or Revuz and Yor (2004).

Exercises

A.1 Consider a fair die toss, X which results in a value contained in the set $(\{1,2,3,4,5,6\})$. Let $\mathcal{F} = \sigma(\{1,3,5\},\{2,4,6\})$. This is an information set that corresponds to distinguishing odd/even outcomes. Determine

$$\mathbb{E}[X|\mathcal{F}].$$

Compute

$$\mathbb{E}[(\mathbb{E}[X|\mathcal{F}] - X)^2],$$

and

$$\mathbb{E}[\mathbb{E}[X|\mathcal{F}]].$$

A.2 Consider two consecutive coin tosses with a fair coin ($\mathbb{P}(H) = \mathbb{P}(T) = \frac{1}{2}$). Let the set of outcomes be given by $\Omega = \{\{HH\}, \{HT\}, \{TH\}, \{TT\}\}$. Consider random variables $S(0)$, $S(1)$, and $S(2)$ defined by

$$S(n) = \text{number of heads in the first n coin tosses}$$

for $n = 0, 1, 2$.

(a) Find the information sets $\mathcal{G}_n = \sigma(S(n))$ for $n = 0, 1, 2$.

(b) For a random variable X given by $X(HH) = 3$, $X(HT) = 1$, $X(TH) = 2$, $X(TT) = 0$, compute $\mathbb{E}[X|\mathcal{G}_n]$ for $n = 0, 1, 2$.

(c) Determine $\mathbb{E}[\mathbb{E}[X|\mathcal{G}_2]|\mathcal{G}_1]$. Is it equal to $\mathbb{E}[X|\mathcal{G}_1]$? Why?
Hint: For the tower property $\mathbb{E}[\mathbb{E}[X|\mathcal{G}_2]|\mathcal{G}_1] = \mathbb{E}[X|\mathcal{G}_1]$ to hold in general, we must have $\mathcal{G}_1 \subseteq \mathcal{G}_2$. Check if the condition $\mathcal{G}_1 \subseteq \mathcal{G}_2$ holds.

A.3 Determine $\mathbb{E}\left[\int_0^T W(t)dW(t)\right]^2$ from Ito's isometry.

A.4 Let $W(t)$ be a standard Brownian motion. Using Ito's formula, determine $dW(t)^4$, and compute $\mathbb{E}[W(T)^4]$.

A.5 Let

$$S(t) = \sigma S(t)dW(t).$$

Use Ito's formula to determine

(a)

$$dS(t)^\alpha,$$

(b)

$$d\log(S(t)).$$

A.6 Define

$$Z(t) = \exp(\sigma W(t) - \tfrac{1}{2}\sigma^2 t).$$

Show that

$$Z(t) \cdot (W(t) - \sigma t)$$

is a martingale.
Hint: Apply the product rule, show that the resulting formula has a zero dt term.

Solutions to Selected Exercises

1.2:

$$dS_M(t) = \sigma S_M(t)dW^M(t) - a(t)S_M(t)dt.$$

1.3:
The self-financing condition is given by

$$(d\Delta^Y(t)) + (d\Delta^X(t)) \cdot dX_Y(t) + (d\Delta^X(t)) \cdot X_Y(t) = 0.$$

(a) We have $\Delta^Y(t) = [\max_{0 \leq s \leq t} X_Y(s)]$, $\Delta^X(t) = 0$. Therefore

$$\begin{aligned}(d\Delta^Y(t)) + (d\Delta^X(t)) \cdot dX_Y(t) + (d\Delta^X(t)) \cdot X_Y(t)&\\ = (d\Delta^Y(t)) = d\left[\max_{0 \leq s \leq t} X_Y(s)\right] \neq 0.&\end{aligned}$$

This portfolio is not self-financing.
(b) We have $\Delta^Y(t) = \left[\frac{1}{t}\int_0^t X_Y(s)ds\right]$, $\Delta^X(t) = 0$. Therefore

$$\begin{aligned}&(d\Delta^Y(t)) + (d\Delta^X(t)) \cdot dX_Y(t) + (d\Delta^X(t)) \cdot X_Y(t)\\ &= (d\Delta^Y(t)) = d\left[\tfrac{1}{t}\int_0^t X_Y(s)ds\right]\\ &= \left[\int_0^t X_Y(s)ds\right] d\left(\tfrac{1}{t}\right) + \tfrac{1}{t}d\left[\int_0^t X_Y(s)ds\right]\\ &= -\left[\tfrac{1}{t^2}\int_0^t X_Y(s)ds\right]dt + \tfrac{1}{t}X_Y(t)dt \neq 0.\end{aligned}$$

1.5:
We want to prove that

$$X_Y(t)dN(d_+) + dX_Y(t)dN(d_-) - KdN(d_-) = 0.$$

Note that we can write

$$N(d_+) = f(t, X_Y(t))$$

for a function $f(t,x)$ given by

$$f(t,x) = N\left(\tfrac{1}{\sigma\sqrt{T-t}} \cdot \log(\tfrac{x}{K}) + \tfrac{1}{2}\sigma\sqrt{T-t}\right),$$

and

$$N(d_-) = g(t, X_Y(t))$$

for a function $g(t,x)$ given by

$$g(t,x) = N\left(\tfrac{1}{\sigma\sqrt{T-t}} \cdot \log(\tfrac{x}{K}) - \tfrac{1}{2}\sigma\sqrt{T-t}\right).$$

Using Ito's formula, we can write

$$\begin{aligned}
dN(d_+) &= df(t, X_Y(t)) \\
&= f_t(t, X_Y(t))dt + f_x(t, X_Y(t))dX_Y(t) + \tfrac{1}{2}f_{xx}(t, X_Y(t))d^2X_Y(t) \\
&= \left[f_t(t, X_Y(t)) + \tfrac{1}{2}f_{xx}(t, X_Y(t))\sigma^2 X_Y^2(t)\right]dt + f_x(t, X_Y(t))dX_Y(t).
\end{aligned}$$

The partial derivatives are given by

$$\begin{aligned}
f_t(t,x) &= \phi(d_+) \cdot \left[\tfrac{1}{2}\tfrac{1}{\sigma\sqrt{(T-t)^3}} \cdot \log(\tfrac{x}{K}) - \tfrac{1}{4}\sigma\tfrac{1}{\sqrt{T-t}}\right], \\
f_x(t,x) &= \phi(d_+) \cdot \tfrac{1}{\sigma\sqrt{T-t}} \cdot \tfrac{1}{x}, \\
f_{xx}(t,x) &= \phi'(d_+) \cdot \left[\tfrac{1}{\sigma\sqrt{T-t}} \cdot \tfrac{1}{x}\right]^2 + \phi(d_+) \cdot \left[-\tfrac{1}{\sigma\sqrt{T-t}} \cdot \tfrac{1}{x^2}\right],
\end{aligned}$$

where $\phi(x) = \frac{1}{\sqrt{2\pi}} \cdot e^{-\frac{x^2}{2}}$ is the density of the standard normal variable. Similarly, we can write

$$\begin{aligned}
dN(d_-) &= dg(t, X_Y(t)) \\
&= g_t(t, X_Y(t))dt + g_x(t, X_Y(t))dX_Y(t) + \tfrac{1}{2}g_{xx}(t, X_Y(t))d^2X_Y(t) \\
&= \left[g_t(t, X_Y(t)) + \tfrac{1}{2}g_{xx}(t, X_Y(t))\sigma^2 X_Y^2(t)\right]dt + g_x(t, X_Y(t))dX_Y(t),
\end{aligned}$$

where

$$\begin{aligned}
g_t(t,x) &= \phi(d_-) \cdot \left[\tfrac{1}{2}\tfrac{1}{\sigma\sqrt{(T-t)^3}} \cdot \log(\tfrac{x}{K}) + \tfrac{1}{4}\sigma\tfrac{1}{\sqrt{T-t}}\right], \\
g_x(t,x) &= \phi(d_-) \cdot \tfrac{1}{\sigma\sqrt{T-t}} \cdot \tfrac{1}{x}, \\
g_{xx}(t,x) &= \phi'(d_-) \cdot \left[\tfrac{1}{\sigma\sqrt{T-t}} \cdot \tfrac{1}{x}\right]^2 + \phi(d_-) \cdot \left[-\tfrac{1}{\sigma\sqrt{T-t}} \cdot \tfrac{1}{x^2}\right].
\end{aligned}$$

Thus we can expand the expression

$$X_Y(t)dN(d_+) + dX_Y(t)dN(d_-) - KdN(d_-)$$

and get

$$\begin{aligned}
&X_Y(t)dN(d_+) + dX_Y(t)dN(d_-) - KdN(d_-) = \\
&= X_Y(t)\left[(f_t(t, X_Y(t)) + \tfrac{1}{2}f_{xx}(t, X_Y(t))\sigma^2 X_Y^2(t))dt + f_x(t, X_Y(t))dX_Y(t))\right] \\
&\qquad + f_x(t, X_Y(t))\sigma^2 X_Y^2(t)dt \\
&\quad - K\left[(g_t(t, X_Y(t)) + \tfrac{1}{2}g_{xx}(t, X_Y(t))\sigma^2 X_Y^2(t))dt + g_x(t, X_Y(t))dX_Y(t))\right].
\end{aligned}$$

Let us check both the $dX_Y(t)$ and dt terms in the above equality. The $dX_Y(t)$ term is

$$X_Y(t) \cdot f_x(t, X_Y(t)) - K \cdot g_x(t, X_Y(t)),$$

or after substitution for f_x and g_x,

$$X_Y(t) \cdot \phi(d_+) \cdot \tfrac{1}{\sigma\sqrt{T-t}} \cdot \tfrac{1}{X_Y(t)} - K \cdot \phi(d_-) \cdot \tfrac{1}{\sigma\sqrt{T-t}} \cdot \tfrac{1}{X_Y(t)} = \\ = \tfrac{1}{\sigma\sqrt{T-t}} \cdot \tfrac{1}{X_Y(t)} \cdot [X_Y(t) \cdot \phi(d_+) - K \cdot \phi(d_-)].$$

The reader should verify that we indeed have

$$X_Y(t) \cdot \phi(d_+) - K \cdot \phi(d_-) = 0,$$

so the $dX_Y(t)$ term is zero. The dt term is given by

$$X_Y(t)\left[f_t(t, X_Y(t)) + \tfrac{1}{2} f_{xx}(t, X_Y(t))\sigma^2 X_Y^2(t)\right] + \\ + f_x(t, X_Y(t))\sigma^2 X_Y^2(t) - K\left[g_t(t, X_Y(t)) + \tfrac{1}{2} g_{xx}(t, X_Y(t))\sigma^2 X_Y^2(t)\right].$$

After substitution for the partial derivatives of f and g, we get

$$X_Y(t) \cdot \phi(d_+) \cdot \left[\tfrac{1}{2}\tfrac{1}{\sigma\sqrt{(T-t)^3}} \cdot \log\left(\tfrac{X_Y(t)}{K}\right) - \tfrac{1}{4}\sigma\tfrac{1}{\sqrt{T-t}}\right] \\ + X_Y(t)\left[\phi'(d_+)\left[\tfrac{1}{\sigma\sqrt{T-t}} \cdot \tfrac{1}{X_Y(t)}\right]^2 + \phi(d_+)\left[-\tfrac{1}{\sigma\sqrt{T-t}} \cdot \tfrac{1}{X_Y^2(t)}\right]\right]\sigma^2 X_Y^2(t) \\ + \phi(d_+) \cdot \tfrac{1}{\sigma\sqrt{T-t}} \cdot \tfrac{1}{X_Y(t)}\sigma^2 X_Y^2(t) \\ - K \cdot \phi(d_-) \cdot \left[\tfrac{1}{2}\tfrac{1}{\sigma\sqrt{(T-t)^3}} \cdot \log\left(\tfrac{X_Y(t)}{K}\right) + \tfrac{1}{4}\sigma\tfrac{1}{\sqrt{T-t}}\right] \\ - K \cdot \left[\phi'(d_-) \cdot \left[\tfrac{1}{\sigma\sqrt{T-t}} \cdot \tfrac{1}{X_Y(t)}\right]^2 + \phi(d_-) \cdot \left[-\tfrac{1}{\sigma\sqrt{T-t}} \cdot \tfrac{1}{X_Y^2(t)}\right]\right]\sigma^2 X_Y^2(t).$$

Note that since $X_Y(t) \cdot \phi(d_+) - K \cdot \phi(d_-) = 0$, two pairs of these terms in the above expression cancel out. More specifically,

$$X_Y(t) \cdot \phi(d_+) \cdot \tfrac{1}{2}\tfrac{1}{\sigma\sqrt{(T-t)^3}} \cdot \log\left(\tfrac{X_Y(t)}{K}\right) \\ - K \cdot \phi(d_-) \cdot \tfrac{1}{2}\tfrac{1}{\sigma\sqrt{(T-t)^3}} \cdot \log\left(\tfrac{X_Y(t)}{K}\right) = 0,$$

and

$$X_Y(t) \cdot \phi(d_+) \cdot \left[-\tfrac{1}{\sigma\sqrt{T-t}} \cdot \tfrac{1}{X_Y^2(t)}\right]\sigma^2 X_Y^2(t) \\ - K \cdot \phi(d_-) \cdot \left[-\tfrac{1}{\sigma\sqrt{T-t}} \cdot \tfrac{1}{X_Y^2(t)}\right]\sigma^2 X_Y^2(t) = 0.$$

This slightly simplifies the dt term, which after additional substitution becomes $\phi'(x) = -x \cdot \phi(x)$,

$$X_Y(t) \cdot \phi(d_+) \cdot \left[-\tfrac{1}{4}\sigma \tfrac{1}{\sqrt{T-t}}\right] - X_Y(t) \cdot d_+ \cdot \phi(d_+) \cdot \tfrac{1}{T-t} + \\ + \phi(d_+) \cdot \tfrac{1}{\sqrt{T-t}} \cdot \sigma X_Y(t) - K \cdot \phi(d_-) \cdot \left[\tfrac{1}{4}\sigma \tfrac{1}{\sqrt{T-t}}\right] + K \cdot d_- \cdot \phi(d_-) \cdot \tfrac{1}{T-t}.$$

Since $K \cdot \phi(d_-) = X_Y(t) \cdot \phi(d_+)$, we can replace these terms in the above expression to get

$$X_Y(t) \cdot \phi(d_+) \cdot \tfrac{\sigma}{\sqrt{T-t}}\left[-\tfrac{1}{4} - \tfrac{d_+}{\sigma\sqrt{T-t}} + 1 - \tfrac{1}{4} + \tfrac{d_-}{\sigma\sqrt{T-t}}\right].$$

But since $-\frac{d_+}{\sigma\sqrt{T-t}} + \frac{d_-}{\sigma\sqrt{T-t}} = -\frac{1}{2}$, the above term sums to zero. Thus we proved that both dt and $dX_Y(t)$ terms of $X_Y(t)dN(d_+) + dX_Y(t)dN(d_-) - KdN(d_-)$ are zero, and therefore the trading strategy is self-financing.

2.1:
(a)

$$V_Y(0) = \mathbb{E}^Y\left[V_Y(1)\right] = \mathbb{E}^Y\left[\mathbb{I}(\omega = H)\right] = \mathbb{P}^Y(H),$$
$$V_0 = \mathbb{P}^Y(H) \cdot Y_0 = \frac{1-d}{u-d} \cdot Y_0.$$

(b) The hedging portfolio has the form

$$P_0 = \Delta^X(0) \cdot X_0 + \Delta^Y(0) \cdot Y_0,$$

where

$$\Delta^X(0) = \frac{V_Y(1,H) - V_Y(1,T)}{X_Y(1,H) - X_Y(1,T)} = \frac{1-0}{u \cdot X_Y(0) - d \cdot X_Y(0)} = \frac{Y_X(0)}{u-d},$$

and

$$\Delta^Y(0) = \frac{V_X(1,H) - V_X(1,T)}{Y_X(1,H) - Y_X(1,T)} = \frac{\frac{1}{u} \cdot Y_X(0) - 0}{\frac{1}{u} \cdot Y_X(0) - \frac{1}{d} \cdot Y_X(0)} = -\frac{d}{u-d}.$$

Therefore the hedging portfolio is

$$P_0 = \frac{Y_X(0)}{u-d} \cdot X_0 - \frac{d}{u-d} \cdot Y_0.$$

Indeed,

$$P_Y(0) = \frac{Y_X(0)}{u-d} \cdot X_Y(0) - \frac{d}{u-d} = \frac{1-d}{u-d},$$
$$P_Y(1,H) = \frac{Y_X(0)}{u-d} \cdot X_Y(1,H) - \frac{d}{u-d} = \frac{u}{u-d} - \frac{d}{u-d} = 1,$$
$$P_Y(1,T) = \frac{Y_X(0)}{u-d} \cdot X_Y(1,T) - \frac{d}{u-d} = \frac{d}{u-d} - \frac{d}{u-d} = 0.$$

(c)

$$U_X(0) = \mathbb{E}^X\left[U_X(1)\right] = \mathbb{E}^X\left[\mathbb{I}(\omega = H)\right] = \mathbb{P}^X(H),$$

$$U_0 = \mathbb{P}^X(H) \cdot X_0 = u \cdot \frac{1-d}{u-d} \cdot X_0.$$

(d) The hedging portfolio has the form

$$P_0 = \Delta^X(0) \cdot X_0 + \Delta^Y(0) \cdot Y_0,$$

where

$$\Delta^X(0) = \frac{U_Y(1,H) - U_Y(1,T)}{X_Y(1,H) - X_Y(1,T)} = \frac{u \cdot X_Y(0) - 0}{u \cdot X_Y(0) - d \cdot X_Y(0)} = \frac{u}{u-d},$$

and

$$\Delta^Y(0) = \frac{U_X(1,H) - U_X(1,T)}{Y_X(1,H) - Y_X(1,T)} = \frac{1-0}{\frac{1}{u} \cdot Y_X(0) - \frac{1}{d} \cdot Y_X(0)} = -\frac{ud \cdot X_Y(0)}{u-d}.$$

Therefore the hedging portfolio is

$$P_0 = \frac{u}{u-d} \cdot X_0 - \frac{ud \cdot X_Y(0)}{u-d} \cdot Y_0.$$

Indeed,

$$\begin{aligned}
P_X(0) &= \frac{u}{u-d} - \frac{ud \cdot X_Y(0)}{u-d} \cdot Y_X(0) = u \cdot \frac{1-d}{u-d}, \\
P_X(1,H) &= \frac{u}{u-d} - \frac{ud \cdot X_Y(0)}{u-d} \cdot Y_X(1,H) = \frac{u}{u-d} - \frac{d}{u-d} = 1, \\
P_X(1,T) &= \frac{u}{u-d} - \frac{ud \cdot X_Y(0)}{u-d} \cdot Y_X(1,T) = \frac{u}{u-d} - \frac{u}{u-d} = 0.
\end{aligned}$$

2.4:
(a) At the terminal time $n = 2$, we have

$$V_\$(2, \omega) = \max(5, S_\$(2, \omega)).$$

Thus

$$V_\$(2, HH) = 16, \qquad V_\$(2, HT) = V_\$(2, TH) = 5, \qquad V_\$(2, TT) = 5.$$

At time $n = 1$, we have to compare the intrinsic value and the continuation value of the contract. This gives us

$$V_\$(1, \omega) = \max\left(\max(5, S_\$(1, \omega)), \tfrac{1}{1+r} \cdot \mathbb{E}^T[V_\$(2)](\omega)\right).$$

The T-forward measure is given by

$$p^T = \mathbb{P}^T(H) = \frac{1 + r - d}{u - d}, \qquad \text{and} \qquad q^T = \mathbb{P}^T(T) = \frac{u - (1 + r)}{u - d}.$$

For the particular choice of the parameters $u = 2$, $d = \frac{1}{2}$, $r = \frac{1}{4}$, we get $p^T = q^T = \frac{1}{2}$. We conclude that

$$V_\$(1, H) = \max\left(\max(5, 8), \tfrac{1}{1+\frac{1}{4}} \cdot [\tfrac{1}{2} \cdot 16 + \tfrac{1}{2} \cdot 5]\right) = \max(8, \tfrac{42}{5}) = \tfrac{42}{5}.$$

The continuation value is larger than the intrinsic value, and thus it is better to continue. Similarly,

$$V_\$(1, T) = \max\left(\max(5, 2), \tfrac{1}{1+\frac{1}{4}} \cdot [\tfrac{1}{2} \cdot 5 + \tfrac{1}{2} \cdot 5]\right) = \max(5, 4) = 5.$$

The intrinsic value is larger than the continuation value, and thus it is better to stop. Finally,

$$V_\$(0) = \max\left(\max(5, 4), \tfrac{1}{1+\frac{1}{4}} \cdot [\tfrac{1}{2} \cdot \tfrac{42}{5} + \tfrac{1}{2} \cdot 5]\right) = \max(5, \tfrac{134}{25}) = \tfrac{134}{25}.$$

The continuation value is larger than the intrinsic value, and thus it is better to continue. The optimal stopping strategy is $\tau^*(HH) = \tau^*(HT) = 2$, and $\tau^*(TH) = \tau^*(TT) = 1$.
(b) The hedging portfolio is given by

$$\Delta^S(0) = \frac{V_\$(1, H) - V_\$(1, T)}{S_\$(1, H) - S_\$(1, T)} = \frac{\frac{42}{5} - 5}{8 - 2} = \frac{17}{30},$$

$$\Delta^M(0) = \frac{V_S(1, H) - V_S(1, T)}{\$_S(1, H) - \$_S(1, T)} \cdot \frac{1}{1 + r} = \frac{\frac{21}{20} - \frac{5}{2}}{\frac{1}{8} - \frac{1}{2}} \cdot \frac{4}{5} = \frac{232}{75},$$

$$\Delta^S(1,H) = \frac{V_\$(2,HH) - V_\$(2,HT)}{S_\$(2,HH) - S_\$(2,HT)} = \frac{16-5}{16-4} = \frac{11}{12},$$
$$\Delta^M(1,H) = \frac{V_S(2,HH) - V_S(2,HT)}{\$_S(2,HH) - \$_S(2,HT)} \cdot \frac{1}{1+r} = \frac{1-\frac{5}{4}}{\frac{1}{16}-\frac{1}{8}} \cdot \frac{4}{5} = \frac{16}{5}.$$

The contract should be exercised at time 1 when $\omega = T$, and thus no hedging is necessary on that outcome.
(c) At the terminal time $n = 2$, we have

$$V_\$(2,\omega) = \min(5, S_\$(2,\omega)).$$

Thus

$$V_\$(2,HH) = 5, \qquad V_\$(2,HT) = V_\$(2,TH) = 4, \qquad V_\$(2,TT) = 1.$$

At time $n = 1$, we have to compare the intrinsic value and the continuation value of the contract. This gives us

$$V_\$(1,\omega) = \max\left(\min(5, S_\$(1,\omega)), \tfrac{1}{1+r} \cdot \mathbb{E}^T[V_\$(2)](\omega)\right).$$

We conclude that

$$V_\$(1,H) = \max\left(\min(5,8), \tfrac{1}{1+\frac{1}{4}} \cdot [\tfrac{1}{2} \cdot 5 + \tfrac{1}{2} \cdot 4]\right) = \max(5, \tfrac{18}{5}) = 5.$$

The intrinsic value is larger than the continuation value, and thus it is better to stop. Similarly,

$$V_\$(1,T) = \max\left(\min(5,2), \tfrac{1}{1+\frac{1}{4}} \cdot [\tfrac{1}{2} \cdot 4 + \tfrac{1}{2} \cdot 1]\right) = \max(2,2) = 2.$$

The intrinsic value is the same as the continuation value, and thus it makes no difference to continue or to stop. Finally,

$$V_\$(0) = \max\left(\min(5,4), \tfrac{1}{1+\frac{1}{4}} \cdot [\tfrac{1}{2} \cdot 5 + \tfrac{1}{2} \cdot 2]\right) = \max(4, \tfrac{14}{5}) = 4.$$

The intrinsic value is larger than the continuation value, and thus it is better to stop. The contract becomes trivial; it should be exercised immediately.

2.5:
(a)

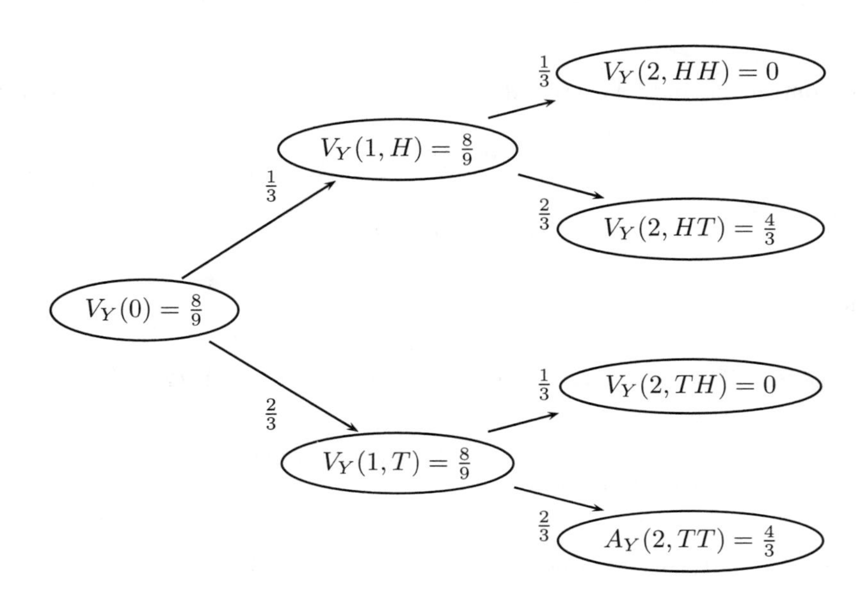

$V_Y(0) = \frac{8}{9}$
$\frac{1}{3}$
$V_Y(1,H) = \frac{8}{9}$
$\frac{2}{3}$
$V_Y(1,T) = \frac{8}{9}$
$\frac{1}{3}$
$V_Y(2,HH) = 0$
$\frac{2}{3}$
$V_Y(2,HT) = \frac{4}{3}$
$\frac{1}{3}$
$V_Y(2,TH) = 0$
$\frac{2}{3}$
$A_Y(2,TT) = \frac{4}{3}$

(b)

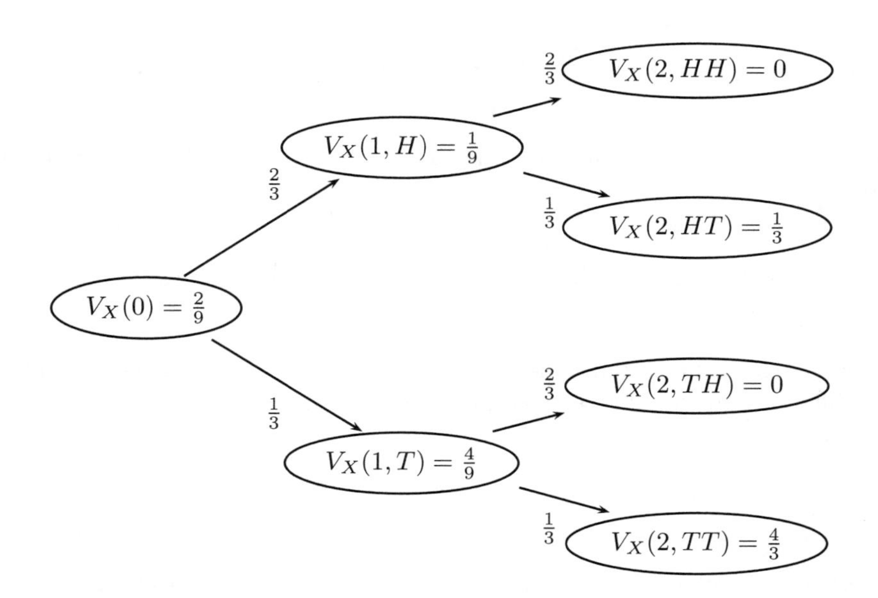

$V_X(0) = \frac{2}{9}$
$\frac{2}{3}$
$V_X(1,H) = \frac{1}{9}$
$\frac{1}{3}$
$V_X(1,T) = \frac{4}{9}$
$\frac{2}{3}$
$V_X(2,HH) = 0$
$\frac{1}{3}$
$V_X(2,HT) = \frac{1}{3}$
$\frac{2}{3}$
$V_X(2,TH) = 0$
$\frac{1}{3}$
$V_X(2,TT) = \frac{4}{3}$

(c)

$$\Delta^X(0) = \frac{V_Y(1,H) - V_Y(1,T)}{X_Y(1,H) - X_Y(1,T)} = \frac{\frac{8}{9} - \frac{8}{9}}{8-2} = 0,$$
$$\Delta^Y(0) = \frac{V_X(1,H) - V_X(1,T)}{Y_X(1,H) - Y_X(1,T)} = \frac{\frac{1}{9} - \frac{4}{9}}{\frac{1}{8} - \frac{1}{2}} = \frac{8}{3},$$

$$\Delta^X(1,H) = \frac{V_Y(2,HH) - V_Y(2,HT)}{X_Y(2,HH) - X_Y(2,HT)} = \frac{0 - \frac{4}{3}}{16-4} = -\frac{1}{9},$$
$$\Delta^Y(1,H) = \frac{V_X(2,HH) - V_X(2,HT)}{Y_X(2,HH) - Y_X(2,HT)} = \frac{0 - \frac{1}{3}}{\frac{1}{16} - \frac{1}{4}} = \frac{16}{9},$$

$$\Delta^X(1,T) = \frac{V_Y(2,TH) - V_Y(2,TT)}{X_Y(2,TH) - X_Y(2,TT)} = \frac{0 - \frac{4}{3}}{4-1} = -\frac{4}{9},$$
$$\Delta^Y(1,T) = \frac{V_X(2,TH) - V_X(2,TT)}{Y_X(2,TH) - Y_X(2,TT)} = \frac{0 - \frac{4}{3}}{\frac{1}{4} - 1} = \frac{16}{9}.$$

(d) The probability measure $\mathbb{P}^A$ is given by

$$\mathbb{P}^A(\omega) = \frac{A_Y(2,\omega)}{A_Y(0)} \cdot \mathbb{P}^Y(\omega).$$

Thus we need values $A_Y(2,\omega)$ for $\omega = HH, HT, TH, TT$. From the definition of A

$$A_Y(2) = \tfrac{1}{3}\left(X_Y(0) + X_Y(1) + X_Y(2)\right),$$

we get

$$A_Y(2,HH) = \tfrac{1}{3}\left(X_Y(0) + X_Y(1,H) + X_Y(2,HH)\right) = \tfrac{1}{3}\left(1 + u + u^2\right) \cdot X_Y(0),$$
$$A_Y(2,HT) = \tfrac{1}{3}\left(X_Y(0) + X_Y(1,H) + X_Y(2,HT)\right) = \tfrac{1}{3}\left(1 + u + u \cdot d\right) \cdot X_Y(0),$$
$$A_Y(2,TH) = \tfrac{1}{3}\left(X_Y(0) + X_Y(1,T) + X_Y(2,TH)\right) = \tfrac{1}{3}\left(1 + d + d \cdot u\right) \cdot X_Y(0),$$
$$A_Y(2,TT) = \tfrac{1}{3}\left(X_Y(0) + X_Y(1,T) + X_Y(2,TT)\right) = \tfrac{1}{3}\left(1 + d + d^2\right) \cdot X_Y(0).$$

Furthermore, from the martingale property of A_Y, we get

$$A_Y(1) = \mathbb{E}_1^Y[A_Y(2)] = \mathbb{E}_1^Y[\tfrac{1}{3}\left(X_Y(0) + X_Y(1) + X_Y(2)\right)] = \tfrac{1}{3} \cdot X_Y(0) + \tfrac{2}{3} \cdot X_Y(1),$$

and therefore

$$A_Y(1,H) = X_Y(1,H) = \left(\tfrac{1}{3} + \tfrac{2}{3} \cdot u\right) \cdot X_Y(0),$$
$$A_Y(1,T) = X_Y(1,T) = \left(\tfrac{1}{3} + \tfrac{2}{3} \cdot d\right) \cdot X_Y(0).$$

Similarly,

$$A_Y(0) = \mathbb{E}^Y[A_Y(2)] = \mathbb{E}^Y[\tfrac{1}{3}\left(X_Y(0) + X_Y(1) + X_Y(2)\right)] = X_Y(0).$$

Therefore

$$\mathbb{P}^A(HH) = \frac{A_Y(2,HH)}{A_Y(0)} \cdot \mathbb{P}^Y(HH) = \tfrac{1}{3}(1+u+u^2) \cdot \left(\frac{1-d}{u-d}\right)^2,$$

$$\mathbb{P}^A(HT) = \frac{A_Y(2,HT)}{A_Y(0)} \cdot \mathbb{P}^Y(HT) = \tfrac{1}{3}(1+u+u \cdot d) \cdot \left(\frac{1-d}{u-d}\right) \cdot \left(\frac{u-1}{u-d}\right),$$

$$\mathbb{P}^A(TH) = \frac{A_Y(2,TH)}{A_Y(0)} \cdot \mathbb{P}^Y(TH) = \tfrac{1}{3}(1+d+d \cdot u) \cdot \left(\frac{1-d}{u-d}\right) \cdot \left(\frac{u-1}{u-d}\right),$$

$$\mathbb{P}^A(TT) = \frac{A_Y(2,TT)}{A_Y(0)} \cdot \mathbb{P}^Y(TT) = \tfrac{1}{3}(1+d+d^2) \cdot \left(\frac{u-1}{u-d}\right)^2.$$

Similarly,

$$\mathbb{P}^A(H) = \frac{A_Y(1,H)}{A_Y(0)} \cdot \mathbb{P}^Y(H) = \tfrac{1}{3}\,(1+2u) \cdot \left(\frac{1-d}{u-d}\right),$$

$$\mathbb{P}^A(T) = \frac{A_Y(1,T)}{A_Y(0)} \cdot \mathbb{P}^Y(T) = \tfrac{1}{3}\,(1+2d) \cdot \left(\frac{u-1}{u-d}\right).$$

This gives us conditional probabilities

$$\mathbb{P}^A(HH|H) = \frac{\mathbb{P}^A(HH)}{\mathbb{P}^A(H)} = \left(\frac{1+u+u^2}{1+2u}\right) \cdot \left(\frac{1-d}{u-d}\right),$$

$$\mathbb{P}^A(HT|H) = \frac{\mathbb{P}^A(HT)}{\mathbb{P}^A(H)} = \left(\frac{1+u+u \cdot d}{1+2u}\right) \cdot \left(\frac{u-1}{u-d}\right),$$

$$\mathbb{P}^A(TH|T) = \frac{\mathbb{P}^A(TH)}{\mathbb{P}^A(T)} = \left(\frac{1+d+d \cdot u}{1+2d}\right) \cdot \left(\frac{1-d}{u-d}\right),$$

$$\mathbb{P}^A(TT|T) = \frac{\mathbb{P}^A(TT)}{\mathbb{P}^A(T)} = \left(\frac{1+d+d^2}{1+2d}\right) \cdot \left(\frac{u-1}{u-d}\right).$$

(e) For the choice of the parameters $u = 2$ and $d = \frac{1}{2}$ we get

$$\mathbb{P}^A(H) = \tfrac{1}{3}\,(1+2u) \cdot \left(\frac{1-d}{u-d}\right) = \frac{5}{9},$$

$$\mathbb{P}^A(T) = \tfrac{1}{3}\,(1+2d) \cdot \left(\frac{u-1}{u-d}\right) = \frac{4}{9},$$

and

$$\mathbb{P}^A(HH|H) = \left(\frac{1+u+u^2}{1+2u}\right) \cdot \left(\frac{1-d}{u-d}\right) = \frac{7}{15},$$

$$\mathbb{P}^A(HT|H) = \left(\frac{1+u+u \cdot d}{1+2u}\right) \cdot \left(\frac{u-1}{u-d}\right) = \frac{8}{15},$$

$$\mathbb{P}^A(TH|T) = \left(\frac{1+d+d \cdot u}{1+2d}\right) \cdot \left(\frac{1-d}{u-d}\right) = \frac{5}{12},$$

$$\mathbb{P}^A(TT|T) = \left(\frac{1+d+d^2}{1+2d}\right) \cdot \left(\frac{u-1}{u-d}\right) = \frac{7}{12}.$$

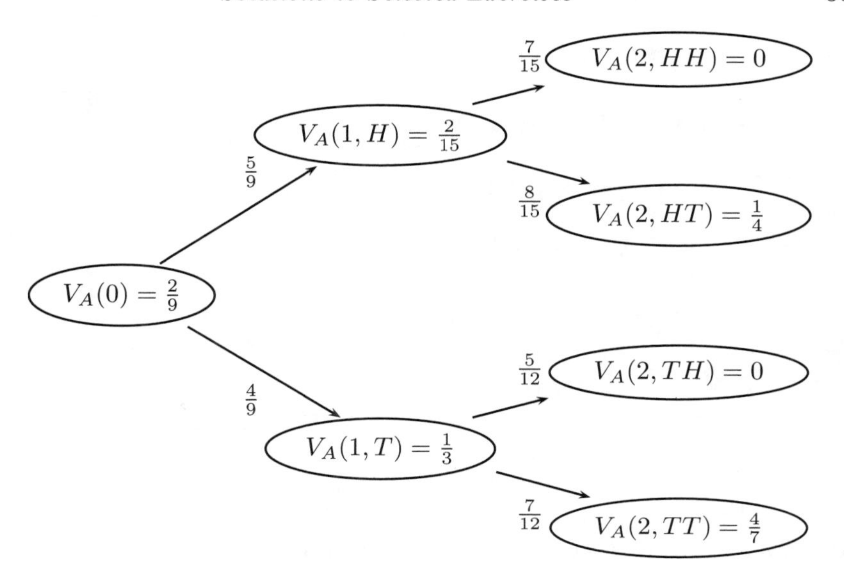

3.1:
(a) From $f^X(x) = f^Y(\frac{1}{x}) \cdot x$ we get

$$[f^X(x)]' = f^Y(\tfrac{1}{x}) - \tfrac{1}{x} \cdot [f^Y(\tfrac{1}{x})]',$$

and

$$[f^X(x)]'' = -\tfrac{1}{x^2} \cdot [f^Y(\tfrac{1}{x})]' + \tfrac{1}{x^2} \cdot [f^Y(\tfrac{1}{x})]' + \tfrac{1}{x^3} \cdot [f^Y(\tfrac{1}{x})]'' = \tfrac{1}{x^3} \cdot [f^Y(\tfrac{1}{x})]'' \geq 0.$$

3.2:
Let us compute

$$X_Y(t) \cdot \tfrac{1}{\sqrt{2\pi}} \cdot e^{-\frac{d_+^2}{2}} - K \cdot \tfrac{1}{\sqrt{2\pi}} \cdot e^{-\frac{d_-^2}{2}}.$$

Note that

$$\begin{aligned} d_\pm^2 &= \left(\tfrac{1}{\sigma\sqrt{T-t}} \cdot \log\left(\tfrac{X_Y(t)}{K}\right) \pm \tfrac{1}{2}\sigma\sqrt{T-t}\right)^2 \\ &= \tfrac{1}{\sigma^2(T-t)} \cdot \left[\log\left(\tfrac{X_Y(t)}{K}\right)\right]^2 + \tfrac{1}{4}\sigma^2(T-t) \pm \log\left(\tfrac{X_Y(t)}{K}\right). \end{aligned}$$

Thus

$$\begin{aligned}
& X_Y(t) \cdot \tfrac{1}{\sqrt{2\pi}} \cdot e^{-\frac{d_+^2}{2}} - K \cdot \tfrac{1}{\sqrt{2\pi}} \cdot e^{-\frac{d_-^2}{2}} \\
= \quad & \tfrac{1}{\sqrt{2\pi}} \cdot \exp\left(\tfrac{1}{2\sigma^2(T-t)} \left[\log\left(\tfrac{X_Y(t)}{K}\right)\right]^2 + \tfrac{1}{8}\sigma^2(T-t) \right) \\
& \times \left[X_Y(t) \cdot \exp\left(-\tfrac{1}{2}\log\left(\tfrac{X_Y(t)}{K}\right)\right) - K \cdot \exp\left(\tfrac{1}{2}\log\left(\tfrac{X_Y(t)}{K}\right)\right)\right] \\
= \quad & \tfrac{1}{\sqrt{2\pi}} \cdot \exp\left(\tfrac{1}{2\sigma^2(T-t)} \left[\log\left(\tfrac{X_Y(t)}{K}\right)\right]^2 + \tfrac{1}{8}\sigma^2(T-t) \right) \\
& \times \left[X_Y(t) \cdot \sqrt{\tfrac{K}{X_Y(t)}} - K \cdot \sqrt{\tfrac{X_Y(t)}{K}} \right] \\
= \quad & 0.
\end{aligned}$$

3.4:

(a) We have seen that the price $V_Y(t)$ is given by

$$V_Y(t) = N(d_-),$$

where

$$d_\pm = \tfrac{1}{\sigma\sqrt{T-t}} \cdot \log\left(\tfrac{X_Y(t)}{K}\right) \pm \tfrac{1}{2}\sigma\sqrt{T-t}.$$

Thus

$$\Delta^X(t) = \frac{\partial V_Y(t)}{\partial X_Y(t)} = \phi(d_-) \cdot \tfrac{1}{\sigma\sqrt{T-t}} \cdot Y_X(t),$$

and

$$\Delta^Y(t) = V_Y(t) - \Delta^X(t) \cdot X_Y(t) = N(d_-) - \phi(d_-) \cdot \tfrac{1}{\sigma\sqrt{T-t}}.$$

The hedging portfolio takes the following form

$$P_t^V = \left[\phi(d_-) \cdot \tfrac{1}{\sigma\sqrt{T-t}} \cdot Y_X(t)\right] \cdot X + \left[N(d_-) - \phi(d_-) \cdot \tfrac{1}{\sigma\sqrt{T-t}}\right] \cdot Y.$$

(b) The $U_X(t)$ price is given by

$$U_X(t) = N(d_+).$$

Therefore

$$\Delta^Y(t) = \frac{\partial U_X(t)}{\partial Y_X(t)} = -\phi(d_+) \cdot \tfrac{1}{\sigma\sqrt{T-t}} \cdot X_Y(t),$$

and

$$\Delta^X(t) = U_X(t) - \Delta^Y(t) \cdot Y_X(t) = N(d_+) + \phi(d_+) \cdot \tfrac{1}{\sigma\sqrt{T-t}}.$$

The hedging portfolio is given by

$$P_t^U = \left[N(d_+) + \phi(d_+) \cdot \tfrac{1}{\sigma\sqrt{T-t}}\right] \cdot X + \left[-\phi(d_+) \cdot \tfrac{1}{\sigma\sqrt{T-t}} \cdot X_Y(t)\right] \cdot Y.$$

(c) The hedge of the contract $W = U - K \cdot V$ is given by

$$\begin{aligned} P_t &= P^U(t) - K \cdot P^V(t) \\ &= \left[N(d_+) + \phi(d_+) \cdot \tfrac{1}{\sigma\sqrt{T-t}} - K \cdot \phi(d_-) \cdot \tfrac{1}{\sigma\sqrt{T-t}} \cdot Y_X(t)\right] \cdot X \\ &\quad + \left[-\phi(d_+) \cdot \tfrac{1}{\sigma\sqrt{T-t}} \cdot X_Y(t) - K \cdot N(d_-) + K \cdot \phi(d_-) \cdot \tfrac{1}{\sigma\sqrt{T-t}}\right] \cdot Y \\ &= [N(d_+)] \cdot X + [-K \cdot N(d_-)] \cdot Y. \end{aligned}$$

We have used the identity

$$X_Y(t) \cdot \phi(d_+) - K \cdot \phi(d_-) = 0$$

that appears in Exercise 3.2.

4.1:
Let us determine the evolution of $dB^T_\$(t)$ first. We have

$$B^T_\$(t) = \exp\left(-\textstyle\int_t^T f(t,u)du\right),$$

and thus according to Ito's formula

$$dB^T_\$(t) = B^T_\$(t) d\left[-\textstyle\int_t^T f(t,u)du\right] + \tfrac{1}{2} B^T_\$(t) d^2 \left[-\textstyle\int_t^T f(t,u)du\right].$$

We also have

$$\begin{aligned} d\left[-\textstyle\int_t^T f(t,u)du\right] &= f(t,t)dt - \int_t^T df(t,u)du \\ &= r(t)dt + \int_t^T [\sigma dW^M(t)]du \\ &= r(t)dt + \sigma(T-t)dW^M(t), \end{aligned}$$

implying

$$dB^T_\$(t) = [r(t) + \tfrac{1}{2}\sigma^2(T-t)^2] B^T_\$(t)dt + \sigma(T-t)B^T_\$(t)dW^M(t).$$

Therefore

$$dB^T_M(t) = \tfrac{1}{2}\sigma^2(T-t)^2 B^T_M(t)dt + \sigma(T-t)B^T_M(t)dW^M(t).$$

The bond B^T price with respect to the money market has positive drift, and thus arbitrage is possible. It cannot be locked by trading in only two assets since there is still a noise term $dW^M(t)$, but it can be locked by trading in three assets, two bonds B^{T_1}, B^{T_2}, and the money market M. Let's find a portfolio of the form $B^{T_2} + \Delta^{T_1}(t)B^{T_1}$ whose price with respect to the money market M has a positive dt term and no noise term.

$$\begin{aligned} d\left(B^{T_2}_M(t) + \Delta^{T_1}(t)B^{T_1}_M(t)\right) \\ &= \left[\tfrac{1}{2}\sigma^2(T_2-t)^2 B^{T_2}_M(t) + \Delta^{T_1}(t)\tfrac{1}{2}\sigma^2(T_1-t)^2 B^{T_1}_M(t)\right] dt \\ &\quad + \left[\sigma(T_2-t)B^{T_2}_M(t) + \Delta^{T_1}(t)\sigma(T_1-t)B^{T_1}_M(t)\right] dW^M(t). \end{aligned}$$

When

$$\Delta^{T_1}(t) = -\frac{T_2-t}{T_1-t}\cdot\frac{B^{T_2}_M(t)}{B^{T_1}_M(t)} = -\frac{T_2-t}{T_1-t}\cdot B^{T_2}_{B^{T_1}}(t),$$

the noise term $dW^M(t)$ cancels, and we get

$$d\left(B^{T_2}_M(t) + \Delta^{T_1}(t)B^{T_1}_M(t)\right) = \tfrac{1}{2}\sigma^2(T_2-t)(T_2-T_2)B^{T_2}_M(t)dt.$$

The dt term is positive when $T_2 > T_1$. Should the arbitrage portfolio start with

$$P_0 = B^{T_2} + \Delta^{T_1}(0)\cdot B^{T_1} + \Delta^M(0)\cdot M = 0,$$

we must have

$$\Delta^M(0) = (\tfrac{T_2}{T_1} - 1)\cdot B^{T_2}_M(0).$$

Therefore a portfolio with the following positions

$$\Delta(t) = (\Delta^{T_2}(t), \Delta^{T_1}(t), \Delta^M(t)) = \left(1, -\frac{T_2-t}{T_1-t}\cdot B^{T_2}_{B^{T_1}}(t), \left(\frac{T_2}{T_1}-1\right)\cdot B^{T_2}_M(0)\right)$$

is an arbitrage portfolio.

5.1:
(a)

$$\begin{aligned} V_Y(t) &= \mathbb{E}^Y_t[V_Y(T)] \\ &= \mathbb{E}^Y_t[X_Y(T)]^\alpha = \mathbb{E}^Y_t\left[[X_Y(t)]^\alpha\cdot\left[\exp(\sigma W^Y(T-t) - \tfrac{1}{2}\sigma^2(T-t))\right]^\alpha\right] \\ &= [X_Y(t)]^\alpha\cdot\exp(-\tfrac{1}{2}\alpha\sigma^2(T-t))\cdot\mathbb{E}^Y_t[\exp(\sigma\alpha W^Y(T-t))] \\ &= [X_Y(t)]^\alpha\cdot\exp(\tfrac{1}{2}\alpha(\alpha-1)\sigma^2(T-t)). \end{aligned}$$

Therefore

$$u^Y(t,x) = x^\alpha\cdot\exp(\tfrac{1}{2}\alpha(\alpha-1)\sigma^2(T-t)).$$

(b) If we substitute $u^Y(t,x) = h(t)\cdot x^\alpha$ into the partial differential equation $u^Y_t(t,x) + \frac{1}{2}\sigma^2x^2u^Y_{xx}(t,x) = 0$, we get

$$h'(t)x^\alpha + \tfrac{1}{2}\sigma^2x^2g(t)\cdot\alpha(\alpha-1)\cdot x^{\alpha-2} = 0,$$

or after dividing by x^α

$$h'(t) + \tfrac{1}{2}\sigma^2\alpha(\alpha-1)\cdot h(t) = 0,$$

which is an ordinary differential equation for $h(t)$ with a terminal condition $h(T)=1$. The solution is given by

$$h(t) = \exp(\tfrac{1}{2}\alpha(\alpha-1)\sigma^2(T-t)),$$

and thus u^Y is given by

$$u^Y(t,x) = x^\alpha\cdot\exp(\tfrac{1}{2}\alpha(\alpha-1)\sigma^2(T-t)),$$

which confirms the result from (a).

(c) The hedging positions are given by

$$\Delta^X(t) = u^Y_x(t,X_Y(t)) = \alpha[X_Y(t)]^{\alpha-1}\cdot\exp(\tfrac{1}{2}\alpha(\alpha-1)\sigma^2(T-t)),$$

and

$$\begin{aligned}\Delta^Y(t) &= u^Y(t,X_Y(t)) - X_Y(t)\cdot u^Y_x(t,X_Y(t))\\ &= (1-\alpha)[X_Y(t)]^\alpha\cdot\exp(\tfrac{1}{2}\alpha(\alpha-1)\sigma^2(T-t)).\end{aligned}$$

5.2:

First, in order to have $L^{1-\alpha}R^\alpha = X$ on the barrier, we must have $L^{1-\alpha}R^\alpha_Y(t) = X_Y(t) = Le^{r(T-t)}$. Solving for α from

$$L^{1-\alpha}R^\alpha_Y(t) = L^{1-\alpha}[X_Y(t)]^\alpha\exp(\tfrac{1}{2}\alpha(\alpha-1)\sigma^2(T-t)) = X_Y(t)$$

leads to $\alpha = -\frac{2r}{\sigma^2}$. Similarly, on the barrier we have

$$\begin{aligned}\mathbb{P}^X_t(X_Y(T)\geq K)\cdot X &= \mathbb{P}^X_t(\tfrac{1}{K}\geq Y_X(T))\cdot X\\ &= \mathbb{P}^X_t\left(\left(\tfrac{1}{K}\right)^{1-\alpha}\geq [Y_X(T)]^{1-\alpha}\right)\cdot X\\ &= \mathbb{P}^X_t\left(\left(\tfrac{1}{K}\right)^{1-\alpha}\geq R^\alpha_X(T)\right)\cdot X\\ &= \mathbb{P}^{(\alpha)}_t\left(\left(\tfrac{1}{K}\right)^{1-\alpha}\geq R^\alpha_X(t)\cdot\frac{X_{R^\alpha}(T)}{X_{R^\alpha}(t)}\right)\cdot L^{1-\alpha}R^\alpha\\ &= \mathbb{P}^{(\alpha)}_t\left(\left(\tfrac{L^2}{K}\right)^{1-\alpha}\geq X_{R^\alpha}(T)\right)\cdot L^{1-\alpha}R^\alpha\\ &= \mathbb{P}^{(\alpha)}_t\left(R^\alpha_X(T)\geq\left(\tfrac{K}{L^2}\right)^{1-\alpha}\right)\cdot L^{1-\alpha}R^\alpha\\ &= \mathbb{P}^{(\alpha)}_t\left([Y_X(T)]^{1-\alpha}\geq\left(\tfrac{K}{L^2}\right)^{1-\alpha}\right)\cdot L^{1-\alpha}R^\alpha\\ &= \mathbb{P}^{(\alpha)}_t\left(\tfrac{L^2}{K}\geq X_Y(T)\right)\cdot L^{1-\alpha}R^\alpha.\end{aligned}$$

The important points in this problem are to realize that the two relevant measures are just $\mathbb{P}^X$ and $\mathbb{P}^{(\alpha)}$, and an observation that $R_X^{\alpha}(T) = [Y_X(T)]^{1-\alpha}$.

5.3:

$$\mathbb{P}^{(\alpha)}(X_Y(T) \geq K) = N\left(\frac{1}{\sigma\sqrt{T-t}} \cdot \log(\frac{X_Y(t)}{K}) + (\alpha - \frac{1}{2})\sigma\sqrt{T-t}\right).$$

5.4:
The price of V is given by the Black-Scholes formula

$$V_t = \mathbb{P}_t^{(2)}(R_X^2(T) \geq K) \cdot R_t^2 - K\mathbb{P}^X(R_X^2(T) \geq K) \cdot X_t.$$

Let us determine $\mathbb{P}_t^{(2)}(R_X^2(T) \geq K)$ and $\mathbb{P}_t^X(R_X^2(T) \geq K)$. The second probability $\mathbb{P}_t^X(R_X^2(T) \geq K)$ is simply

$$\begin{aligned}\mathbb{P}_t^X(R_X^2(T) \geq K) &= \mathbb{P}_t^X([X_Y(T)]^2 \cdot Y_X(T) \geq K) = \mathbb{P}_t^X(X_Y(T) \geq K) \\ &= N(\frac{1}{\sigma\sqrt{T-t}} \cdot \log(\frac{X_Y(t)}{K}) + \frac{1}{2}\sigma\sqrt{T-t}).\end{aligned}$$

The first probability $\mathbb{P}_t^{(2)}(R_X^2(T) \geq K)$ is given by

$$\begin{aligned}\mathbb{P}_t^{(2)}(R_X^2(T) \geq K) &= \mathbb{P}_t^{(2)}(X_Y(T) \geq K) = \mathbb{P}_t^{(2)}([X_Y(T)]^2 \geq K^2) \\ &= \mathbb{P}_t^{(2)}(R_Y^2(T) \geq K^2) = \mathbb{P}_t^{(2)}(\tfrac{1}{K^2} \geq Y_{R^2}(T)) \\ &= \mathbb{P}_t^{(2)}(\tfrac{1}{K^2} \geq Y_{R^2}(t) \cdot \exp(2\sigma W^{(2)}(T-t) - 2\sigma^2(T-t))) \\ &= \mathbb{P}_t^{(2)}\left(\frac{1}{\sigma\sqrt{T-t}} \cdot \log(\frac{X_Y(t)}{K}) + \frac{3}{2}\sigma\sqrt{T-t} \geq \frac{W^{(2)}(T-t)}{\sqrt{T-t}}\right) \\ &= N\left(\frac{1}{\sigma\sqrt{T-t}} \cdot \log(\frac{X_Y(t)}{K}) + \frac{3}{2}\sigma\sqrt{T-t}\right).\end{aligned}$$

The hedging portfolio P is given by

$$\begin{aligned}P_t &= [N(\frac{1}{\sigma\sqrt{T-t}} \cdot \log(\frac{X_Y(t)}{K}) + \frac{3}{2}\sigma\sqrt{T-t})] \cdot R_t^2 \\ &\qquad + [-K \cdot N(\frac{1}{\sigma\sqrt{T-t}} \cdot \log(\frac{X_Y(t)}{K}) + \frac{1}{2}\sigma\sqrt{T-t})] \cdot X_t \\ &= \Big[N(\frac{1}{\sigma\sqrt{T-t}} \cdot \log(\frac{X_Y(t)}{K}) + \frac{3}{2}\sigma\sqrt{T-t}) \cdot [2X_Y(t)e^{\sigma^2(T-t)}] \\ &\qquad -K \cdot N(\frac{1}{\sigma\sqrt{T-t}} \cdot \log(\frac{X_Y(t)}{K}) + \frac{1}{2}\sigma\sqrt{T-t})\Big] \cdot X_t \\ &\quad + \Big[-N(\frac{1}{\sigma\sqrt{T-t}} \cdot \log(\frac{X_Y(t)}{K}) + \frac{3}{2}\sigma\sqrt{T-t}) \cdot [X_Y(t)]^2 e^{\sigma^2(T-t)}\Big] \cdot Y_t.\end{aligned}$$

We have expressed the hedge for R^2 in terms of the assets X and Y using the explicit formulas from Exercise 5.1.

9.1:
The solutions 1, x, and y represent contracts with payoffs Y_T, X_T, and A_T.

10.2:
Let $f(x) = \log(x)$. Then $f'(x) = \frac{1}{x}$, and from Ito's formula we get

$$\begin{aligned}
\log(X_Y(t)) &= \log(X_Y(0)) + \int_0^t \frac{1}{X_Y(s-)} \cdot (e^\gamma - 1) \cdot X_Y(s-) d(N(s) - \lambda^Y s) \\
&\quad + \sum_{0 \le s \le t} \left[\log(X_Y(s)) - \log(X_Y(s-)) - \frac{1}{X_Y(s-)} \Delta X_Y(s) \right] \\
&= \log(X_Y(0)) + \int_0^t (e^\gamma - 1) d(N(s) - \lambda^Y s) \\
&\quad + \int_0^t [\gamma - (e^\gamma - 1)] \, dN(s) \\
&= \log(X_Y(0)) + \int_0^t \gamma dN(s) + \int_0^t (e^\gamma - 1) d(-\lambda^Y s).
\end{aligned}$$

Rewriting these dynamics in differential form, we get

$$d \log(X_Y(t)) = \gamma dN(t) - (e^\gamma - 1)\lambda^Y dt.$$

10.3:

$$d[X_Y(t)]^\alpha = [e^{\alpha\gamma} - 1] \cdot [X_Y(t-)]^\alpha d\left(N(t) - \frac{\alpha(e^\gamma - 1)}{e^{\alpha\gamma} - 1} \lambda^Y t \right).$$

10.4:
(a)

$$\begin{aligned}
V_Y(t) &= \mathbb{E}_t^Y [V_Y(T)] = \mathbb{E}_t^Y [\mathbb{I}(N(T) = 0)] \\
&= \mathbb{P}_t^Y (N(T) = 0) = \mathbb{P}_t^Y ((N(T) - N(t) = 0) \cdot \mathbb{I}(N(t) = 0)) \\
&= \exp\left(-\lambda^Y (T - t)\right) \cdot \mathbb{I}(N(t) = 0).
\end{aligned}$$

(b)

$$\begin{aligned}
V_X(t) &= \mathbb{E}_t^X [V_X(T)] = \mathbb{E}_t^X [Y_X(T)\mathbb{I}(N(T) = 0)] \\
&= \mathbb{E}_t^X \left[Y_X(t) \cdot \exp\left(-(e^{-\gamma} - 1)\lambda^X (T - t)\right) \cdot \mathbb{I}(N(T) = 0) \right] \\
&= Y_X(t) \cdot \exp\left(-(e^{-\gamma} - 1)\lambda^X (T - t)\right) \cdot \mathbb{P}_t^X (N(T) = 0) \\
&= Y_X(t) \cdot \exp\left(-e^{-\gamma} \cdot \lambda^X (T - t)\right) \cdot \mathbb{I}(N(t) = 0).
\end{aligned}$$

Note that using the change of numeraire formula we have

$$\begin{aligned}
V_Y(t) &= V_X(t) \cdot X_Y(t) \\
&= Y_X(t) \cdot \exp\left(-e^{-\gamma}\lambda^X (T - t)\right) \cdot \mathbb{I}(N(t) = 0) \cdot X_Y(t) \\
&= \exp\left(-e^{-\gamma}\lambda^X (T - t)\right) \cdot \mathbb{I}(N(t) = 0) \\
&= \exp\left(-\lambda^Y (T - t)\right) \cdot \mathbb{I}(N(t) = 0),
\end{aligned}$$

so the two prices are indeed consistent.

(c) The hedging portfolio $P_t = \Delta^X(t)\cdot X + \Delta^Y(t)\cdot Y$ takes the following form:

$$P_t = \left[\frac{u^Y(t, e^\gamma X_Y(t)) - u^Y(t, X_Y(t))}{(e^\gamma - 1)X_Y(t)}\right]\cdot X + \left[\frac{u^X(t, e^{-\gamma}Y_X(t)) - u^X(t, Y_X(t))}{(e^{-\gamma} - 1)Y_X(t)}\right]\cdot Y.$$

The price of the contract is zero after the first jump, so the values of $u^Y(t, e^\gamma X_Y(t))$ and $u^X(t, e^{-\gamma}Y_X(t))$ are both zero. Thus

$$\Delta^X(t) = \left[\frac{u^Y(t, e^\gamma X_Y(t)) - u^Y(t, X_Y(t))}{(e^\gamma - 1)X_Y(t)}\right] = -\frac{\exp\left(-\lambda^Y(T-t)\right)\cdot \mathbb{I}(N(t)=0)}{(e^\gamma - 1)X_Y(t)},$$

and

$$\begin{aligned}\Delta^Y(t) &= \frac{u^X(t, e^{-\gamma}Y_X(t)) - u^X(t, Y_X(t))}{(e^{-\gamma} - 1)Y_X(t)}\\ &= -\frac{Y_X(t)\cdot \exp\left(-e^{-\gamma}\lambda^X(T-t)\right)\cdot \mathbb{I}(N(t)=0)}{(e^{-\gamma} - 1)Y_X(t)}\\ &= -\frac{\exp\left(-\lambda^Y\cdot(T-t)\right)\cdot \mathbb{I}(N(t)=0)}{(e^{-\gamma} - 1)}.\end{aligned}$$

Thus the hedging portfolio has the form

$$P_t = -\frac{e^{-\lambda^Y(T-t)}\cdot \mathbb{I}(N(t)=0)}{(e^\gamma - 1)X_Y(t)}\cdot X + e^\gamma\cdot\frac{e^{-\lambda^Y\cdot(T-t)}\cdot \mathbb{I}(N(t)=0)}{(e^\gamma - 1)}\cdot Y.$$

(d) Let $N(t) = m \le k$. Then

$$\begin{aligned}V_Y(t) &= \mathbb{E}_t^Y[V_Y(T)] = \mathbb{E}_t^Y[\mathbb{I}(N(T)=k)]\\ &= \mathbb{P}_t^Y(N(T)=k) = \mathbb{P}_t^Y(N(T)-N(t)=k-m)\\ &= \exp\left(-\lambda^Y(T-t)\right)\cdot\frac{\left[\lambda^Y(T-t)\right]^{k-m}}{(k-m)!},\end{aligned}$$

$$V_X(t) = V_Y(t)\cdot Y_X(t) = Y_X(t)\cdot\exp\left(-\lambda^Y(T-t)\right)\cdot\frac{\left[\lambda^Y(T-t)\right]^{k-m}}{(k-m)!}.$$

For the hedging portfolio we need the post-jump prices $u^Y(t, e^\gamma X_Y(t-))$ and $u^X(t, e^{-\gamma}Y_X(t-))$. When $k > m$, the number of jumps left to reach $N(T) = k$ is reduced by one (now $N(t) = m+1$ after the jump), and so we have the same formula as above with $k-m-1$ instead of $k-m$. Therefore

$$\begin{aligned}
\Delta^X(t) &= \left[\frac{u^Y\left(t, e^{\gamma} X_Y(t)\right) - u^Y\left(t, X_Y(t)\right)}{(e^{\gamma}-1)X_Y(t)}\right] \\
&= \frac{e^{-\lambda^Y(T-t)} \cdot \frac{\left[\lambda^Y(T-t)\right]^{k-m-1}}{(k-m-1)!} - e^{-\lambda^Y(T-t)} \cdot \frac{\left[\lambda^Y(T-t)\right]^{k-m}}{(k-m)!}}{(e^{\gamma}-1)X_Y(t)} \\
&= e^{-\lambda^Y(T-t)} \cdot \frac{\left[\lambda^Y(T-t)\right]^{k-m-1}}{(k-m)!} \cdot \frac{k-m-\lambda^Y(T-t)}{(e^{\gamma}-1)X_Y(t)},
\end{aligned}$$

and

$$\begin{aligned}
\Delta^Y(t) &= \frac{u^X\left(t, e^{-\gamma} Y_X(t)\right) - u^X\left(t, Y_X(t)\right)}{(e^{-\gamma}-1)\, Y_X(t)} \\
&= \frac{Y_X(t)\left[e^{-\lambda^Y(T-t)} \cdot \frac{\left[\lambda^Y(T-t)\right]^{k-m-1}}{(k-m-1)!} - e^{-\lambda^Y(T-t)} \cdot \frac{\left[\lambda^Y(T-t)\right]^{k-m}}{(k-m)!}\right]}{(e^{-\gamma}-1)\, Y_X(t)} \\
&= e^{-\lambda^Y(T-t)} \cdot \frac{\left[\lambda^Y(T-t)\right]^{k-m-1}}{(k-m)!} \cdot \frac{k-m-\lambda^Y(T-t)}{(e^{-\gamma}-1)}.
\end{aligned}$$

10.7:
(a)

$$dA_Y(t) = \int_{-\infty}^{\infty} \bar{\Delta}^X(t-) \cdot (e^x - 1) \cdot X_Y(t-)\Big(\mu^Y(dx, dt) - \nu^Y(dx)dt\Big),$$

where $\bar{\Delta}^X(t) = \int_t^T \mu(ds)$.
(b)

$$\begin{aligned}
&u_t^Y(t,x,y) + \int_{-\infty}^{\infty} \nu^Y(d\xi)\Bigg[u^Y\left(t, e^{\xi}x, \bar{\Delta}^X(t)(e^{\xi}-1)x + y\right) \\
&- u^Y(t,x,y) - u_x^Y(t,x,y)\cdot(e^{\xi}-1)x - u_y^Y(t,x,y)\cdot\bar{\Delta}^X(t)(e^{\xi}-1)x\Bigg] = 0.
\end{aligned}$$

(c)

$$dA_X(t) = \int_{-\infty}^{\infty} \left[A_X(t-) - \bar{\Delta}^X(t-)\right] \cdot (e^x - 1) \cdot \Big(\mu^X(dx, dt) - \nu^X(dx)dt\Big),$$

$$\begin{aligned}
&u_t^X(t,x,y) + \int_{-\infty}^{\infty} \nu^X(d\xi)\Bigg[u^X\left(t, e^{\xi}x, e^{\xi}y - (e^{\xi}-1)\bar{\Delta}^X(t)\right) \\
&- u^X(t,x,y) - u_x^X(t,x,y)\cdot(e^{\xi}-1)x - u_y^X(t,x,y)\cdot(y - \bar{\Delta}^X(t))(e^{\xi}-1)\Bigg] = 0.
\end{aligned}$$

The reduced equation becomes

$$
\begin{aligned}
u_t^X(t,y) + \int_{-\infty}^{\infty} \nu^X(d\xi) \Big[u^X\left(t, e^{\xi} y - (e^{\xi}-1)\bar{\Delta}^X(t)\right) \\
- u^X(t,y) - u_y^X(t,x,y) \cdot (y - \bar{\Delta}^X(t))(e^{\xi}-1) \Big] = 0.
\end{aligned}
$$

References

Arrow, K. J. and G. Debreu (1954). Existence of an equilibrium for a competitive economy. *Econometrica 22*(3), 265–290.

Bachelier, L. (1900). Théorie de la spéculation. *Ann. Sci. École Norm. Sup. 17*, 21–86.

Barone-Adesi, G. and R. Whaley (1987). Efficient analytic approximation of American option values. *Journal of Finance 42*(2), 301–320.

Baxter, M. and A. Rennie (1996). *Financial Calculus*. Cambridge University Press.

Bayraktar, E. and H. Xing (2010). Pricing Asian options for jump diffusions. *Mathematical Finance*.

Bensoussan, A. (1984). On the theory of option pricing. *Acta Applicandae Mathematicae 2*(2), 139–158.

Bjork, T. (2004). *Arbitrage Theory in Continuous Time* (2 ed.). Oxford University Press, USA.

Black, F. (1976). The pricing of commodity contracts. *Journal of Financial Economics 3*(1-2), 167–179.

Black, F. and M. Scholes (1973). The pricing of options and corporate liabilities. *Journal of Political Economy 81*(3), 637–654.

Boyarchenko, S. and S. Levendorskii (2002). Perpetual American options under Levy processes. *SIAM Journal on Control and Optimization 40*(6), 1663–1696.

Boyle, P. (1977). Options: A Monte Carlo approach. *Journal of Financial Economics 4*(3), 323–338.

Boyle, P., M. Broadie, and P. Glasserman (1997). Monte Carlo methods for security pricing. *Journal of Economic Dynamics & Control 21*(8–9), 1267–1321.

Boyle, P. P. (1988). A lattice framework for option pricing with two state variables. *The Journal of Financial and Quantitative Analysis 23*(1), 1–12.

Boyle, P. P., J. Evnine, and S. Gibbs (1989). Numerical evaluation of multivariate contingent claims. *The Review of Financial Studies 2*(2), 241–250.

Brace, A., D. Gatarek, and M. Musiela (1997). The market model of interest rate dynamics. *Mathematical Finance 7*(2), 127–155.

Brekke, K. (1997). The numeraire matters in cost-benefit analysis. *Journal of Public Economics 64*(1), 117–123.

Brigo, D. and F. Mercurio (2006). *Interest Rate Models – Theory and Practice: With Smile, Inflation and Credit* (2 ed.). Springer.

Broadie, M. and J. Detemple (1996). American option valuation: New bounds, approximations, and a comparison of existing methods. *Review of Financial Studies 9*(4), 1211–1250.

Broadie, M. and J. Detemple (1997). The valuation of American options on multiple assets. *Mathematical Finance 7*(3), 241–286.

Broadie, M. and P. Glasserman (1996). Estimating security price derivatives using simulation. *Management Science 42*(2), 269–285.

Broadie, M. and P. Glasserman (1997). Pricing American-style securities using simulation. *Journal of Economic Dynamics & Control 21*(8-9), 1323–1352.

Broadie, M., P. Glasserman, and S. Kou (1997). A continuity correction for discrete barrier options. *Mathematical Finance 7*(4), 325–349.

Brown, H., D. Hobson, and L. Rogers (2001). Robust hedging of barrier options. *Mathematical Finance 11*(3), 285–314.

Buchen, P. and O. Konstandatos (2005). A new method of pricing lookback options. *Mathematical Finance 15*(2), 245–259.

Cameron, R. H. and W. T. Martin (1944). Transformation of Wiener integrals under translations. *Ann. Math 45*, 386–396.

Carr, P. (1998). Randomization and the American put. *Review of Financial Studies 11*(3), 597–626.

Carr, P. and J. Bowie (1994). Static simplicity. *Risk 7*, 44–50.

Carr, P. and A. Chou (1997). Breaking barriers. *Risk 10*, 139–145.

Carr, P., K. Ellis, and V. Gupta (1998). Static hedging of exotic options. *Journal of Finance 53*(3), 1165–1190.

Carr, P., H. Geman, and D. Madan (2001). Pricing and hedging in incomplete markets. *Journal of Financial Economics 62*(1), 131–167.

Carr, P., H. Geman, D. B. Madan, and M. Yor (2002). The fine structure of asset returns: an empirical investigation. *Journal of Business 75*(2), 305–332.

Carr, P. and R. Lee (2009). Put-call symmetry: Extensions and applications. *Mathematical Finance 19*(4), 523–560.

Carr, P. P., R. Jarrow, and R. Myneni (1992). Alternative characterizations of American put options. *Mathematical Finance 2*(2), 87–106.

Cerny, A. (2009). *Mathematical Techniques in Finance: Tools for Incomplete Markets* (2 ed.). Princeton University Press.

Cherubini, U., E. Luciano, and W. Vecchiato (2004). *Copula Methods in Finance.* John Wiley.

Clement, E., D. Lamberton, and P. Protter (2002). An analysis of a least squares regression method for American option pricing. *Finance and Stochastics 6*(4), 449–471.

Cont, R. and P. Tankov (2003). *Financial Modelling with Jump Processes.* Chapman & Hall.

Cont, R. and E. Voltchkova (2005a). A finite difference scheme for option pricing in jump diffusion and exponential levy models. *SIAM Journal on Numerical Analysis 43*(4), 1596–1626.

Cont, R. and E. Voltchkova (2005b). Integro-differential equations for option prices in exponential Levy models. *Finance and Stochastics 9*(3), 299–325.

Cox, J. C., J. E. Ingersoll, and S. A. Ross (1985). A theory of the term structure of interest rates. *Econometrica 53*(2), 385–407.

Cox, J. C., S. A. Ross, and M. Rubinstein (1979). Option pricing: A simplified approach. *Journal of Financial Economics 7*(3), 229–263.

Cox, J. C. and M. Rubinstein (1985). *Options Markets.* Prentice Hall.

Curran, M. (1994). Valuing Asian and portfolio options by conditioning on the geometric mean price. *Management Science 40*(12), 1705–1711.

Dana, R. A. and M. Jeanblanc (2007). *Financial Markets in Continuous Time* (2 ed.). Springer.

Delbaen, F. and W. Schachermayer (1996). The fundamental theorem of asset pricing for unbounded stochastic processes. *Mathematische Annalen 312*, 215–250.

Delbaen, F. and W. Schachermayer (2006). *The Mathematics of Arbitrage.* Springer Finance. Springer-Verlag.

D'Halluin, Y., P. Forsyth, and G. Labahn (2005). A semi-Lagrangian approach for American-Asian options under jump diffusion. *SIAM Journal on Scientific Computing 27*(1), 315–345.

Duffie, D. (2001). *Dynamic Asset Pricing Theory* (3 ed.). Princeton University Press, Princeton, New Jersey.

Duffy, D. J. (2006). *Finite Difference Methods in Financial Engineering: A*

Partial Differential Equation Approach. Wiley.

Dufresne, D. (2000). Laguerre series for Asian and other options. *Mathematical Finance 10*(4), 407–428.

Eberlein, E. and A. Papapantoleon (2005). Equivalence of floating and fixed strike Asian and lookback options. *Stochastic Processes and Their Applications 115*(1), 31–40.

El Karoui, N., M. Jeanblanc-Picque, and S. Shreve (1998). Robustness of the Black and Scholes formula. *Mathematical Finance 8*(2), 93–126.

Fajardo, J. and E. Mordecki (2006). Symmetry and duality in Levy markets. *Quantitative Finance 6*(3), 219–227.

Figlewski, S. and B. Gao (1999). The Adaptive Mesh Model: A new approach to efficient option pricing. *Journal of Financial Economics 53*(3), 313–351.

Filipovic, D. (2008). Optimal numeraires for risk measures. *Mathematical Finance 18*(2), 333–336.

Filipovic, D. (2009). *Term-Structure Models: A Graduate Course.* Springer Verlag.

Flemming, J., S. Turnovsky, and M. Kemp (1977). Choice of numeraire and certainty price in general equilibrium-models of price uncertainty. *Review of Economic Studies 44*(3), 573–583.

Fouque, J. P. and C. H. Han (2003). Pricing Asian options with stochastic volatility. *Quantitative Finance 3*(5), 353–362.

Fouque, J. P., G. Papanicolaou, and R. Sircar (2000). *Derivatives in Financial Markets with Stochastic Volatility.* Cambridge University Press.

Garman, M. and S. Kohlhagen (1983). Foreign currency option values. *Journal of International Money and Finance 2*(3), 231–237.

Gatheral, J. (2006). *The Volatility Surface: A Practitioner's Guide.* Wiley Finance. Wiley.

Geman, H., N. El Karoui, and J. C. Rochet (1995). Changes of numeraire, changes of probability measure and option pricing. *Journal of Applied Probability 32*, 443–458.

Geman, H. and M. Yor (1993). Bessel processes, Asian options, and perpetuities. *Mathematical Finance 3*(4), 349–375.

Gerber, H. and E. Shiu (1996). Martingale approach to pricing perpetual American options on two stocks. *Mathematical Finance 6*(3), 303–322.

Geske, R. and H. Johnson (1984). The American put option valued analytically. *Journal of Finance 39*(5), 1511–1524.

Girsanov, I. V. (1960). On transforming a certain class of stochastic processes by absolutely continuous substitution of measures. *Theory Probability Applications 5*, 285–301.

Glasserman, P. (2003). *Monte Carlo Methods in Financial Engineering.* Springer.

Goldman, M. B., H. B. Sossin, and M. A. Gatto (1979). Path dependent options: "buy at the low, sell at the high". *Journal of Finance 34*(5), 1111–1127.

Gourieroux, C., J. Laurent, and H. Pham (1998). Mean-variance hedging and numeraire. *Mathematical Finance 8*(3), 179–200.

Harrison, J. M. and D. Kreps (1979). Martingales and arbitrage in multiperiod securities markets. *Journal of Economic Theory 20*, 381–408.

Harrison, J. M. and S. Pliska (1981). Martingales and stochastic integrals in the theory of continuos trading. *Stochastic Processes and Their Applications 15*, 313–316.

Haug, E. G. (1997). *The Complete Guide to Option Pricing Formulas.* McGraw-Hill.

Haugh, M. and L. Kogan (2004). Pricing American options: A duality approach. *Operations Research 52*(2), 258–270.

Heath, D., R. Jarrow, and A. Morton (1992). Bond pricing and the term structure of interest rates: A new methodology for contingent claims valuation. *Econometrica 60*(1), 77–105.

Henderson, V. and R. Wojakowski (2002). On the equivalence of floating- and fixed-strike Asian options. *Journal of Applied Probability 39*(2), 391–394.

Heston, S. L. (1993). A closed-form solution for options with stochastic volatility with applications to bond and currency options. *The Review of Financial Studies 6*(2), 327–343.

Hiriart-Urruty, J.-B. and C. Lemarechal (1993). *Convex Analysis and Minimization Algorithms I: Fundamentals.* Springer.

Hobson, D. (1998). Robust hedging of the lookback option. *Finance and Stochastics 2*(4), 329–347.

Hoogland, J. and C. D. D. Neumann (2000). Asians and cash dividends: exploiting symmetries in pricing theory. *Working Paper*.

Hoogland, J. and C. D. D. Neumann (2001a). Local scale invariance and contingent claim pricing I. *International Journal of Theoretical and Applied Finance 4*(1), 1–21.

Hoogland, J. and C. D. D. Neumann (2001b). Local scale invariance and contingent claim pricing II: Path-dependent contingent claims. *International Journal of Theoretical and Applied Finance 4*(1), 23–43.

Hull, J. and A. White (1990). Pricing interest-rate-derivative securities. *Review of Financial Studies 3*(4), 573–592.

Hull, J. C. (2008). *Options, Futures, and Other Derivatives* (7 ed.). Prentice Hall.

Ito, K. (1944). Stochastic integral. *Proceedings of the Imperial Academy, Tokyo 20*, 519–524.

Jaeckel, P. (2002). *Monte Carlo Methods in Finance.* Wiley.

Jaillet, P., D. Lamberton, and B. Lapeyre (1990). Variational-inequalities and the pricing of American options. *Acta Applicandae Mathematicae 21*(3), 263–289.

James, J. and N. Webber (2000). *Interest Rate Modelling: Financial Engineering.* John Wiley & Sons.

Jamshidian, F. (1989). An exact bond option formula. *The Journal of Finance 44*(1), 205–209.

Jamshidian, F. (1997). LIBOR and swap market models and measures. *Finance and Stochastics 1*(4), 293–330.

Jamshidian, F. (2007). Exchange options. *Working Paper*, 1–22.

Jeanblanc, M., M. Yor, and M. Chesney (2009). *Mathematical Methods for Financial Markets.* Springer Finance. Springer-Verlag.

Johansson, P. (1998). Does the choice of numeraire matter in cost-benefit analysis? *Journal of Public Economics 70*(3), 489–493.

Johnson, H. (1987). Options on the maximum or the minimum of several assets. *Journal of Financial and Quantitative Analysis 22*(03), 277–283.

Joshi, M. S. (2008). *The Concepts and Practice of Mathematical Finance.* Cambridge University Press.

Karatzas, I. (1988). On the pricing of American options. *Applied Mathematics and Optimization 17*(1), 37–60.

Karatzas, I. and C. Kardaras (2007). The numeraire portfolio in semimartingale financial models. *Finance and Stochastics 11*(4), 447–493.

Karatzas, I. and S. E. Shreve (1991). *Brownian Motion and Stochastic Calculus* (2 ed.). Springer.

Karatzas, I. and S. E. Shreve (2001). *Methods of Mathematical Finance.* Springer.

Korn, R., E. Korn, and G. Kroisandt (2010). *Monte Carlo Methods and Models in Finance and Insurance.* Chapman & Hall.

Kou, S. (2002). A jump-diffusion model for option pricing. *Management Science 48*(8), 1086–1101.

Kunitomo, N. and M. Ikeda (1992). Pricing options with curved boundaries. *Mathematical Finance 2*(4), 275–298.

Lawler, G. F. (2006). *Introduction to Stochastic Processes.* Chapman & Hall.

Levendorskii, S. (2004). Early exercise boundary and option prices in Levy driven models. *Quantitative Finance 4*(5), 525–547.

Lewis, A. L. (2000). *Option Valuation Under Stochastic Volatility.* Finance Press.

Linetsky, V. (2004). Spectral expansions for Asian (average price) options. *Operations Research 52*(6), 856–867.

Long, J. (1990). The numeraire portfolio. *Journal of Financial Economics 26*(1), 29–69.

Longstaff, F. and E. Schwartz (2001). Valuing American options by simulation: A simple least-squares approach. *Review of Financial Studies 14*(1), 113–147.

Magdon-Ismail, M. and A. Atiya (2004). Maximum drawdown. *Risk 17*(10), 99–102.

Magdon-Ismail, M., A. Atiya, A. Pratap, and Y. Abu-Mostafa (2004). On the maximum drawdown of a Brownian motion. *Journal of Applied Probability 41*, 147–161.

Margrabe, W. (1976). A theory of forward and futures prices. *Wharton School of Business, Preprint.*

Margrabe, W. (1978). The value of an option to exchange one asset for another. *The Journal of Finance 33*(1), 177–186.

McKean, H. P. (1965). A free boundary problem for the heat equation arising from a problem in mathematical economics. *Industrial Management Review 6*(2), 32–39.

Merton, R. (1973). Theory of rational option pricing. *Bell Journal of Economics and Management Science 4*, 141–183.

Merton, R. C. (1969). Lifetime portfolio selection under uncertainty: The continuous-time case. *The Review of Economics and Statistics 51*(3), 247–257.

Merton, R. C. (1976). Option pricing when underlying stock returns are discontinuous. *Journal of Financial Economics 3*, 125–144.

Merton, R. C. (1992). *Continuous-Time Finance.* Wiley-Blackwell.

Mikosch, T. (1999). *Elementary Stochastic Calculus With Finance in View.* World Scientific Publishing Company.

Milevsky, M. and S. Posner (1998). Asian options, the sum of lognormals, and the reciprocal gamma distribution. *Journal of Financial and Quantitative Analysis 33*(3), 409–422.

Musiela, M. and M. Rutkowski (2008). *Martingale Methods in Financial Modelling* (2 ed.). Springer.

Neftci, S. N. (2008). *Principles of Financial Engineering* (2 ed.). Academic Press.

Nielsen, J. and K. Sandmann (2003). Pricing bounds on Asian options. *Journal of Financial and Quantitative Analysis 38*(2), 449–473.

Oksendal, B. (2007). *Stochastic Differential Equations: An Introduction with Applications* (6 ed.). Springer.

Papell, D. and H. Theodoridis (2001). The choice of numeraire currency in panel tests of purchasing power parity. *Journal of Money, Credit and Banking 33*(3), 790–803.

Pellser, A. (2000). *Efficient Methods for Valuing Interest Rate Derivatives.* Springer Verlag.

Pham, H. (1997). Optimal stopping, free boundary, and American option in a jump-diffusion model. *Applied Mathematics and Optimization 35*(2), 145–164.

Platen, E. (2004). A class of complete benchmark models with intensity-based jumps. *Journal of Applied Probability 41*(1), 19–34.

Platen, E. (2006). A benchmark approach to finance. *Mathematical Finance 16*(1), 131–151.

Pospisil, L. and J. Vecer (2008). PDE methods for the maximum drawdown. *Journal of Computational Finance 2*(12), 59–76.

Pospisil, L. and J. Vecer (2010). Portfolio sensitivity to changes in the maximum and the maximum drawdown. *Quantitative Finance 10*(6), 617–627.

Protter, P. E. (2005). *Stochastic Integration and Differential Equations* (2 ed.). Springer.

Rebonato, R. (2004). *Volatility and Correlation: The Perfect Hedger and the Fox.* Wiley.

Revuz, D. and M. Yor (2004). *Continuous Martingales and Brownian Motion.*

Springer.

Rogers, L. (2002). Monte Carlo valuation of American options. *Mathematical Finance 12*(3), 271–286.

Rogers, L. and Z. Shi (1995). The value of an Asian option. *Journal of Applied Probability 32*(4), 1077–1088.

Rubinstein, M. and E. Reiner (1991). Breaking down barriers. *Risk 4*(9), 28–35.

Sadr, A. (2009). *Interest Rate Swaps and Their Derivatives: A Practitioner's Guide (Wiley Finance).* Wiley.

Samuelson, P. A. (1965). Proof that properly anticipated prices fluctuate randomly. *Industrial Management Review 6*(2), 41–49.

Samuelson, P. A. (1973). Mathematics of speculative price. *SIAM Review 15*(1), 1–42.

Schoutens, W. (2003). *Levy Processes in Finance: Pricing Financial Derivatives.* Wiley.

Schroder, M. (1999). Changes of numeraire for pricing futures, forwards, and options. *Review of Financial Studies 12*(5), 1143–1163.

Shiryaev, A. N. (1999). *Essentials of Stochastic Finance: Facts, Models, Theory.* World Scientific Publishing Company.

Shreve, S. E. (2004a). *Stochastic Calculus for Finance. I.* Springer Finance. Springer-Verlag.

Shreve, S. E. (2004b). *Stochastic Calculus for Finance. II.* Springer Finance. Springer-Verlag.

Stulz, R. M. (1982). Options on the minimum or the maximum of two risky assets : Analysis and applications. *Journal of Financial Economics 10*(2), 161 – 185.

Tavella, D. and C. Randall (2000). *Pricing Financial Instruments: The Finite Difference Method.* Wiley.

Tsitsiklis, J. and B. Van Roy (2001). Regression methods for pricing complex American-style options. *Neural Networks, IEEE Transactions on 12*(4), 694–703.

Vasicek, O. (1977). An equilibrium characterization of the term structure. *Journal of Financial Economics 5*(2), 177–188.

Vecer, J. (2001). A new PDE approach for pricing arithmetic average Asian options. *Journal of Computational Finance 4*(4), 105–113.

Vecer, J. (2002). Unified pricing of Asian options. *Risk 15*(6), 113–116.

Vecer, J. (2006). Maximum drawdown and directional trading. *Risk 19*(12), 88–92.

Vecer, J. and M. Xu (2004). Pricing Asian options in a semimartingale model. *Quantitative Finance 4*(2), 170–175.

Williams, D. (1991). *Probability with Martingales.* Cambridge University Press.

Wilmott, P. (2006). *Paul Wilmott on Quantitative Finance 3 Volume Set* (2 ed.). Wiley.

Wystup, U. (2008). Foreign exchange symmetries. *Working Paper*, 1–16.

Zhang, X. (1997). Numerical analysis of American option pricing in a jump-diffusion model. *Mathematics of Operations Research 22*(3), 668–690.

Zhu, S.-P. (2006). An exact and explicit solution for the valuation of American put options. *Quantitative Finance 6*(3), 229–242.

Zvan, R., P. Forsyth, and K. Vetzal (1998). Penalty methods for American options with stochastic volatility. *Journal of Computational and Applied Mathematics 91*(2), 199–218.

Zvan, R., K. Vetzal, and P. Forsyth (2000). PDE methods for pricing barrier options. *Journal of Economic Dynamics & Control 24*(11-12), 1563–1590.

Index